Sixth Form
Pure Mathematics

VOLUME TWO

Sixth Form
Pure Mathematics
VOLUME TWO

C. PLUMPTON
Queen Mary College, London

W. A. TOMKYS
Belle Vue Boys Grammar School
Bradford

PERGAMON PRESS
OXFORD · LONDON · EDINBURGH · NEW YORK
TORONTO · SYDNEY · PARIS · BRAUNSCHWEIG

Pergamon Press Ltd., Headington Hill Hall, Oxford
4 & 5 Fitzroy Square, London W.1
Pergamon Press (Scotland) Ltd., 2 & 3 Teviot Place, Edinburgh 1
Pergamon Press Inc., 44-01 21st Street, Long Island City, New York 11101
Pergamon of Canada, Ltd., 6 Adelaide Street East, Toronto, Ontario
Pergamon Press (Aust.) Pty. Ltd., 20-22 Margaret Street, Sydney, N.S.W.
Pergamon Press S.A.R.L., 24 rue des Écoles, Paris 5e
Vieweg & Sohn GmbH, Burgplatz 1, Braunschweig

First published 1963
Reprinted (with corrections) 1967

Library of Congress Catalog Card Number 61-10010

Printed in Great Britain by
Bookprint Ltd., London and Crawley
957/63

CONTENTS

CHAPTER XI

CHAPTER XII

CHAPTER XIII

CHAPTER XIV

CHAPTER XV

Definitions. Loci in polar coordinates. Curve sketching in polar coordinates. The lengths of chords of polar curves which are drawn through the pole. Transformations from polar to cartesian equations and the reverse process. Areas in polar coordinates. The length of an arc in polar coordinates. Volumes of revolution and areas of surfaces of revolution in polar coordinates. The angle between the tangent and the radius vector. The tangential polar equation—Curvature.

CHAPTER XVI

The number system. Definition of complex number. The cube roots of unity. Conjugate pairs of complex roots. The geometry of complex numbers. The polar coordinate form of a complex number—Modulus and Argument. Products and quotients. De Moivre's Theorem. The exponential form of a complex number. Exponential values of sine and cosine.

CHAPTER XVII

Formation of differential equations. The solution of a differential equation. First order differential equations with variables separable. Homogeneous equations. The law of natural growth. Linear equations of the first order. Bernoulli's equation $\dfrac{dy}{dx} + Py = Qy^n$. Equations of higher orders. Linear equations of the second order with constant coefficients. The complementary function. The particular integral.

CHAPTER XVIII

Graphical methods. The number of real roots of an equation. The approximate value of a small root of a polynomial equation. Newton's method for obtaining a closer approximation to a real root of an equation.

CHAPTER XIX

Rules of manipulation. Fundamental inequalities. The calculus applied to inequalities.

CHAPTER XX

The straight line. Line pairs. The circle. The radical axis. Coaxal circles. Conic sections. Note on the general equation of the second degree. The chord of contact of tangents drawn from an external point to a conic— Pole and polar. The equation of a pair of tangents drawn from an external point to a conic. Conjugate diameters. The polar equation of a conic.

CHAPTER XXI

The coordinate system. The distance between two points. The coordinates of a point which divides the line joining two given points in a given ratio. The equation of a plane. The equations of a line—Direction Cosines. The angle between two directions. The intersection of three planes.

PREFACE

In this volume the pupil is introduced to inverse trigonometric functions, to hyperbolic and inverse hyperbolic functions and to a new range of mathematical methods including the use of determinants, the manipulation of inequalities, the solution of easy differential equations and the use of approximate numerical methods. Complex numbers are defined and the various ways of representing them, and of manipulating them, are considered.

Polar coordinates, curvature, an elementary study of lengths of curves and areas of surfaces of revolution, a more mature discussion of two-dimensional coordinate geometry than was possible in Volume I, and an elementary introduction to the methods of three dimensional coordinate geometry comprise the geometrical content of the book.

Throughout, the authors have tried to preserve the concentric style which they used in Volume I and the many worked examples and exercises in each chapter are designed or chosen to provide a continuous reminder of the work of the preceding chapters.

The aim of the book has been to provide an adequate course for mathematical pupils at Grammar Schools and a useful introductory course for Science and Engineering students in their first year at University or Technical College or engaged in private study. Except for Pure Geometry, the two volumes cover almost all of the syllabuses for Advanced Pure Mathematics of the nine Examining Boards.

The authors wish to thank the authorities of the University of London, the Cambridge Syndicate, the Oxford and Cambridge Joint Board and the Northern Joint Board for permission to include questions (marked L., C., O.C., and N. respectively) from papers set by them.

C. Plumpton
W. A. Tomkys

CHAPTER XI

LINEAR EQUATIONS AND DETERMINANTS

11.1 The solution of linear simultaneous equations in two unknowns

If

$$a_1 x + b_1 y + c_1 = 0,$$
$$a_2 x + b_2 y + c_2 = 0, \tag{11.1}$$

where none of $a_1, a_2, b_1, b_2, c_1, c_2$ is zero, then

$$b_2(a_1 x + b_1 y + c_1) - b_1(a_2 x + b_2 y + c_2) = 0.$$
$$\therefore x(b_2 a_1 - b_1 a_2) = b_1 c_2 - b_2 c_1.$$
$$\therefore x = (b_1 c_2 - b_2 c_1)/(b_2 a_1 - b_1 a_2) \tag{11.2}$$

unless $a_1 b_2 - b_1 a_2 = 0$. Also

$$a_2(a_1 x + b_1 y + c_1) - a_1(a_2 x + b_2 y + c_2) = 0.$$
$$\therefore y = -(a_1 c_2 - a_2 c_1)/(a_1 b_2 - a_2 b_1) \tag{11.3}$$

unless $a_1 b_2 - a_2 b_1 = 0$. These results can be shown by substitution to satisfy the original equations (11.1). We write the solution in the form

$$\frac{x}{b_1 c_2 - b_2 c_1} = \frac{-y}{a_1 c_2 - a_2 c_1} = \frac{1}{a_1 b_2 - a_2 b_1} \tag{11.4}$$

and, at this stage, it is best remembered in the form

$$\frac{x}{\underset{b_2 \searrow\nwarrow c_2}{b_1 \nwarrow\nearrow c_1}} = \frac{-y}{\underset{a_2 \nearrow\searrow c_2}{a_1 \nwarrow\nearrow c_1}} = \frac{1}{\underset{a_2 \nearrow\searrow b_2}{a_1 \nwarrow\nearrow b_1}} \tag{11.5}$$

where the arrows indicate cross-multiplication and each *set* of coefficients in the denominator is obtained by "covering-up" the column in the equations corresponding to x, y and the constant respectively.

(a) If $a_1 b_2 - a_2 b_1 = 0$, eqn (11.2) is meaningless unless also $b_1 c_2 - b_2 c_1 = 0$ and eqn (11.3) is meaningless unless also $a_2 c_1 - a_1 c_2 = 0$. In fact, if $a_1 b_2 - a_2 b_1 = 0$ and one or both of $(b_1 c_2 - b_2 c_1)$, $(c_1 a_2 - c_2 a_1)$ are non-zero, equations (11.1) have no solutions.

(b) If $(a_1b_2 - a_2b_1) = (b_1c_2 - b_2c_1) = (a_2c_1 - a_1c_2) = 0$, then $a_1/a_2 = b_1/b_2 = c_1/c_2$ and the original equations (11.1) can assume identical forms if one of them is divided by a numerical factor. In this case eqns (11.1) have an infinity of pairs of solutions in which x can take any arbitrary value and $y = (-a_1x - c_1)/b_1$.

(c) If $a_1 = 0$, $b_1 \neq 0$, $a_2 \neq 0$, then $y = -c_1/b_1$.

$$\therefore a_2x - b_2c_1/b_1 + c_2 = 0.$$
$$\therefore x = -(b_1c_2 - b_2c_1)/a_2b_1$$

and this result agrees with the one we have already obtained.

Similar results are obtained in other cases in which one of a_1, a_2, or one of b_1, b_2 is zero. If $a_1 = a_2 = 0$ or if $b_1 = b_2 = 0$, one equation becomes trivial or untrue.

If one of a_1, a_2 is zero and one of b_1, b_2 is zero, the solution is evident and in each case it is included in the general solution obtained above.

Geometrical Interpretation. The eqns (11.1) each represent a straight line. The lines intersect at one point [Fig. 101 (i)] if $a_1b_2 - a_2b_1 \neq 0$. The lines are parallel [Fig. 101 (ii)] if $a_1b_2 - a_2b_1 = 0$ and $a_1c_2 - a_2c_1 \neq 0$. The lines are coincident [Fig. 101 (iii)] if $a_1b_2 - a_2b_1 = 0$ and $a_1c_2 - a_2c_1 = 0$.

Examples. (i) Solve the equations

$$23x - 17y - 5 = 0,$$
$$19x + 21y + 18 = 0.$$

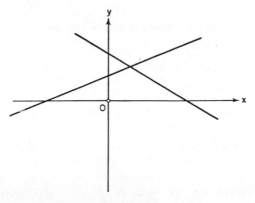

FIG. 101 (i).

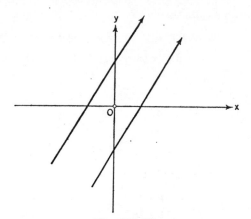

FIG. 101 (ii).

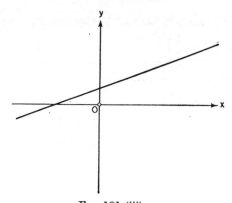

FIG. 101 (iii).

$$\frac{x}{-17 \times 18 + 5 \times 21} = \frac{-y}{23 \times 18 + 5 \times 19} = \frac{1}{23 \times 21 + 19 \times 17}.$$

$$\therefore x = -201/806, \; y = -509/806.$$

(ii) Solve the equations

$$(a + 1)x - \qquad 3y = 4a + 3,$$
$$2x - (a + 1)y = 5a - 1,$$

and find the values of a for which the lines represented are parallel. Discuss the case $a = -1$.

We write the equations as

$$(a + 1)x - \qquad 3y - (4a + 3) = 0,$$
$$2x - (a + 1)y - (5a - 1) = 0.$$

Then

$$\frac{x}{3(5a - 1) - (a + 1)(4a + 3)} = \frac{-y}{-(a + 1)(5a + 1) + 2(4a + 3)} =$$

$$= \frac{1}{-(a + 1)^2 + 6}$$

unless $(a + 1)^2 = 6$, i.e., unless $a = -1 \pm \sqrt{6}$.

$$\therefore\ x = \frac{4a^2 - 8a + 6}{a^2 + 2a - 5}, \quad y = -\frac{5a^2 - 2a - 5}{a^2 + 2a - 5}$$

$$\text{unless } a = -1 \pm \sqrt{6}$$

If $a = -1$, the lines are parallel to the x and y axes respectively and they intersect at $\left(-3, \dfrac{1}{3}\right)$.

Two important cases of elimination.

(i) If $a_1 b_2 - a_2 b_1 \neq 0$, the equations

$$a_1 x^2 + b_1 x + c_1 = 0,$$
$$a_2 x^2 + b_2 x + c_2 = 0,$$

have a common root α (say) if

$$a_1 \alpha^2 + b_1 \alpha + c_1 = 0,$$
$$a_2 \alpha^2 + b_2 \alpha + c_2 = 0,$$

i.e., if $\qquad \dfrac{\alpha^2}{b_1 c_2 - b_2 c_1} = \dfrac{-\alpha}{a_1 c_2 - a_2 c_1} = \dfrac{1}{a_1 b_2 - a_2 b_1}$,

i.e., if $\qquad \left(\dfrac{a_1 c_2 - a_2 c_1}{a_1 b_2 - a_2 b_1}\right)^2 = \dfrac{b_1 c_2 - b_2 c_1}{a_1 b_2 - a_2 b_1}$,

i.e., $\qquad (a_1 c_2 - a_2 c_1)^2 = (a_1 b_2 - a_2 b_1)(b_1 c_2 - b_2 c_1).$

(ii) If $\qquad a_1 \cos \theta + b_1 \sin \theta + c_1 = 0,$
$$a_2 \cos \theta + b_2 \sin \theta + c_2 = 0,$$

then

$$\frac{\cos \theta}{b_1 c_2 - b_2 c_1} = \frac{-\sin \theta}{a_1 c_2 - a_2 c_1} = \frac{1}{a_1 b_2 - a_2 b_1}.$$

Hence the identity $\sin^2\theta + \cos^2\theta = 1$ gives

$$\left(\frac{b_1 c_2 - b_2 c_1}{a_1 b_2 - a_2 b_1}\right)^2 + \left(\frac{a_1 c_2 - a_2 c_1}{a_1 b_2 - a_2 b_1}\right)^2 = 1.$$

Examples. (i) Show that the result of eliminating θ between the equations

$$x = (4 - k)\cos\theta + 2\sin\theta, \; y = 2\cos\theta + (1 - k)\sin\theta,$$

in which k is constant, is generally an equation of the second degree in x and y. Show, however, that there are two values of k for which the eliminant is of the first degree in x and y, and find the eliminant in all cases. (N.)

We have
$$2\sin\theta + (4 - k)\cos\theta - x = 0,$$
$$(1 - k)\sin\theta + \qquad 2\cos\theta - y = 0.$$

$$\therefore \; \frac{\sin\theta}{-(4 - k)y + 2x} = \frac{-\cos\theta}{-2y + x(1 - k)} = \frac{1}{4 - (4 - k)(1 - k)}.$$

if $4 - (4 - k)(1 - k) \neq 0$, i.e., if $k^2 - 5k \neq 0$, which is so provided $k \neq 0$ and $k \neq 5$. Hence in general

$$\{-2y + x(1 - k)\}^2 + \{2x - y(4 - k)\}^2 = \{4 - (4 - k)(1 - k)\}^2$$

which reduces to

$$x^2(k^2 - 2k + 5) + xy(8k - 20) + y^2(k^2 - 8k + 20) = \{k(k - 5)\}^2.$$

This is the *eliminant* in the general case and it is an equation of the second degree in x and y. This solution is not valid in the cases $k = 0$ and $k = 5$ which must now be considered separately.

If $k = 5$, the original equations become

$$x = -\cos\theta + 2\sin\theta, \; y = 2\cos\theta - 4\sin\theta$$

and the eliminant is $y = -2x$.

If $k = 0$, the original equations become

$$x = 4\cos\theta + 2\sin\theta, \; y = 2\cos\theta + \sin\theta$$

and the eliminant is $x = 2y$.

(ii) As the parameter k varies, the equation $\dfrac{x^2}{a + k} + \dfrac{y^2}{k} = 1$ represents a variable conic S_k. Prove that there is just one value of k for which S_k touches the line

$$x\cos\alpha + y\sin\alpha = p.$$

If now this line moves parallel to itself, prove that the locus of the point of contact is

$$x^2 - 2xy \cot 2\alpha - y^2 = a^2. \quad \text{(N.)}$$

In this example the word *parameter* is used to mean *the variable element in the equation of a family of curves*. Hitherto we have used it to mean *the variable element in the coordinates of a family of points*.

In Vol. I we showed that $y = mx + c$ is a tangent to $x^2/a^2 \pm y^2/b^2 = 1$ if $c^2 = a^2m^2 \pm b^2$. Hence $y = -\cot\alpha + p \csc\alpha$ is a tangent to

$$\frac{x^2}{a^2 + k} + \frac{y^2}{k} = 1 \text{ if}$$

$$p^2 \csc^2\alpha = (a^2 + k)\cot^2\alpha + k,$$

i.e., if

$$k(1 + \cot^2\alpha) = p^2 \csc^2\alpha - a^2 \cot^2\alpha,$$

which reduces to

$$k = p^2 - a^2 \cos^2\alpha.$$

This is a unique value of k since a, p, α are constants. Hence there is just one value of k for which S_k touches the line $x \cos\alpha + y \sin\alpha = p$.

If the line touches the conic at (x_1, y_1) then, by eqn (3.9) of Vol. I, the equation of the line is

$$\frac{x x_1}{a^2 + k} + \frac{y y_1}{k} = 1.$$

But the equation of the line is

$$x \cos\alpha + y \sin\alpha = p$$

and hence by comparing coefficients

$$\frac{x_1}{(a^2 + k) \cos\alpha} = \frac{y_1}{k \sin\alpha} = \frac{1}{p}.$$

But

$$k = p^2 - a^2 \cos^2\alpha.$$

$$\therefore px_1 = (a^2 + p^2 - a^2 \cos^2\alpha)\cos\alpha = p^2 \cos\alpha + a^2 \sin^2\alpha \cos\alpha.$$

$$\therefore p^2 \cos\alpha - px_1 + a^2 \sin^2\alpha \cos\alpha = 0.$$

Similarly $p^2 \sin\alpha - py_1 - a^2 \cos^2\alpha \sin\alpha = 0.$

$$\therefore \frac{p^2}{a^2 x_1 \cos^2\alpha \sin\alpha + a^2 y_1 \sin^2\alpha \cos\alpha} = \frac{-p}{-a^2 \cos^3\alpha \sin\alpha - a \sin^3\alpha \cos\alpha} =$$

$$= \frac{1}{-y_1 \cos\alpha + x_1 \sin\alpha}.$$

$$\therefore p^2 = \frac{\frac{1}{2}a^2 \sin 2\alpha \, (x_1 \cos \alpha + y_1 \sin \alpha)}{(x_1 \sin \alpha - y_1 \cos \alpha)}$$

and
$$p = \frac{\frac{1}{2}a^2 \sin 2\alpha \, (\sin^2 \alpha + \cos^2 \alpha)}{(x_1 \sin \alpha - y_1 \cos \alpha)} = \frac{\frac{1}{2}a^2 \sin 2\alpha}{x_1 \sin \alpha - y_1 \cos \alpha} \, .$$

$$\therefore \frac{\frac{1}{2}a^2 \sin 2\alpha \, (x_1 \cos \alpha + y_1 \sin \alpha)}{(x_1 \sin \alpha - y_1 \cos \alpha)} = \left(\frac{\frac{1}{2}a^2 \sin 2\alpha}{x_1 \sin \alpha - y_1 \cos \alpha} \right)^2 .$$

$$\therefore (x_1^2 - y_1^2) \sin \alpha \cos \alpha - x_1 y_1 (\cos^2 \alpha - \sin^2 \alpha) = \tfrac{1}{2}a^2 \sin 2\alpha.$$

$$\therefore \sin 2\alpha \left\{ \tfrac{1}{2}(x_1^2 - y_1^2) - x_1 y_1 \cot 2\alpha \right\} = \tfrac{1}{2}a^2 \sin 2\alpha.$$

$$\therefore x_1^2 - y_1^2 - 2\,x_1 y_1 \cot 2\alpha = a^2.$$

Hence the locus of the point of contact is

$$x^2 - 2xy \cot 2\alpha - y^2 = a^2.$$

Exercises 11.1

Solve each of the following pairs of simultaneous equations

1. $2x - 5y - 7 = 0,$
   $2x + 9y - 11 = 0.$

2. $2x + 5y + 8 = 0,$
   $13x - 2y + 7 = 0.$

3. $14x + 9y = 16,$
   $11x - 17y = 1.$

4. $ax + (a - 1)y - ab = 0,$
   $(a - 1)x + ay + 3ab = 0.$

5. $a(x - b) + 2b(y - a) = ab,$
   $3a(x - 2b) - b(y + a) = -2ab.$

6. $\dfrac{x}{a} + \dfrac{y}{b} = 2a,$
   $x + y = a^2 + b^2.$

7. Solve the equations
$$ax + 3y = 1,$$
$$2x + (a + 1)y = 1$$

and discuss their geometrical interpretation in the special cases $a = -3$ and $a = 2$.

8. Show that there is one value of a for which
$$x + (a + 1)y = a,$$
$$ax + 2y = a + 6$$

represent two parallel lines and one value of a for which the equations represent coincident lines. Find the coordinates of the point of intersection of the lines represented by these equations in the case in which they are perpendicular.

9. If the equations
$$2x^2 - kax - 15a^2 = 0,$$
$$2x^2 + 3kax - 5a^2 = 0$$

have a common root, find the possible values of k and in each case solve the equations.

10. Eliminate t between $x = 2t^2 + 3t - 1$, $y = t^2 - t + 1$.

11. Eliminate t between $x = 2t/(1 - t^2)$, $\quad y = (1 + t^2)/2t$.

12. Show that, if a parabola $y = x^2 + bx + c$ passes through $(1,1)$ and one of the intersections of $y = x^2 - 3x + 1$ with the x-axis, then $b = (-2 + \sqrt{5})$ and $c = (2 - \sqrt{5})$, or $b = (-2 - \sqrt{5})$ and $c = 2 + \sqrt{5}$.

13. Eliminate θ between

$$x = \cos\theta - \sin\theta,$$
$$y = 2\cos\theta + 3\sin\theta.$$

14. Eliminate θ between

$$x\cos\theta - y\sin\theta = 1,$$
$$2x\sin\theta + y\cos\theta = 3.$$

15. Eliminate θ between

$$x\sec\theta + y\tan\theta = 1,$$
$$2x\sec\theta - 2y\tan\theta = -1.$$

16. The straight lines $x\cos\theta + y\sin\theta - p_1 = 0$ and $3x\cos\theta + 4y\sin\theta - p_2 = 0$ intersect at R. Find, in terms of p_1 and p_2, the locus of R as θ varies and consider the particular cases (i) $p_1 = p_2 = 0$, (ii) $p_2 = 3p_1$, (iii) $p_2 = 4p_1$.

11.2 Second order determinants

The form of result (11.5) of § 11.1 is a convenient one, and it is now written as

$$\frac{x}{\begin{vmatrix} b_1 & c_1 \\ b_2 & c_2 \end{vmatrix}} = \frac{-y}{\begin{vmatrix} a_1 & c_1 \\ a_2 & c_2 \end{vmatrix}} = \frac{1}{\begin{vmatrix} a_1 & b_1 \\ a_2 & b_2 \end{vmatrix}}.$$

The *set* of numbers written as $\begin{vmatrix} b_1 & c_1 \\ b_2 & c_2 \end{vmatrix}$ is defined to mean $b_1 c_2 - b_2 c_1$ and is called a *determinant*. The *elements* b_1 and b_2 of this set comprise a *column* and the elements b_1 and c_1 comprise a *row*. We shall denote the columns taken in order from left to right as $\varkappa_1$, $\varkappa_2$ and the rows taken in order from top to bottom as ϱ_1, ϱ_2. Because such sets of numbers are of frequent occurrence (or of frequent use) in mathematical theory it is helpful to be able to manipulate them without first having to expand them. There is, perhaps, little need to manipulate two-rowed determinants as such, but later in this chapter (§ 11.4) we shall consider the manipulation of three-rowed and n-rowed determinants, making use of the methods we now discuss concerning two-rowed determinants.

11.3 Rules of manipulation for second order determinants

(a) A determinant is unaltered if the rows are written as columns and the columns as rows.

For

$$\begin{vmatrix} a_1 & b_1 \\ a_2 & b_2 \end{vmatrix} = a_1 b_2 - b_1 a_2$$

and

$$\begin{vmatrix} a_1 & a_2 \\ b_1 & b_2 \end{vmatrix} = a_1 b_2 - b_1 a_2 .$$

$$\therefore \begin{vmatrix} a_1 & b_1 \\ a_2 & b_2 \end{vmatrix} = \begin{vmatrix} a_1 & a_2 \\ b_1 & b_2 \end{vmatrix} .$$

(b) If two rows, or two columns, of a determinant are interchanged, the determinant changes sign.

For

$$\begin{vmatrix} b_1 & a_1 \\ b_2 & a_2 \end{vmatrix} = b_1 a_2 - a_1 b_2 .$$

$$\therefore \begin{vmatrix} b_1 & a_1 \\ b_2 & a_2 \end{vmatrix} = - \begin{vmatrix} a_1 & b_1 \\ a_2 & b_2 \end{vmatrix} .$$

Similarly

$$\begin{vmatrix} a_2 & b_2 \\ a_1 & b_1 \end{vmatrix} = - \begin{vmatrix} a_1 & b_1 \\ a_2 & b_2 \end{vmatrix} .$$

(c) If two rows or two columns of a determinant are identical, the determinant is equal to zero. This is immediately evident in the case of a two-rowed determinant. It is also a corollary of (a) and (b).

(d) If the elements of a row or of a column have a common factor, then that factor is a factor of the whole determinant.

For

$$\begin{vmatrix} k a_1 & k b_1 \\ a_2 & b_2 \end{vmatrix} = k a_1 b_2 - k b_1 a_2 = k (a_1 b_2 - b_1 a_2) = k \begin{vmatrix} a_1 & b_1 \\ a_2 & b_2 \end{vmatrix} .$$

Similarly a common factor can be extracted from a column.

(e) Determinants can be *added*, if they have one row or one column in common, in this way:

$$\begin{vmatrix} a_1 & a_2 \\ b_1 & b_2 \end{vmatrix} + \begin{vmatrix} a_1 & a_2 \\ c_1 & c_2 \end{vmatrix} = \begin{vmatrix} a_1 & a_2 \\ b_1 + c_1 & b_2 + c_2 \end{vmatrix} .$$

This follows from the direct expansion of the determinants.

(f) Any multiple of a row may be added to any other row without altering the value of the determinant (and similarly for columns).

For
$$\begin{vmatrix} a_1 + kb_1 & b_1 \\ a_2 + kb_2 & b_2 \end{vmatrix} = \begin{vmatrix} a_1 & b_1 \\ a_2 & b_2 \end{vmatrix} + \begin{vmatrix} kb_1 & b_1 \\ kb_2 & b_2 \end{vmatrix} = \begin{vmatrix} a_1 & b_1 \\ a_2 & b_2 \end{vmatrix} + k \begin{vmatrix} b_1 & b_1 \\ b_2 & b_2 \end{vmatrix} = \begin{vmatrix} a_1 & b_1 \\ a_2 & b_2 \end{vmatrix}.$$

This last rule, especially, affords a means of shortening manipulative work with determinants.

Examples.

(i) $\begin{vmatrix} 160 & 63 \\ 80 & 64 \end{vmatrix} = 80 \begin{vmatrix} 2 & 63 \\ 1 & 64 \end{vmatrix} \underset{(\varrho_2 - \varrho_1)}{=} 80 \begin{vmatrix} 2 & 63 \\ -1 & 1 \end{vmatrix} = 80\,(2 + 63) = 5200.$

(ii) $\begin{vmatrix} 79 & 58 \\ 23 & 17 \end{vmatrix} \underset{(\varrho_1 - 3\varrho_2)}{=} \begin{vmatrix} 10 & 7 \\ 23 & 17 \end{vmatrix} \underset{(\varkappa_1 - \varkappa_2)}{=} \begin{vmatrix} 3 & 7 \\ 6 & 17 \end{vmatrix} = 3 \begin{vmatrix} 1 & 7 \\ 2 & 17 \end{vmatrix}$

$$= 3\,(17 - 14) = 9.$$

(iii) Factorize $\Delta = \begin{vmatrix} x^2 + y^2 & y^2 + z^2 \\ xy & zy \end{vmatrix}.$

$$\underset{(\varkappa_1 - \varkappa_2)}{\Delta =} \begin{vmatrix} x^2 - z^2 & y^2 + z^2 \\ xy - zy & zy \end{vmatrix} = (x - z) \begin{vmatrix} x + z & y^2 + z^2 \\ y & zy \end{vmatrix}$$

$$= y\,(x - z) \begin{vmatrix} x + z & y^2 + z^2 \\ 1 & z \end{vmatrix} \underset{(\varkappa_2 - z\varkappa_1)}{=} y\,(x - z) \begin{vmatrix} x + z & y^2 - xz \\ 1 & 0 \end{vmatrix}$$

$$= - y\,(x - z)\,(y^2 - xz).$$

Exercises 11.3

Evaluate each of the following:

1. $\begin{vmatrix} 64 & 23 \\ 16 & 5 \end{vmatrix}$ 2. $\begin{vmatrix} 51 & 15 \\ 47 & 11 \end{vmatrix}$ 3. $\begin{vmatrix} 859 & -83 \\ -215 & 21 \end{vmatrix}$ 4. $\begin{vmatrix} 73 & 17 \\ -15 & -3 \end{vmatrix}$ 5. $\begin{vmatrix} 221 & 16 \\ 51 & 3 \end{vmatrix}.$

6. Evaluate $\begin{vmatrix} 5 & 9 \\ 2 & 5 \end{vmatrix} + \begin{vmatrix} 5 & 9 \\ -3 & -4 \end{vmatrix}.$

7. Evaluate $\begin{vmatrix} -161 & 2 \\ 133 & -5 \end{vmatrix} + \begin{vmatrix} -161 & 17 \\ 133 & 7 \end{vmatrix} + \begin{vmatrix} -161 & -19 \\ 133 & -2 \end{vmatrix}.$

8. Without expanding the separate determinants, prove that

$$\begin{vmatrix} x - 1 & x + 1 \\ 2x + 3 & -x - 2 \end{vmatrix} + \begin{vmatrix} x - 1 & x - 3 \\ x + 1 & x + 2 \end{vmatrix} + \begin{vmatrix} 3x & x - 1 \\ 0 & x + 1 \end{vmatrix} = 0.$$

9. Factorize $\begin{vmatrix} a + x + 1 & b + x + 1 \\ a + y + 1 & b + y + 1 \end{vmatrix}.$

10. Factorize $\begin{vmatrix} a^2 - ab & ab - b^2 \\ b & a \end{vmatrix}.$

11.4 Third order determinants

Consider the simultaneous equations

$$a_1 x + b_1 y + c_1 = 0, \qquad (1)$$

$$a_2 x + b_2 y + c_2 = 0, \qquad (2)$$

$$a_3 x + b_3 y + c_3 = 0. \qquad (3)$$

The solution of (2) and (3) is

$$\frac{x}{\begin{vmatrix} b_2 & c_2 \\ b_3 & c_3 \end{vmatrix}} = \frac{-y}{\begin{vmatrix} a_2 & c_2 \\ a_3 & c_3 \end{vmatrix}} = \frac{1}{\begin{vmatrix} a_2 & b_2 \\ a_3 & b_3 \end{vmatrix}}$$

if $\begin{vmatrix} a_2 & b_2 \\ a_3 & b_3 \end{vmatrix} \neq 0$. Assuming this condition to be satisfied, this solution is also a solution of (1) if

$$a_1 \begin{vmatrix} b_2 & c_2 \\ b_3 & c_3 \end{vmatrix} - b_1 \begin{vmatrix} a_2 & c_2 \\ a_3 & c_3 \end{vmatrix} + c_1 \begin{vmatrix} a_2 & b_2 \\ a_3 & b_3 \end{vmatrix} = 0.$$

This condition is written thus:

$$\begin{vmatrix} a_1 & b_1 & c_1 \\ a_2 & b_2 & c_2 \\ a_3 & b_3 & c_3 \end{vmatrix} = 0$$

and the left hand side is called a determinant of the third order.

If $\begin{vmatrix} a_2 & b_2 \\ a_3 & b_3 \end{vmatrix} = 0$ and $\begin{vmatrix} a_2 & c_2 \\ a_3 & c_3 \end{vmatrix} \neq 0$, there are no finite solutions of equations (2) and (3). If $\begin{vmatrix} a_2 & b_2 \\ a_3 & b_3 \end{vmatrix} = \begin{vmatrix} a_2 & c_2 \\ a_3 & c_3 \end{vmatrix} = 0$, equations (2) and (3) are equivalent and there will be a unique solution of the three simultaneous equations unless $\begin{vmatrix} a_1 & b_1 \\ a_2 & b_2 \end{vmatrix} = 0$.

Rules for the manipulation of three–rowed determinants. The general statements of these rules are the same as those given for two–rowed determinants in § 11.3. Particular statements and verifications follow here. Proofs in other cases follow lines similar to those given in § 11.3.

(a) $\begin{vmatrix} a_1 & b_1 & c_1 \\ a_2 & b_2 & c_2 \\ a_3 & b_3 & c_3 \end{vmatrix} = \begin{vmatrix} a_1 & a_2 & a_3 \\ b_1 & b_2 & b_3 \\ c_1 & c_2 & c_3 \end{vmatrix}.$

This is verified by direct expansion.

(b) $\begin{vmatrix} a_1 & b_1 & c_1 \\ a_2 & b_2 & c_2 \\ a_3 & b_3 & c_3 \end{vmatrix} = - \begin{vmatrix} a_1 & c_1 & b_1 \\ a_2 & c_2 & b_2 \\ a_3 & c_3 & b_3 \end{vmatrix}.$

For $\begin{vmatrix} a_1 & b_1 & c_1 \\ a_2 & b_2 & c_2 \\ a_3 & b_3 & c_3 \end{vmatrix} = a_1 \begin{vmatrix} b_2 & c_2 \\ b_3 & c_3 \end{vmatrix} - b_1 \begin{vmatrix} a_2 & c_2 \\ a_3 & c_3 \end{vmatrix} + c_1 \begin{vmatrix} a_2 & b_2 \\ a_3 & b_3 \end{vmatrix}$

$$= - a_1 \begin{vmatrix} c_2 & b_2 \\ c_3 & b_3 \end{vmatrix} + c_1 \begin{vmatrix} a_2 & b_2 \\ a_3 & b_3 \end{vmatrix} - b_1 \begin{vmatrix} a_2 & c_2 \\ a_3 & c_3 \end{vmatrix} = - \begin{vmatrix} a_1 & c_1 & b_1 \\ a_2 & c_2 & b_2 \\ a_3 & c_3 & b_3 \end{vmatrix}.$$

(c) $\begin{vmatrix} a_1 & a_1 & b_1 \\ a_2 & a_2 & b_2 \\ a_3 & a_3 & b_3 \end{vmatrix} = 0.$

For interchanging x_1 and x_2 gives

$$\begin{vmatrix} a_1 & a_1 & b_1 \\ a_2 & a_2 & b_2 \\ a_3 & a_3 & b_3 \end{vmatrix} = - \begin{vmatrix} a_1 & a_1 & b_1 \\ a_2 & a_2 & b_2 \\ a_3 & a_3 & b_3 \end{vmatrix}.$$

(d) $\begin{vmatrix} k a_1 & b_1 & c_1 \\ k a_2 & b_2 & c_2 \\ k a_3 & b_3 & c_3 \end{vmatrix} = k \begin{vmatrix} a_1 & b_1 & c_1 \\ a_2 & b_2 & c_2 \\ a_3 & b_3 & c_3 \end{vmatrix}.$

This is easily verified by expanding the determinants. It is a corollary of (c) and (d) that

$$\begin{vmatrix} k a_1 & a_1 & b_1 \\ k a_2 & a_2 & b_2 \\ k a_3 & a_3 & b_3 \end{vmatrix} = 0.$$

(e) $\begin{vmatrix} a_1 + x_1 & b_1 & c_1 \\ a_2 + x_2 & b_2 & c_2 \\ a_3 + x_3 & b_3 & c_3 \end{vmatrix} = \begin{vmatrix} a_1 & b_1 & c_1 \\ a_2 & b_2 & c_2 \\ a_3 & b_3 & c_3 \end{vmatrix} + \begin{vmatrix} x_1 & b_1 & c_1 \\ x_2 & b_2 & c_2 \\ x_3 & b_3 & c_3 \end{vmatrix}.$

For the L.H.S. $= (a_1 + x_1) \begin{vmatrix} b_2 & c_2 \\ b_3 & c_3 \end{vmatrix} - (a_2 + x_2) \begin{vmatrix} b_1 & c_1 \\ b_3 & c_3 \end{vmatrix} + (a_3 + x_3) \begin{vmatrix} b_1 & c_1 \\ b_2 & c_2 \end{vmatrix}.$

[Note the expansion from the elements of the first column instead of, as hitherto, from the elements of the first row. This process is possible because of rule (a) above.]

$$\therefore \text{L.H.S.} = a_1 \begin{vmatrix} b_2 & c_2 \\ b_3 & c_3 \end{vmatrix} - a_2 \begin{vmatrix} b_1 & c_1 \\ b_3 & c_3 \end{vmatrix} + a_3 \begin{vmatrix} b_1 & c_1 \\ b_2 & c_2 \end{vmatrix} + x_1 \begin{vmatrix} b_2 & c_2 \\ b_3 & c_3 \end{vmatrix}$$

$$- x_2 \begin{vmatrix} b_1 & c_1 \\ b_3 & c_3 \end{vmatrix} + x_3 \begin{vmatrix} b_1 & c_1 \\ b_2 & c_2 \end{vmatrix}$$

$$= \begin{vmatrix} a_1 & b_1 & c_1 \\ a_2 & b_2 & c_2 \\ a_3 & b_3 & c_3 \end{vmatrix} + \begin{vmatrix} x_1 & b_1 & c_1 \\ x_2 & b_2 & c_2 \\ x_3 & b_3 & c_3 \end{vmatrix}.$$

(f) $$\begin{vmatrix} a_1 & b_1 & c_1 \\ a_2 & b_2 & c_2 \\ a_3 & b_3 & c_3 \end{vmatrix} = \begin{vmatrix} a_1 + k b_1 & b_1 & c_1 \\ a_2 + k b_2 & b_2 & c_2 \\ a_3 + k b_3 & b_3 & c_3 \end{vmatrix}.$$

For the R.H.S. $$= \begin{vmatrix} a_1 & b_1 & c_1 \\ a_2 & b_2 & c_2 \\ a_3 & b_3 & c_3 \end{vmatrix} + k \begin{vmatrix} b_1 & b_1 & c_1 \\ b_2 & b_2 & c_2 \\ b_3 & b_3 & c_3 \end{vmatrix} = \begin{vmatrix} a_1 & b_1 & c_1 \\ a_2 & b_2 & c_2 \\ a_3 & b_3 & c_3 \end{vmatrix}.$$

Examples. (i) $$\begin{vmatrix} 133 & 38 & 95 \\ -15 & 8 & 20 \\ -3 & -4 & -11 \end{vmatrix} \underset{\text{rule (d)}}{=} 19 \begin{vmatrix} 7 & 2 & 5 \\ -15 & 8 & 20 \\ -3 & -4 & -11 \end{vmatrix}$$

$$\underset{\substack{(\varrho_2 - 4\varrho_1) \\ \text{rule (f)}}}{=} 19 \begin{vmatrix} 7 & 2 & 5 \\ -43 & 0 & 0 \\ -3 & -4 & -11 \end{vmatrix} \underset{\text{rule (b)}}{=} -19 \begin{vmatrix} -43 & 0 & 0 \\ 7 & 2 & 5 \\ -3 & -4 & -11 \end{vmatrix}$$

$$= -19 \times \left\{ -43 \left(-22 + 20 \right) \right\} = -1634.$$

(ii) Find the values of a for which the equations

$$(5 + a)x - 4y + 5 = 0,$$
$$x + (2 - a)y - 1 = 0,$$
$$x - ay = 0$$

have a unique consistent solution.

The equations have a unique consistent solution if

$$\begin{vmatrix} (5 + a) & -4 & 5 \\ 1 & (2 - a) & -1 \\ 1 & -a & 0 \end{vmatrix} = 0, \tag{1}$$

provided that $$\begin{vmatrix} 5 + a & -4 \\ 1 & 2 - a \end{vmatrix} \neq 0, \tag{2}$$

i.e., if,
$(\varrho_1 + 5\varrho_2)$

$$\begin{vmatrix} 10+a & 6-5a & 0 \\ 1 & 2-a & -1 \\ 1 & -a & 0 \end{vmatrix} = 0.$$

$$\therefore -10a - a^2 - 6 + 5a = 0,$$

or
$$a^2 + 5a + 6 = 0,$$

i.e.
$$a = -3 \text{ or } a = -2$$

since these values of a satisfy condition (2).

Hence, the equations have a unique consistent solution when $a = -3$ and when $a = -2$.

n^{th} *order determinants.* The definition of a third order determinant is extended to define an nth order determinant

$$\Delta = \begin{vmatrix} a_{11} & a_{12} & . & . & . & a_{1n} \\ a_{21} & a_{22} & . & . & . & a_{2n} \\ . & . & . & . & . & . \\ . & . & . & . & . & . \\ . & . & . & . & . & . \\ a_{n1} & a_{n2} & . & . & . & a_{nn} \end{vmatrix}$$

$$= \sum_{r=1}^{n} (-1)^{r+1} a_{1r} \begin{vmatrix} a_{21} & \cdots & a_{2,r-1} & a_{2,r+1} & \cdots & a_{2n} \\ a_{31} & \cdots & a_{3,r-1} & a_{3,r+1} & \cdots & a_{3n} \\ . & \cdots & . & . & \cdots & . \\ . & \cdots & . & . & \cdots & . \\ . & \cdots & . & . & \cdots & . \\ a_{n1} & \cdots & a_{n,r-1} & a_{n,r+1} & \cdots & a_{nn} \end{vmatrix}.$$

In this expansion, each of the elements of the first row is multiplied by the determinant which remains when the row and column containing that element is removed. The terms corresponding to the odd-ordered elements are positive and those corresponding to the even-ordered elements are negative.

It is convenient at this stage to shorten the notation and to define Δ as $\sum_{r=1}^{n} a_{1r} A_{1r}$, where $A_{1r} = (-1)^{r+1} M_{1r}$ and M_{1r} is the $(n-1)^{\text{th}}$ order determinant obtained by striking out the first row and the rth column of Δ. M_{1r} and A_{1r} are called the minor and cofactor respectively of a_{1r}.

An nth order determinant is thus defined in terms of $(n-1)^{\text{th}}$ order determinants and so, ultimately, in terms of second order determinants. The rules for the manipulation of third order determinants can be shown to apply to the manipulation of nth order determinants.

Exercises 11.4

Evaluate

1. $\begin{vmatrix} 1 & 3 & 7 \\ 2 & 4 & 9 \\ 3 & 5 & 9 \end{vmatrix}$ 2. $\begin{vmatrix} -18 & 1 & 9 \\ 5 & -6 & -3 \\ 3 & -2 & 6 \end{vmatrix}$ 3. $\begin{vmatrix} 2 & -3 & 5 \\ 10 & -15 & 24 \\ 8 & -12 & 21 \end{vmatrix}$ 4. $\begin{vmatrix} -4 & 12 & 5 \\ -5 & 18 & 6 \\ -6 & 27 & 7 \end{vmatrix}$

5. $\begin{vmatrix} 8 & -2 & -4 \\ 7 & 1 & -2 \\ 6 & 4 & 0 \end{vmatrix}$ 6. $\begin{vmatrix} 21 & 5 & -6 \\ 10 & 3 & -3 \\ 25 & 7 & -8 \end{vmatrix}$ 7. $\begin{vmatrix} 17 & 13 & 12 \\ 7 & 7 & 7 \\ 5 & 11 & 6 \end{vmatrix}$ 8. $\begin{vmatrix} 26 & 16 & 7 \\ -1 & 7 & 4 \\ 80 & 71 & 32 \end{vmatrix}$

9. $\begin{vmatrix} 1 & x & x^2+1 \\ 1 & y & xy \\ 1 & -(x+y) & -x^2 \end{vmatrix}$ 10. $\begin{vmatrix} a & 2b & -c \\ -3a & 4b & -4c \\ -a & 6b & -6c \end{vmatrix}$.

11. Eliminate x and y between
$$x/a + y/b = 1; \; x/b - y/a = 1; \; x + y = a + b.$$

12. Find two values of a for each of which the equations
$$ax + y + \cdot \; a^{\cdot} = 0,$$
$$3x - ay + (a + 6) = 0,$$
$$(a + 3)x + (a + 2)y + 5 = 0$$
have a consistent solution, and find the solution of the equations in each case.

11.5 Factorization of determinants

The Remainder Theorem states that if the polynomial $f(x)$ is divided by $(x - a)$ the remainder is $f(a)$.

The Factor Theorem states that if $f(a) = 0$ then $(x - a)$ is a factor of the polynomial $f(x)$.

Each of these theorems follows from the assumption of the existence of the identity $f(x) \equiv (x - a)\varphi(x) + R$ where R is independent of x. It is often possible to use the factor theorem to find factors of a determinant without first expanding the determinant. This method can sometimes be combined with deductions made from the symmetry of the determinant to be factorized.

Examples. (i) Factorize $\Delta = \begin{vmatrix} bc & ca & ab \\ a^2 & b^2 & c^2 \\ 1 & 1 & 1 \end{vmatrix}$.

(a) Δ is symmetrical in a, b and c because the cyclic interchange $a \to b$, $b \to c$, $c \to a$ gives the determinant

$$\begin{vmatrix} ca & ab & bc \\ b^2 & c^2 & a^2 \\ 1 & 1 & 1 \end{vmatrix}$$

and this determinant, being the result of two interchanges of columns from Δ [rule (b)], is equal to Δ.

The factors as a whole of Δ are therefore symmetrical in a, b and c.

(b) The substitution $a = b$ in Δ gives the determinant

$$\begin{vmatrix} bc & cb & b^2 \\ b^2 & b^2 & c^2 \\ 1 & 1 & 1 \end{vmatrix}$$

and this determinant is equal to zero because the first and second columns are identical [rule (c)].

$\therefore (a - b)$ and, by symmetry, $(b - c)$ and $(c - a)$ are factors of Δ.

(c) Δ is of the fourth degree in a, b and c, and therefore Δ has a fourth linear factor which must be a symmetrical one. The only such factor is $(a + b + c)$.

$\therefore \Delta \equiv k(a - b)(b - c)(c - a)(a + b + c)$, where k is constant.

Inspection of Δ shows that the term $+ b^3 c$ occurs in the expansion. Hence $k = -1$ and

$$\Delta = - (a - b)(b - c)(c - a)(a + b + c).$$

(ii) Factorize $\Delta = \begin{vmatrix} x^2 & x & 1 \\ 1 & x^2 & x \\ x & 1 & x^2 \end{vmatrix}$.

Taking $\varkappa_1 + \varkappa_2 + \varkappa_3$ and extracting the common factor $(x^2 + x + 1)$ from the first column gives

$$\Delta = (x^2 + x + 1) \begin{vmatrix} 1 & x & 1 \\ 1 & x^2 & x \\ 1 & 1 & x^2 \end{vmatrix} .$$

The factor theorem shows $(x - 1)$ to be a factor of Δ and hence, taking $\varkappa_2 - \varkappa_3$ and extracting the factor $(x - 1)$,

$$\Delta = (x^2 + x + 1)\,(x - 1)\begin{vmatrix} 1 & 1 & 1 \\ 1 & x & x \\ 1 & -(x+1) & x^2 \end{vmatrix}.$$

The factor theorem now shows $(x - 1)$ to be a factor and hence, taking $\varrho_1 - \varrho_2$ and extracting the factor $(x - 1)$,

$$\Delta = (x^2 + x + 1)\,(x - 1)^2\begin{vmatrix} 0 & -1 & -1 \\ 1 & x & x \\ 1 & -(x+1) & x^2 \end{vmatrix}.$$

$$\therefore\ \Delta = (x^2 + x + 1)(x - 1)^2(x^2 - x + x + 1 + x)$$

$$= (x^2 + x + 1)^2(x - 1)^2.$$

(iii) Solve the equation $\begin{vmatrix} 1 - x & 1 & 1 \\ 3 - x & 2 - x & 3 - x \\ 2 - x & 2 & 6 - x \end{vmatrix} = 0.$

$(\varkappa_1 + 2\varkappa_2 - \varkappa_3)$ gives $\begin{vmatrix} 2 - x & 1 & 1 \\ 4 - 2x & 2 - x & 3 - x \\ 0 & 2 & 6 - x \end{vmatrix} = 0.$

$$\therefore (2 - x)\begin{vmatrix} 1 & 1 & 1 \\ 2 & 2 - x & 3 - x \\ 0 & 2 & 6 - x \end{vmatrix} = 0.$$

$(\varkappa_3 - \varkappa_2,\ \varkappa_2 - \varkappa_1)$ give $(2 - x)\begin{vmatrix} 1 & 0 & 0 \\ 2 - x & 1 \\ 0 & 2 & 4 - x \end{vmatrix} = 0.$

$$\therefore (2 - x)\,(x^2 - 4x - 2) = 0.$$

$$\therefore x = 2 \text{ or } 2 + \sqrt{6} \text{ or } 2 - \sqrt{6}.$$

Exercises 11.5

Factorize each of the following determinants:

1. $\begin{vmatrix} x^2 & y^2 & z^2 \\ x & y & z \\ 1 & 1 & 1 \end{vmatrix}$ 2. $\begin{vmatrix} x^3 & y^3 & 1 \\ x & y & 1 \\ 1 & 1 & 1 \end{vmatrix}$ 3. $\begin{vmatrix} x^3 & y^3 & z^3 \\ x^2 & y^2 & z^2 \\ 1 & 1 & 1 \end{vmatrix}$ 4. $\begin{vmatrix} ab & bc & ca \\ c & a & b \\ 1 & 1 & 1 \end{vmatrix}$ 5. $\begin{vmatrix} a & b & c \\ c & a & b \\ b & c & a \end{vmatrix}$

6. $\begin{vmatrix} x & xy & x^2 \\ y & yz & y^2 \\ z & zx & z^2 \end{vmatrix}$

7. $\begin{vmatrix} a-x & 2a & 2a \\ 2(x-c) & x-2c-a & 2(x-c) \\ 2c & 2c & c-x \end{vmatrix}$

8. $\begin{vmatrix} -x & x+y & x+z \\ x+y & -y & y+z \\ y+z & x+z & -z \end{vmatrix}$

9. $\begin{vmatrix} c & a-b & -c \\ b-2c & -2a-b & a-c \\ c & a+b & c \end{vmatrix}.$

10. Solve the equation $\begin{vmatrix} x & 2 & 3 \\ 3 & 3+x & 5 \\ 7 & 8 & 8+x \end{vmatrix} = 0.$

11. Solve the equation $\begin{vmatrix} x+5 & 3 & 2 \\ 2 & x+3 & 5 \\ 1 & 2 & x+7 \end{vmatrix} = 0.$

12. Show that the equation $\begin{vmatrix} x+1 & 2 & 5 \\ 3 & -x & 1 \\ 1 & -2 & x+1 \end{vmatrix} = 0$

has only one real root, and find that root.

13. Solve the equation $\begin{vmatrix} x & 2 & 3 \\ 4 & 1 & x \\ 1 & 2+x & 2 \end{vmatrix} = 0.$

11.6 Geometrical interpretation

The equations
$$a_1 x + b_1 y + c_1 = 0,$$
$$a_2 x + b_2 y + c_2 = 0,$$
$$a_3 x + b_3 y + c_3 = 0$$

are the equations of straight lines and the condition that they have
a consistent solution is the condition that the lines are concurrent,
i.e., the lines are concurrent if

$$\begin{vmatrix} a_1 & b_1 & c_1 \\ a_2 & b_2 & c_2 \\ a_3 & b_3 & c_3 \end{vmatrix} = 0.$$

Special cases occur when

(i) $\begin{vmatrix} a_1 & b_1 \\ a_2 & b_2 \end{vmatrix} = 0,$ $\begin{vmatrix} a_2 & b_2 \\ a_3 & b_3 \end{vmatrix} = 0,$ $\begin{vmatrix} b_1 & c_1 \\ b_2 & c_2 \end{vmatrix} \neq 0,$ $\begin{vmatrix} b_2 & c_2 \\ b_3 & c_3 \end{vmatrix} \neq 0;$

in this case the lines are parallel;

(ii) $\begin{vmatrix} a_1 & b_1 \\ a_2 & b_2 \end{vmatrix} = \begin{vmatrix} a_2 & b_2 \\ a_3 & b_3 \end{vmatrix} = \begin{vmatrix} b_1 & c_1 \\ b_2 & c_2 \end{vmatrix} = 0, \quad \begin{vmatrix} b_2 & c_2 \\ b_3 & c_3 \end{vmatrix} \neq 0;$

in this case two of the lines are coincident, and parallel to the third line;

(iii) $\begin{vmatrix} a_1 & b_1 \\ a_2 & b_2 \end{vmatrix} = \begin{vmatrix} a_2 & b_2 \\ a_3 & b_3 \end{vmatrix} = \begin{vmatrix} b_1 & c_1 \\ b_2 & c_2 \end{vmatrix} = \begin{vmatrix} b_2 & c_2 \\ b_3 & c_3 \end{vmatrix} = 0;$

in this case the three lines are coincident.

Example. Prove that the normals at the points of parameters t_1, t_2, t_3 on the parabola $x = at^2$, $y = 2at$ are concurrent if $t_1 + t_2 + t_3 = 0$.

The equation of the normal at $(at^2, 2at)$ on the parabola is

$$tx + y - (2at + at^3) = 0.$$

Hence the normals at t_1, t_2, t_3 are concurrent if

$$\begin{vmatrix} t_1 & 1 & -(2at_1 + at_1^3) \\ t_2 & 1 & -(2at_2 + at_2^3) \\ t_3 & 1 & -(2at_3 + at_3^3) \end{vmatrix} = 0,$$

i.e., if $\quad \begin{vmatrix} t_1 & 1 & -2at_1 \\ t_2 & 1 & -2at_2 \\ t_3 & 1 & -2at_3 \end{vmatrix} + \begin{vmatrix} t_1 & 1 & -at_1^3 \\ t_2 & 1 & -at_2^3 \\ t_3 & 1 & -at_3^3 \end{vmatrix} = 0.$

The first of these determinants is equal to zero so that the normals are concurrent if $\quad -a \begin{vmatrix} t_1 & 1 & t_1^3 \\ t_2 & 1 & t_2^3 \\ t_3 & 1 & t_3^3 \end{vmatrix} = 0,$

i.e., if $\quad (t_1 - t_2)(t_2 - t_3)(t_3 - t_1)(t_1 + t_2 + t_3) = 0$. [Exercises 11.5 Question 2.] Hence, since no two of t_1, t_2, t_3 are equal, the normals are concurrent if $t_1 + t_2 + t_3 = 0$.

11.7 The equation of a straight line through two given points

The equation $\qquad \begin{vmatrix} x & y & 1 \\ x_1 & y_1 & 1 \\ x_2 & y_2 & 1 \end{vmatrix} = 0 \qquad\qquad (11.6)$

is the equation of a straight line because it is of the form $Ax + By + C = 0$. Substitution of x_1 for x and y_1 for y in the determinant gives a determinant with two rows the same. The point (x_1, y_1) is therefore on the line and similarly (x_2, y_2) is on the line. Equation (11.6) is therefore the equation of the straight line joining (x_1, y_1), (x_2, y_2).

Examples. (i) Prove that the medians of a triangle are concurrent.

Choose coordinate axes and scales in relation to the triangle ABC so that A is $(0, 0)$, B is $(2, 0)$ and C is $(2h, 2k)$.

Then P, Q, R, the mid-points of BC, CA, AB respectively, are $(h + 1, k)$, (h, k) and $(1, 0)$.

The equation of AP is $kx - (h + 1)y = 0$.

The equation of BQ is
$$\begin{vmatrix} x & y & 1 \\ 2 & 0 & 1 \\ h & k & 1 \end{vmatrix} = 0,$$

i.e., $kx + (2 - h)y - 2k = 0$.

The equation of CR is
$$\begin{vmatrix} x & y & 1 \\ 2h & 2k & 1 \\ 1 & 0 & 1 \end{vmatrix} = 0,$$

i.e., $2kx + (1 - 2h)y - 2k = 0$.

Hence since
$$\begin{vmatrix} k & -(h+1) & 0 \\ k & (2 - h) & -2k \\ 2k & (1 - 2h) & -2k \end{vmatrix} = 0, \quad (\varrho_1 + \varrho_2 = \varrho_3),$$

the medians are concurrent.

(ii) Find the equation of the chord joining the points t_1, t_2 on the curve $x : y : a = t : 1 : 1 + t^3$.

The equation of the chord is
$$\begin{vmatrix} x & y & 1 \\ \dfrac{a t_1}{1 + t_1^3} & \dfrac{a}{1 + t_1^3} & 1 \\ \dfrac{a t_2}{1 + t_2^3} & \dfrac{a}{1 + t_2^3} & 1 \end{vmatrix} = 0.$$

Multiplication of the last column by a, followed by the division of the last two rows by a, gives the equivalent form
$$\begin{vmatrix} x & y & a \\ \dfrac{t_1}{1 + t_1^3} & \dfrac{1}{1 + t_1^3} & 1 \\ \dfrac{t_2}{1 + t_2^3} & \dfrac{1}{1 + t_2^3} & 1 \end{vmatrix} = 0,$$

i. e., $\begin{vmatrix} x & y & a \\ t_1 & 1 & 1+t_1^3 \\ t_2 & 1 & 1+t_2^3 \end{vmatrix} = 0$ provided that $t_1 \neq -1,\, t_2 \neq -1$.

Therefore $(\varrho_3 - \varrho_2)$ followed by division by $(t_2 - t_1)\, [t_2 \neq t_1]$ gives the equation of the chord,

$$\begin{vmatrix} x & y & a \\ t_1 & 1 & 1+t_1^3 \\ 1 & 0 & t_1^2 + t_2^2 + t_1 t_2 \end{vmatrix} = 0,$$

i.e., $(t_1^2 + t_2^2 + t_1 t_2)x - (t_1^2 t_2 + t_1 t_2^2 - 1)y - a = 0$.

11.8 The equation of a tangent to the curve $x = f(t),\, y = g(t)$

The equation of the tangent at the point t, is obtained from the equation of the chord in the last example by considering the limiting form $t_2 \to t_1$ of the equation of the chord.

The equation of the tangent is therefore

$$\begin{vmatrix} x & y & a \\ t_1 & 1 & 1+t_1^3 \\ 1 & 0 & 3t_1^2 \end{vmatrix} = 0 \qquad\qquad (1)$$

which reduces to, $(3\varrho_2 - t_1 \varrho_3)$,

$$\begin{vmatrix} x & y & a \\ 2t_1 & 3 & 3 \\ 1 & 0 & 3t_1^2 \end{vmatrix} = 0$$

or $9t_1^2 x - (6t_1^3 - 3)y - 3a = 0$.

In the form (1), the elements of the third row are the derivatives with respect to t_1 of the elements of the second row and this is a special case of a proposition which is true in general.

The tangent to $x = f(t),\, y = g(t)$ at the point t_1 is

$$\begin{vmatrix} x & y & 1 \\ f(t_1) & g(t_1) & 1 \\ f'(t_1) & g'(t_1) & 0 \end{vmatrix} = 0. \qquad\qquad (11.7)$$

For the chord through the points $t_1,\, t_2$ is

$$\begin{vmatrix} x & y & 1 \\ f(t_1) & g(t_1) & 1 \\ f(t_2) & g(t_2) & 1 \end{vmatrix} = 0,$$

i. e., $\begin{vmatrix} x & y & 1 \\ f(t_1) & g(t_1) & 1 \\ f(t_2) - f(t_1) & g(t_2) - g(t_1) & 0 \end{vmatrix} = 0,$

i. e., $\begin{vmatrix} x & y & 1 \\ f(t_1) & g(t_1) & 1 \\ \dfrac{f(t_2) - f(t_1)}{t_2 - t_1} & \dfrac{g(t_2) - g(t_1)}{t_2 - t_1} & 0 \end{vmatrix} = 0.$

The equation of the tangent at the point t_1 is given by this determinant equation in which each term takes its limiting value as $t_2 \to t_1$. Hence the equation of the tangent is as given by equation (11.7).

Exercises 11.8

1. Write down in determinant form, and simplify, the equation of the straight line through each of the following pairs of points:

(i) $(-3, 2)$, $(1, -4)$,   (ii) $(-5, 1)$, $(-2, -3)$,

(iii) $(-4, -2)$, $(2, 1)$,   (iv) $(5, 0)$, $(3, -1)$,

(v) $(3, -1)$, $(0, -5)$.

2. Use determinants to prove that the altitudes of a triangle are concurrent.

3. If the lines
$$(a - 1)x - 3y - 4 = 0,$$
$$ax + y - 17 = 0,$$
$$(a - 2)x - 3y + 1 = 0$$
are concurrent, calculate the value of a.

4. The converse of Ceva's Theorem in Pure Geometry states that if P, Q, R are points on the sides BC, CA, AB respectively of a triangle ABC such that $\dfrac{BP \cdot CQ \cdot AR}{PC \cdot QA \cdot RB} = 1$, then AP, BQ, CR are concurrent. Verify this result in the particular case in which $BP/PC = 2$; $CQ/QA = \dfrac{1}{3}$; and $AR/RB = 3/2$.

5. If two triangles are similar and similarly situated, i.e., if their corresponding sides are parallel, they are said to be *in similitude*. The lines joining corresponding vertices of two such triangles are concurrent at a point called the *centre of similitude*. Verify this by determinant methods in the case of the triangles ABC, PQR where A is $(0, 0)$, B is $(1, 0)$, C is $(a \cos\theta, a \sin\theta)$, P is $(1, 1)$, Q is $(1 + b, 1)$, R is $(1 + ab \cos\theta, 1 + ab \sin\theta)$.

6. Use the methods suggested in question 5 to prove that the lines joining the ends of parallel radii in two fixed circles are concurrent.

7. Write down, in determinant form, the equation of the chord joining the points t_1 and t_2 on each of the following curves,

(i) $x : y : a = t^2 : 2t : 1$, (ii) $x : y : a = t : t^3 : 1$,

(iii) $x : y : a = t^2 : t^3 : 1$, (iv) $x : y : a = t : 1/t : 1$.

8. Write down in determinant form, and simplify, the equation of the tangent at the point t_1 for each of the curves given in Question 7.

11.9 The equation of a line-pair

In Volume I, § 6.4, we showed that the equation

$$ax^2 + 2hxy + by^2 + 2gx + 2fy + c = 0$$

represents a line pair, i.e., two straight lines, if it has a rational solution for y in terms of x. This is so if

$$(hx + f)^2 - b(ax^2 + 2gx + c) \equiv (h^2 - ab)x^2 + 2(hf - bg)x + f^2 - bc$$

is a perfect square, i.e., if

$$(hf - bg)^2 - (h^2 - ab)(f^2 - bc) = 0.$$

This condition reduces to

$$abc + 2fgh - af^2 - bg^2 - ch^2 = 0$$

and it can be expressed in determinant form thus:

$$\begin{vmatrix} a & h & g \\ h & b & f \\ g & f & c \end{vmatrix} = 0. \qquad (11.8)$$

Exercises 11.9

1. Use determinants to determine which of the following are equations of a line pair:

(i) $2x^2 - xy - 3y^2 - x + 4y - 1 = 0$.

(ii) $2x^2 - 3xy - 2y^2 - 2x - y + 5 = 0$.

(iii) $3x^2 + 4xy - 15y^2 - 15x + 25y = 0$.

(iv) $x^2 + xy + 3y^2 - 3x + 8 = 0$.

(v) $x^2 - 2xy - 3y^2 - 2x + 14y - 8 = 0$.

11.10 The equation of a circle through three given points

The equation
$$\begin{vmatrix} x^2 + y^2 & x & y & 1 \\ x_1^2 + y_1^2 & x_1 & y_1 & 1 \\ x_2^2 + y_2^2 & x_2 & y_2 & 1 \\ x_3^2 + y_3^2 & x_3 & y_3 & 1 \end{vmatrix} = 0 \qquad (11.9)$$

is the equation of a circle. The substitution x_1 for x and y_1 for y in the determinant gives a determinant with two identical rows which is therefore equal to zero. Hence the point (x_1, y_1) is on the circle and similarly the points (x_2, y_2) and (x_3, y_3) are on the circle and (11.9) is the equation of the circle through (x_1, y_1), (x_2, y_2), (x_3, y_3).

Example. PQ is a chord joining the points $P(at^2, 2at)$, $Q(as^2, 2as)$ of a parabola and R is the point of intersection of the tangents at P and Q to the parabola. Show that if the circle through P, Q and R meets the x-axis at $M(x_1, 0)$ and $N(x_2, 0)$, then $x_1 + x_2 = a\{(t + s)^2 + 2\}$.

The tangents at P and Q respectively are
$$x - ty + at^2 = 0,$$
$$x - sy + as^2 = 0,$$

and these tangents intersect where.
$$\frac{x}{ats(t - s)} = \frac{-y}{a(s^2 - t^2)} = \frac{1}{t - s},$$

i.e., where $x = ats$, $y = a(t + s)$.
$$\therefore R \text{ is } \{ats, a(t + s)\}.$$

Hence the equation of the circle through P, Q and R is
$$\begin{vmatrix} x^2 + y^2 & x & y & 1 \\ a^2t^4 + 4a^2t^2 & at^2 & 2at & 1 \\ a^2s^4 + 4a^2s^2 & as^2 & 2as & 1 \\ a^2t^2s^2 + a^2(t + s)^2 & ats & a(t + s) & 1 \end{vmatrix} = 0,$$

i.e., $(2\varrho_4 - \varrho_2 - \varrho_3)$,
$$\begin{vmatrix} x^2 + y^2 & x & y & 1 \\ a^2t^4 + 4a^2t^2 & at^2 & 2at & 1 \\ a^2s^4 + 4a^2s^2 & as^2 & 2as & 1 \\ -a^2(t^2 - s^2)^2 - 2a^2(t - s)^2 & -a(t - s)^2 & 0 & 0 \end{vmatrix} = 0,$$

i.e.,
$$
\begin{vmatrix}
x^2 + y^2 & x & y & 1 \\
a^2 t^4 + 4a^2 t^2 & at^2 & 2at & 1 \\
a^2 s^4 + 4a^2 s^2 & as^2 & 2as & 1 \\
a(t+s)^2 + 2a & 1 & 0 & 0
\end{vmatrix} = 0.
$$

This circle meets the x-axis where $y = 0$ and $(x_1 + x_2)$ is equal to $\dfrac{-\text{ the coefficient of } x}{\text{the coefficient of } x^2}$ from the equation obtained by putting $y = 0$ in the equation of the circle.

$$
\therefore x_1 + x_2 = \frac{\begin{vmatrix}
a^2 t^4 + 4a^2 t^2 & 2at & 1 \\
a^2 s^4 + 4a^2 s^2 & 2as & 1 \\
a(t+s)^2 + 2a & 0 & 0
\end{vmatrix}}{\begin{vmatrix}
at^2 & 2at & 1 \\
as^2 & 2as & 1 \\
1 & 0 & 0
\end{vmatrix}} = \left\{(t+s)^2 + 2\right\}a.
$$

Exercises 11.10

Find the equation of the circle through each group of three points named in the following Questions:

1. $(1, 3)$, $(5, -3)$, $(-2, -2)$, 2. $(2, -2)$, $(-11, -3)$, $(1, 0)$,

3. $(-1, 0)$, $(0, 1)$, $(-5, -7)$, 4. $(2a, 2b)$, $(2a, 0)$, $(2a + b, a)$.

5. A circle through $P(at^2, 2at)$ and $Q(as^2, 2as)$ of a parabola also goes through the focus. If the circle cuts the y-axis at M and N prove that

$$OM \cdot ON = a^2 ts(t^2 + s^2 + ts + 3)/(1 + ts).$$

6. Write down the equation of the circle through the points t_1, t_2, t_3 of the curve $x : y : 1 = t^2 : 1/t : 1$ in determinant form and show that it reduces to the form

$$t_1 t_2 t_3 (x^2 + y^2) - x(t_1 t_2 t_3 \Sigma t_1 + 1) - y(\Sigma t_1 t_2 + t_1^2 t_2^2 t_3^2) + (t_1 t_2 t_3 \Sigma t_1 t_2 - \Sigma t_1) = 0.$$

[Use rule (e) and the results of Questions 1, 2 and 3 of Exercises 11.5.]

11.11 The area of a triangle

The equation of the line joining $A(x_1, y_1)$ to $B(x_2, y_2)$ is given by equation (11.6). From this it follows that the length of the perpendicular from $C(x_3, y_3)$ to AB is

$$
\begin{vmatrix}
x_3 & y_3 & 1 \\
x_1 & y_1 & 1 \\
x_2 & y_2 & 1
\end{vmatrix} \Big/ \sqrt{\left\{(x_1 - x_2)^2 + (y_1 - y_2)^2\right\}}
$$

B

numerically. But the length of AB is $\sqrt{\{(x_1 - x_2)^2 + (y_1 - y_2)^2\}}$. Hence the area of $\triangle ABC$ is

$$\tfrac{1}{2} \begin{vmatrix} x_1 & y_1 & 1 \\ x_2 & y_2 & 1 \\ x_3 & y_3 & 1 \end{vmatrix} \tag{11.10}$$

numerically. The sign of the determinant is determined by the order round the triangle in which the vertices are lettered.

Example. Find the area of a triangle inscribed in the semi-cubical parabola $x : y : 1 = t^3 : t^2 : 1$ in terms of the parameters of the vertices and deduce the condition that three points on the curve are collinear.

The area of $\triangle PQR$ where P, Q and R are points on the curve of parameters t_1, t_2, t_3 respectively is given by

$$A = \tfrac{1}{2} \begin{vmatrix} t_1^3 & t_1^2 & 1 \\ t_2^3 & t_2^2 & 1 \\ t_3^3 & t_3^2 & 1 \end{vmatrix} \text{ numerically.}$$

$$\therefore \ A = \tfrac{1}{2} \left| (t_1 - t_2)(t_2 - t_3)(t_3 - t_1) \sum t_1 t_2 \right| \text{ (Exercise 11.5 No. 3).}$$

Since P, Q and R are distinct points, no two of t_1, t_2, t_3 are equal. $\therefore A = 0$ if $\sum t_1 t_2 = 0$, and this is the condition that P, Q and R are collinear.

Exercises 11.11

Calculate the areas of the triangles whose vertices are given in each of the Questions $1-5$.

1. $(3, 0)$, $(-2, -1)$, $(-1, 4)$.

2. $(-1, 2)$, $(3, -4)$, $(-2, 6)$.

3. $(-1, -4)$, $(2, 3)$, $(6, -2)$.

4. $(ct_1, c/t_1)$, $(ct_2, c/t_2)$, $(ct_3, c/t_3)$.

5. $\theta_1 = \alpha$, $\theta_2 = \alpha + 2\pi/3$, $\theta_3 = \alpha + 4\pi/3$ on the ellipse $x = 3\cos\theta$, $y = 2\sin\theta$.

6. Prove that the area of the triangle joining the points (x_1, y_1), (x_2, y_2), (x_3, y_3) on the parabola $y^2 = 4ax$ is

$$|(y_1 - y_2)(y_2 - y_3)(y_3 - y_1)/8a|.$$

7. Prove that the area of the triangle joining the points of parameters t_1, t_2, t_3 on the rectangular hyperbola $xy = c^2$ is

$$\left| \frac{c^2}{2}(t_1 - t_2)(t_2 - t_3)(t_3 - t_1)/t_1 t_2 t_3 \right|$$

and that the tangents at these points form a triangle, whose area is

$$|2c^2(t_1 - t_2)(t_2 - t_3)(t_3 - t_1)/(t_1 + t_2)(t_2 + t_3)(t_3 + t_1)|.$$

8. Normals to the parabola $x = at^2$, $y = 2at$ at the points of parameters t_1, t_2, t_3 intersect at P, Q and R. Prove that the area of the triangle PQR is

$$\left| \frac{a^2}{2} (t_1 - t_2)(t_2 - t_3)(t_3 - t_1)(t_1 + t_2 + t_3)^2 \right|$$

and deduce the condition, in terms of the parameters, that the three normals are concurrent.

11.12 The solution of simultaneous equations

The equations
$$a_1 x + b_1 y + c_1 z = 0,$$
$$a_2 x + b_2 y + c_2 z = 0,$$
$$a_3 x + b_3 y + c_3 z = 0$$
(11.11)

have a solution $x = y = z = 0$.

If one of x, y, z is not zero, each equation can be put into such a form as

$$a_1 \frac{x}{z} + b_1 \frac{y}{z} + c_1 = 0$$

and the three equations in x/z, y/z have consistent solutions if

$$\begin{vmatrix} a_1 & b_1 & c_1 \\ a_2 & b_2 & c_2 \\ a_3 & b_3 & c_3 \end{vmatrix} = 0$$

with the exceptions in the special cases discussed in § 11.4.

In Chapter XXI equations (11.11) will be shown to be the equations of planes through the origin in three-dimensional coordinates.

(i) $(0, 0, 0)$ is a point common to the three planes, which, in general, have no other common point.

(ii) The planes may intersect in a line, whose equation is $x/p = y/q = z/r$ where p, q and r are constants. This is the case discussed above in which the determinant of the coefficients is zero.

(iii) The planes may be coincident. This is the exceptional case which will not be discussed in detail here.

Solution of the equations
$$a_1 x + b_1 y + c_1 z + d_1 = 0,$$
$$a_2 x + b_2 y + c_2 z + d_2 = 0,$$
$$a_3 x + b_3 y + c_3 z + d_3 = 0.$$
(11.12)

Multiply the equations by A_1, A_2, A_3 respectively where A_1, A_2, A_3 are the cofactors of a_1, a_2, a_3 (§ 11.4) in the determinant

$$\Delta = \begin{vmatrix} a_1 & b_1 & c_1 \\ a_2 & b_2 & c_2 \\ a_3 & b_3 & c_3 \end{vmatrix}$$

and add. Then

$$x \Sigma a_1 A_1 + y \Sigma b_1 A_1 + z \Sigma c_1 A_1 + \Sigma d_1 A_1 = 0.$$

But $\Sigma a_1 A_1 = \Delta$ and $\Sigma b_1 A_1 = \begin{vmatrix} b_1 & b_1 & c_1 \\ b_2 & b_2 & c_2 \\ b_3 & b_3 & c_3 \end{vmatrix} = 0.$

Similarly $\Sigma c_1 A_1 = 0.$

$$\therefore x = - \begin{vmatrix} d_1 & b_1 & c_1 \\ d_2 & b_2 & c_2 \\ d_3 & b_3 & c_3 \end{vmatrix} \bigg/ \Delta = - \begin{vmatrix} b_1 & c_1 & d_1 \\ b_2 & c_2 & d_2 \\ b_3 & c_3 & d_3 \end{vmatrix} \bigg/ \Delta.$$

Expressions for y and z can be obtained in the same way giving the complete solution

$$\frac{x}{\begin{vmatrix} b_1 & c_1 & d_1 \\ b_2 & c_2 & d_2 \\ b_3 & c_3 & d_3 \end{vmatrix}} = \frac{-y}{\begin{vmatrix} a_1 & c_1 & d_1 \\ a_2 & c_2 & d_2 \\ a_3 & c_3 & d_3 \end{vmatrix}} = \frac{z}{\begin{vmatrix} a_1 & b_1 & d_1 \\ a_2 & b_2 & d_2 \\ a_3 & b_3 & d_3 \end{vmatrix}} = \frac{-1}{\begin{vmatrix} a_1 & b_1 & c_1 \\ a_2 & b_2 & c_2 \\ a_3 & b_3 & c_3 \end{vmatrix}} \qquad (11.13)$$

the signs having been determined by the number of changes of columns required to put determinants such as $\Sigma b_1 B_1$ in the form used in (11.13).

Example. Solve the equations

$$x + y + z = 1,$$
$$ax + by + cz = 1,$$
$$a^3 x + b^3 y + c^3 z = 1.$$

From (11.13) $x = - \begin{vmatrix} 1 & 1 & -1 \\ b & c & -1 \\ b^3 & c^3 & -1 \end{vmatrix} \bigg/ \begin{vmatrix} 1 & 1 & 1 \\ a & b & c \\ a^3 & b^3 & c^3 \end{vmatrix}$

$$= \frac{(1-b)(b-c)(c-1)(1+b+c)}{(a-b)(b-c)(c-a)(a+b+c)}$$

$$= \frac{(1-b)(c-1)(1+b+c)}{(a-b)(c-a)(a+b+c)}.$$

By symmetry $y = \dfrac{(1 - c)\,(a - 1)\,(a + 1 + c)}{(b - c)\,(a - b)\,(a + b + c)}$,

$$z = \frac{(1 - a)\,(b - 1)\,(a + b + 1)}{(c - a)\,(b - c)\,(a + b + c)}.$$

11.13 Summary

(i) The solution of the equations

$$a_1 x + b_1 y + c_1 = 0,$$
$$a_2 x + b_2 y + c_2 = 0$$

is given in § 11.1.

(ii) The condition that the equations

$$a_1 x + b_1 y + c_1 = 0,$$
$$a_2 x + b_2 y + c_2 = 0,$$
$$a_3 x + b_3 y + c_3 = 0$$

have a consistent solution is obtained in § 11.4 and this provides the condition that three straight lines should be concurrent (§ 11.6).

(iii) The condition that the equations

$$a_1 x + b_1 y + c_1 z = 0,$$
$$a_2 x + b_2 y + c_2 z = 0,$$
$$a_3 x + b_3 y + c_3 z = 0$$

have solutions other than $x = 0$, $y = 0$, $z = 0$ is found in § 11.12 and these solutions, when they exist, are in the form $x : y : z = p : q : r$. Geometrically they concern the intersection of planes.

(iv) The solution of the equations

$$a_1 x + b_1 y + c_1 z + d_1 = 0,$$
$$a_2 x + b_2 y + c_2 z + d_2 = 0,$$
$$a_3 x + b_3 y + c_3 z + d_3 = 0$$

is given in § 11.12.

(v) The deductive process used in arriving at these results can be extended to the general case of the solution of n linear equations in n unknowns.

Exercises 11.13

Solve the equations

1. $\begin{aligned} x + 3y - 4z &= 9, \\ 2x - y + z &= 3, \\ x + 2y - 5z &= 16. \end{aligned}$

2. $\begin{aligned} 2x + y + z &= 11, \\ x + 3y - 4z &= -5, \\ 3x + y + 5z &= 29. \end{aligned}$

3. $\begin{aligned} 3x - 4y + 2z &= 11, \\ 2x + 5y - z &= -5, \\ x - y + 3z &= 8. \end{aligned}$

4. $\begin{aligned} 5x - 4y + z &= 12, \\ x + 7y - z &= -9, \\ 2x + 3y + 3z &= 8. \end{aligned}$

5. $\begin{aligned} 2x - 3y + z &= -14, \\ 3x + 4y - 5z &= 26, \\ x + 2y - 2z &= 12. \end{aligned}$

6. $\begin{aligned} 2x + 5y - z &= 6, \\ 3x - y + 8z &= 8, \\ x + 4y - z &= 2. \end{aligned}$

7. $\begin{aligned} 2x - y + z &= -8, \\ x + 3y - 3z &= 10, \\ 5x + 7y + 2z &= 9. \end{aligned}$

8. $\begin{aligned} 3x - 9y + 2z &= 7, \\ 2x + 7y - 5z &= 11, \\ 5x - y - z &= 16. \end{aligned}$

9. $\begin{aligned} ax + by + cz &= 1, \\ bx + cy + az &= 1, \\ cx + ay + bz &= 1. \end{aligned}$

10. $\begin{aligned} x + y + z &= 1, \\ ax + by + cz &= 1, \\ a^2 x + b^2 y + c^2 z &= 1. \end{aligned}$

11.14 The product of two determinants

If $\Delta_1 = \begin{vmatrix} a_1 & b_1 & c_1 \\ a_2 & b_2 & c_2 \\ a_3 & b_3 & c_3 \end{vmatrix}$, $\Delta_2 = \begin{vmatrix} x_1 & y_1 & z_1 \\ x_2 & y_2 & z_2 \\ x_3 & y_3 & z_3 \end{vmatrix}$,

then

$$\Delta_1 \Delta_2 = \begin{vmatrix} a_1 x_1 + b_1 y_1 + c_1 z_1 & a_1 x_2 + b_1 y_2 + c_1 z_2 & a_1 x_3 + b_1 y_3 + c_1 z_3 \\ a_2 x_1 + b_2 y_1 + c_2 z_1 & a_2 x_2 + b_2 y_2 + c_2 z_2 & a_2 x_3 + b_2 y_3 + c_2 z_3 \\ a_3 x_1 + b_3 y_1 + c_3 z_1 & a_3 x_2 + b_3 y_2 + c_3 z_2 & a_3 x_3 + b_3 y_3 + c_3 z_3 \end{vmatrix}.$$

$$(11.14)$$

This result will not be proved here. It can be verified by expansion of the expression for $\Delta_1 \Delta_2$ into 27 determinants.

11.15 The derivative of a determinant

If the elements of $\Delta = \begin{vmatrix} a_1 & b_1 & c_1 \\ a_2 & b_2 & c_2 \\ a_3 & b_3 & c_3 \end{vmatrix}$ are functions of x, then

$$\frac{\mathrm{d}\Delta}{\mathrm{d}x} = \begin{vmatrix} \mathrm{d}a_1/\mathrm{d}x & b_1 & c_1 \\ \mathrm{d}a_2/\mathrm{d}x & b_2 & c_2 \\ \mathrm{d}a_3/\mathrm{d}x & b_3 & c_3 \end{vmatrix} + \begin{vmatrix} a_1 & \mathrm{d}b_1/\mathrm{d}x & c_1 \\ a_2 & \mathrm{d}b_2/\mathrm{d}x & c_2 \\ a_3 & \mathrm{d}b_3/\mathrm{d}x & c_3 \end{vmatrix} + \begin{vmatrix} a_1 & b_1 & \mathrm{d}c_1/\mathrm{d}x \\ a_2 & b_2 & \mathrm{d}c_2/\mathrm{d}x \\ a_3 & b_3 & \mathrm{d}c_3/\mathrm{d}x \end{vmatrix}. \quad (11.15)$$

This result follows directly from the product rule for differentiation.

Exercises 11.15

In questions 1–3 do not expand the original determinants and express the answers in determinant form.

1.
$$\begin{vmatrix} x & 0 & 1 \\ x^2 & x & 1 \\ 1 & x^2 & 0 \end{vmatrix} \times \begin{vmatrix} 0 & 1 & y \\ y & y^2 & 1 \\ 0 & 1 & y^2 \end{vmatrix}.$$

2.
$$\begin{vmatrix} a & 0 & b \\ 0 & b & c \\ c & a & 0 \end{vmatrix} \times \begin{vmatrix} a & 0 & b \\ 0 & b & 3 \\ 3 & a & 0 \end{vmatrix}.$$

3.
$$\begin{vmatrix} a & f & g \\ h & b & f \\ g & f & c \end{vmatrix} \times \begin{vmatrix} a & 2 & -1 \\ 2 & b & 0 \\ -1 & 0 & c \end{vmatrix}.$$

4. Evaluate $\dfrac{d}{dx} \begin{vmatrix} 1 & 1 & 1 \\ 1 & x & x^2 \\ x & x^2 & x^3 \end{vmatrix}.$

5. Evaluate $\dfrac{d}{dx} \begin{vmatrix} 1 & x & \alpha \\ x & \sin x & \sin \alpha \\ x^2 & \cos x & \cos \alpha \end{vmatrix}.$

Miscellaneous Exercises XI

1. (a) Evaluate the determinant

$$\begin{vmatrix} 4 & 5 & 8 \\ 6 & 10 & 28 \\ 16 & 25 & 64 \end{vmatrix}.$$

(b) Show that the value of the determinant

$$\begin{vmatrix} 1 & x & y \\ 1 & 1+x & 1+x+y \\ 1 & 2+x & 3+2x+y \end{vmatrix}$$

is independent of the values of x and y.

(c) Express the determinant

$$\begin{vmatrix} x^2+6 & x & -x \\ -x & x^2+6 & x \\ x & -x & x^2+6 \end{vmatrix}$$

as a product of three quadratic functions of x.

(N.)

2. The rows of a third order determinant each contain, in different orders, the elements a, b, c. Show that $a + b + c$ is a factor of the determinant.

Indicating clearly the method used for each step in your calculation, evaluate

$$\begin{vmatrix} 7 & 9 & 4 \\ 8 & 14 & 18 \\ 36 & 16 & 28 \end{vmatrix}.$$ (N.)

3. Indicating clearly your method in each case

    (i) evaluate the determinant

$$\begin{vmatrix} 1 & 2 & 3 \\ 6 & 5 & 4 \\ 7 & 8 & 9 \end{vmatrix},$$

    (ii) solve the equation

$$\begin{vmatrix} x & 1 & 1 \\ 2 & x+1 & 2 \\ 3 & 3 & x+2 \end{vmatrix} = 0.$$ (N.)

4. (a) Show that for all values of θ the value of the determinant

$$\begin{vmatrix} 1 & \sin\theta & 1 \\ -\sin\theta & 1 & \sin\theta \\ -1 & -\sin\theta & 1 \end{vmatrix}$$

lies between 2 and 4 inclusive. State one value of θ for which the determinant has the value 2, and one for which it has the value 4.

    (b) Expand the determinant

$$y = \begin{vmatrix} x & x^2 & x^3 \\ a & b & c \\ p & q & r \end{vmatrix}$$

by the first row, and from the expansion find dy/dx. Express dy/dx in the form of a determinant.

Hence, or otherwise, find two values of x for which the determinant

$$y = \begin{vmatrix} x & x^2 & x^3 \\ 1 & 2 & 3 \\ 1 & -2 & 3 \end{vmatrix}$$

has stationary values. (N.)

5. (a) Show that there is only one real value of x which satisfies the equation

$$\begin{vmatrix} x & 2 & 3 \\ 3 & x & 2 \\ 2 & 3 & x \end{vmatrix} = 0$$

and find this value.

    (b) Factorize completely

$$\begin{vmatrix} 1 & 1 & 1 \\ a^2 & b^2 & c^2 \\ bc & ca & ab \end{vmatrix}.$$ (N.)

6. (a) The points A, B, C lie on the curve $y = x(x + 1)$, and their x-coordinates are $(k - 1)$, k, $(k + 1)$ respectively. By evaluating a determinant show that the area of the triangle ABC is independent of the value of k.

(b) Prove that
$$\begin{vmatrix} \cos(x + y) & -\sin(x + y) & \cos 2y \\ \sin x & \cos x & \sin y \\ -\cos x & \sin x & \cos y \end{vmatrix}$$

is equal to $a + b \cos 2y$, where a and b are independent of x and y.   (N.)

7. (a) Express the value of the determinant
$$\begin{vmatrix} 1 & -\cos\theta & -\sin\theta \\ 1 & \cos\theta & \sin\theta \\ 1 & \sin\theta & \cos\theta \end{vmatrix}$$
in terms of $\cos 2\theta$.

(b) Solve the equation
$$\begin{vmatrix} 1 & 1 & 1 \\ x & 2 & 3 \\ x^3 & 8 & 27 \end{vmatrix} = 0.$$ (N.)

8. Express the determinant
$$\begin{vmatrix} 1 & \alpha & \alpha^4 \\ 1 & \beta & \beta^4 \\ 1 & \gamma & \gamma^4 \end{vmatrix}$$

as a product of linear and quadratic factors. Hence or otherwise find all real solutions of the equation
$$\begin{vmatrix} 1 & -1 & 1 \\ 1 & 1 & 1 \\ x & x^2 & x^5 \end{vmatrix} = 0.$$ (N.)

9. (a) Find all the values of θ between $0°$ and $90°$ that satisfy the equation
$$\begin{vmatrix} 1 & \sin\theta & \sin^2\theta \\ 1 & \cos\theta & \cos^2\theta \\ 4 & 2 & 1 \end{vmatrix} = 0.$$

(b) Determine the values of a for which the equations
$$a(y + 1) + 2x = 0,$$
$$a(x + 1) + 2y = 0,$$
$$a(x + y) + 2 = 0$$
are consistent.   (N.)

10. (a) Find the values of k for each of which the quadratic equations
$$x^2 + kx - 6k = 0,$$
$$x^2 - 2x - k = 0$$
have a common root.

(b) Solve the equation
$$\begin{vmatrix} 2x & 7 & 1 \\ 7 & 2x & 1 \\ 2 & 7 & x^2 \end{vmatrix} = 0.$$ (N.)

11. (a) Find the values of k for which the lines

$$2x + ky + 4 = 0, 4x - y - 2k = 0, 3x + y - 1 = 0$$

are concurrent.

(b) Show that

$$\begin{vmatrix} 1+a & a+a^2 & a^2+1 \\ 1+b & b+b^2 & b^2+1 \\ 1+c & c+c^2 & c^2+1 \end{vmatrix} = k(b-c)(c-a)(a-b)$$

where k is a numerical constant, and determine the value of k. (N.)

12. Solve the equation $\begin{vmatrix} x & 2a & a \\ a & x+a & a \\ 2a & 2a & x-a \end{vmatrix} = 0.$ (N.)

13. Find the equation of the circle that passes through the points $(0, 1)$, $(0, 4)$, $(2, 5)$. Show that the axis of x is a tangent to this circle and determine the equation of the other tangent that passes through the origin. (N.)

14. Show that $\begin{vmatrix} t_1^2 & 2t_1 & 1 \\ t_2^2 & 2t_2 & 1 \\ t_1 t_2 & t_1 + t_2 & 1 \end{vmatrix} = (t_2 - t_1)^3.$

The tangents at the points Q and R on the parabola $y^2 = 4ax$ meet at the point P. If S is the mid-point of QR and P moves in such a way that the area of the triangle PQR is constant, show that the distance between P and S is constant. (N.)

15. Prove that there are just two values of the constant a for which the three equations

$$3x - 2y = 8,$$
$$2x - ay = 2a + 1,$$
$$(a - 2)x + y = 4 - a$$

have a common solution and find these values. Solve the equations for each of these values of a. (N.)

16. (a) Find the values of a for which the equations $x + 2y = a$, $x + ay = 2$, $ax + 4y = 4$ are consistent for the determination of x, y.

Give a geometrical interpretation for each value of a.

(b) Factorize $\begin{vmatrix} x & y & z \\ y-z & x+z & y-x \\ y+z & x-z & x-y \end{vmatrix}.$ (N.)

17. (a) Solve the simultaneous equations

$$y = mx, y = n(x - a)$$

giving geometrical illustrations of the different cases that may arise.

(b) Express the determinant

$$\begin{vmatrix} a & a^2 x - 1/x & a^2 x + 1/x \\ b & b^2 x - 1/x & b^2 x + 1/x \\ c & c^2 x - 1/x & c^2 x + 1/x \end{vmatrix}$$

as a product of linear factors. (N.)

18. (a) Solve the equation $\begin{vmatrix} x & 2 & -2 \\ 2 & x & -2 \\ -2 & 2 & x \end{vmatrix} = 0$.

(b) Write down the first and second derivatives with respect to x of the product vy where v and y are functions of x.

By eliminating y and $\dfrac{dy}{dx}$ deduce that, if $u = vy$, then

$$v^3 \frac{d^2 y}{dx^2} = \begin{vmatrix} u & du/dx & d^2u/dx^2 \\ v & dv/dx & d^2v/dx^2 \\ 0 & v & 2dv/dx \end{vmatrix}. \tag{N.}$$

19. If $ad - bc \neq 0$, prove that the simultaneous equations

$$ax + by = 0, \, cx + dy = 0$$

can only be satisfied by $x = 0$, $y = 0$.

If $ad - bc = 0$, show that they have other solutions. Find all the pairs of real values of x and y that satisfy the equations

$$2a(x^2 - 4ay) - x(y^2 - 4ax) = 0,$$
$$y(x^2 - 4ay) - 2a(y^2 - 4ax) = 0. \tag{N.}$$

20. If λ, x, y and z are functions of t, and if

$$X = \lambda x, \, Y = \lambda y, \, Z = \lambda z,$$

prove that

$$\begin{vmatrix} X & dX/dt & d^2X/dt^2 \\ Y & dY/dt & d^2Y/dt^2 \\ Z & dZ/dt & d^2Z/dt^2 \end{vmatrix} = \lambda^3 \begin{vmatrix} x & dx/dt & d^2x/dt^2 \\ y & dy/dt & d^2y/dt^2 \\ z & dz/dt & d^2z/dt^2 \end{vmatrix}$$

If $X = 1/(1 + t^2)^{2/3}$, $Y = t/(1 + t^2)^{2/3}$, $Z = t^2/(1 + t^2)^{2/3}$, show that

$$\begin{vmatrix} X & dX/dt & d^2X/dt^2 \\ Y & dY/dt & d^2Y/dt^2 \\ Z & dZ/dt & d^2Z/dt^2 \end{vmatrix} = \frac{2}{(1 + t^2)^2}. \tag{N.}$$

21. Can the simultaneous equations

$$ax + 2y = c, \, 3x + y = 5$$

be satisfied by finite values of x and y in the three cases (i) $a \neq 6$, (ii) $a = 6$, $c = 10$; (iii) $a = 6$, $c = 8$? Illustrate each case geometrically. Obtain the solution in each case in which the equations can be satisfied by finite values of x and y. (N.)

22. Prove that
$$\begin{vmatrix} 2\cos\theta & 1 & 0 \\ 1 & 2\cos\theta & 1 \\ 0 & 1 & 2\cos\theta \end{vmatrix} = \frac{\sin 4\theta}{\sin \theta}.$$
(N.)

23. Solve the equations
$$x + y + z = 3,$$
$$x - y + 2z = 4,$$
$$kx + 4y + z = 5$$

when k is not equal to 2. Show that, when $k = 2$, there is a solution in which z has any given value c. (N.)

24. Given that no two of a, b, c are equal, solve the simultaneous equations
$$a(b - y) + b(a - x) = (a - x)(b - y),$$
$$a^2(b - y) + b^2(a - x) = c(a - x)(b - y).$$
(N.)

25. (a) Expand the determinant
$$\begin{vmatrix} a & b & c \\ b & c & a \\ c & a & b \end{vmatrix}$$

(i) without factorizing it, (ii) in the form of the product of a linear and a quadratic factor. Hence, or otherwise, solve for x the equation
$$(p - x)^3 + (q - x)^3 = (p + q - 2x)^3.$$

(b) Solve the equation
$$\begin{vmatrix} 3 + x & 4 & 2 + x \\ 2 - 8x^2 & 2x - 8x^2 & 1 - 8x^2 \\ 1 + x & 2 & 3 + x \end{vmatrix} = 0.$$
(N.)

26. (a) Prove that
$$\begin{vmatrix} (n-2)! & (n-1)! & n! \\ (n-1)! & n! & (n+1)! \\ n! & (n+1)! & (n+2)! \end{vmatrix} = 2(n-2)!\,(n-1)!\,n!.$$

(b) Find the values of t for which the equations
$$x - y + 1 = 0,$$
$$x + ty + t = 0,$$
$$x - t^2y + t^2 = 0$$

can be simultaneously satisfied by finite values of x and y. (N.)

27. Find all the values of x which satisfy the equation
$$\begin{vmatrix} x & x^2 & 1 + x^4 \\ a & a^2 & 1 + a^4 \\ 1 & 1 & 2 \end{vmatrix} = 0.$$
(N.)

28. If no two of a, b, c are equal and
$$\begin{vmatrix} a^4 - 1 & a^3 & a \\ b^4 - 1 & b^3 & b \\ c^4 - 1 & c^3 & c \end{vmatrix} = 0,$$
prove that $abc(bc + ca + ab) = a + b + c$. (N.)

29. Prove that
$$\begin{vmatrix} 1 & 1 & 1 \\ \cos\alpha & \cos\beta & \cos\gamma \\ \cos 2\alpha & \cos 2\beta & \cos 2\gamma \end{vmatrix}$$
$$= 2(\cos\beta - \cos\gamma)(\cos\gamma - \cos\alpha)(\cos\alpha - \cos\beta).$$
Prove also that
$$\begin{vmatrix} \cos\alpha & \cos\beta & \cos\gamma \\ \cos 2\alpha & \cos 2\beta & \cos 2\gamma \\ \cos 3\alpha & \cos 3\beta & \cos 3\gamma \end{vmatrix}$$
$$= 4(\cos\beta - \cos\gamma)(\cos\gamma - \cos\alpha)(\cos\alpha - \cos\beta)(\cos\alpha + \cos\beta + \cos\gamma + 2\cos\alpha\cos\beta\cos\gamma).$$
(N.)

30. Prove, without solving the equations, that the three lines
$$x + y - 1 = 0,$$
$$x + (1 - \cos\alpha)\, y - 1 - \cos\alpha = 0,$$
$$x + (1 - \sin\alpha)\, y - 1 - \sin\alpha = 0$$
are concurrent. (O.C.)

31. (i) Find, in terms of a, b, and c the values of x, y, and z which satisfy the simultaneous equations
$$(1 + a)\, x + y + z = 1,$$
$$x + (1 + b)\, y + z = 1,$$
$$x + y + (1 + c)\, z = 1.$$

Show also that no finite values of x, y, and z can be found to satisfy the equations, if a, b and c are the roots of a cubic equation of the form
$$A t^3 + B t^2 + t + 1 = 0.$$

(ii) Prove that the most general solution of the simultaneous equations
$$(1 + a)\, x + y + z = 1,$$
$$x + y + z = 1$$

(obtained by putting $b = c = 0$ in the preceding equations) can be expressed in the form $x = 0$, $y = \frac{1}{2} + p\lambda$, $z = \frac{1}{2} + q\lambda$, and determine the relation connecting p and q. (O.C.)

32. Prove that the three common chords of the circles
$$x^2 + y^2 \quad\ - 6y - \ 1 = 0,$$
$$x^2 + y^2 - \ 2x - 2y + \ 1 = 0,$$
$$x^2 + y^2 + \ 6x - 4y - 21 = 0$$

taken in pairs are concurrent, and find the coordinates of the point of intersection. (O.C.)

33. Prove that the equation of the normal to the hyperbola $xy = c^2$ at the point $(ct,\ ct^{-1})$ is

$$t^3 x - ty + c - ct^4 = 0.$$

Prove that the parameters of those points of $xy = c^2$ which lie on the curve $y^2 - x^2 - gx - fy = 0$ satisfy the equation

$$ct^4 + gt^3 + ft - c = 0,$$

and that the normals to $xy = c^2$ at these points pass through the point $(-g, f)$.

Prove that $t_1 t_2 t_3 t_4 = -1$ and $\Sigma t_1 t_2 = 0$ together form sufficient conditions for the normals to $xy = c^2$ at the points whose parameters are t_1, t_2, t_3, and t_4 to be concurrent. (O.C.)

34. Solve the equations

$$x + \ y - 3z = 0,$$
$$3x + 5y + \ z = -\frac{5}{3},$$
$$2x + 3y + 2z = \frac{1}{6}.$$

Find a necessary and sufficient condition that the equations

$$ax + by + cz = 0,$$
$$bx + cy + az = 0,$$
$$cx + ay + bz = 0$$

may be consistent for values of x, y, z which are not all zero. (O.C.)

35. By means of determinants solve for x, y and z the simultaneous equations

$$x + \ y + 3z = 4,$$
$$x + 2y + 4z = 5,$$
$$x - \ y + 7z = 6.$$ (L.)

36. By eliminating x and y show that there is only one real value of z satisfying the equations

$$zx - 3y - 19 = 0,$$
$$3x + 7y + 3z = 0,$$
$$2x + zy + 11 = 0.$$

Solve the equations for this value of z. (L.)

37. Show that x and $x + y + z$ are factors of the determinant

$$\begin{vmatrix} (y+z)^2 & x^2 & x^2 \\ y^2 & (z+x)^2 & y^2 \\ z^2 & z^2 & (x+y)^2 \end{vmatrix}.$$

Hence, or otherwise, evaluate the determinant as a product of linear factors. (L.)

38. Show that the simultaneous equations

$$x + ky = 2,$$
$$y - kz = 3,$$
$$kx - z = -5,$$

where k is real, have a unique solution provided $k \neq -1$, and that if $k = -1$ the equations are consistent. Find the solution when $k = 1$.   (L.)

39. (i) Solve completely the equation

$$\begin{vmatrix} 4 & x+3 & 2x+6 \\ 2x+5 & 2 & x-4 \\ 2x+7 & 3 & x-2 \end{vmatrix} = 0.$$

(ii) Factorize the determinant

$$\begin{vmatrix} x^2 & yz & x(y+z) \\ y^2 & zx & y(z+x) \\ z^2 & xy & z(x+y) \end{vmatrix}.$$ (L.)

40. (i) Express the determinant

$$\begin{vmatrix} y+z-x, & z+x-y, & x+y-z \\ x, & y, & z \\ x^2, & y^2, & z^2 \end{vmatrix}$$

as a product of linear factors.

(ii) Find the values of k for which the equations

$$(k+1)x + 2y + 3 = 0,$$
$$2x + (k+3)y + 1 = 0,$$
$$3x + y + (k+2) = 0$$

are consistent.   (L.)

41. The two values of y given by the simultaneous equations

$$y = \frac{a_1 x^2 + b_1 x + c_1}{a_2 x^2 + b_2 x + c_2} \quad \text{and} \quad ax^2 + bx + c = 0$$

are equal. Prove that either $b^2 = 4ac$ or

$$\begin{vmatrix} a & b & c \\ a_1 & b_1 & c_1 \\ a_2 & b_2 & c_2 \end{vmatrix} = 0.$$ (L.)

42. Show that

$$\begin{vmatrix} 2 & a+b & a^2+b^2 \\ a+b & a^2+b^2 & a^3+b^3 \\ 1 & c & c^2 \end{vmatrix} = \begin{vmatrix} 1 & 1 & 0 \\ a & b & 0 \\ 0 & 0 & 1 \end{vmatrix} \times \begin{vmatrix} 1 & 1 & 1 \\ a & b & c \\ a^2 & b^2 & c^2 \end{vmatrix}.$$ (L.)

43. (i) Show that $(x + y + z)$ is a factor of the determinant

$$\begin{vmatrix} y + z & x^2 & x \\ z + x & y^2 & y \\ x + y & z^2 & z \end{vmatrix},$$

and find the remaining factors.

(ii) If $2s = a + b + c$, prove that

$$\begin{vmatrix} \sin(s - a) & \sin a & 1 \\ \sin(s - b) & \sin b & 1 \\ \sin(s - c) & \sin c & 1 \end{vmatrix}$$

is equal to $4 \sin s \sin \tfrac{1}{2}(b - c) \sin \tfrac{1}{2}(c - a) \sin \tfrac{1}{2}(a - b)$.   (L.)

44. The three distinct straight lines

$$y = mx - m^2 + m^3,$$

$$y = nx - n^2 + n^3,$$

$$y = px - p^2 + p^3$$

meet in a point. Show that $m + n + p = 1$ and that the common point is

$$x = mn + np + pm,$$

$$y = mnp.$$ (L.)

CHAPTER XII

INVERSE CIRCULAR FUNCTIONS, HYPERBOLIC FUNCTIONS AND INVERSE HYPERBOLIC FUNCTIONS

12.1 Inverse circular functions

We have already used the notation $\text{Sin}^{-1}x$ to denote "the angle whose sine is x". If $y = \text{Sin}^{-1}x$, then $x = \sin y$ and the graph of $y = \text{Sin}^{-1}x$ is shown in Fig. 102 (i).

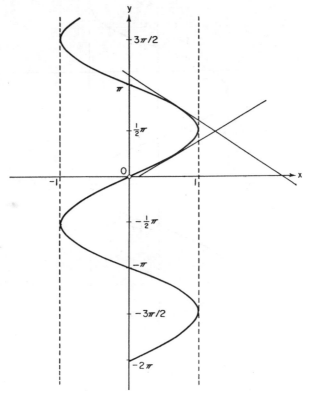

FIG. 102 (i). The graph of $y = \text{Sin}^{-1}x$.

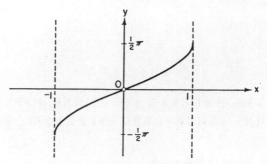

Fig. 102 (ii). The graph of $y = \sin^{-1}x$.

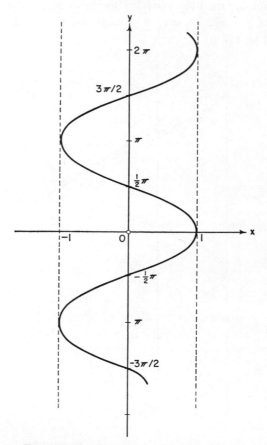

Fig. 102 (iii). The graph of $y = \mathrm{Cos}^{-1}x$.

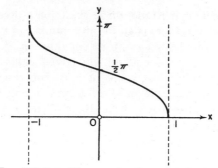

FIG. 102 (iv). The graph of $y = \cos^{-1}x$.

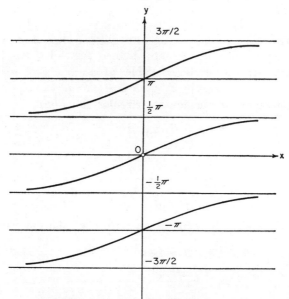

FIG. 102 (v). The graph of $y = \mathrm{Tan}^{-1}x$.

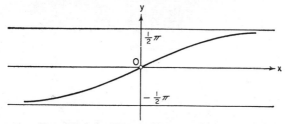

FIG. 102 (vi). The graph of $y = \tan^{-1}x$.

A characteristic of the function is that, for a given value of the independent variable x, there is an infinite number of values of the dependent variable y. $\mathrm{Sin}^{-1}x$ is a *many-valued function of* x. Also, x must lie in the range $-1 \leqq x \leqq 1$ and for any given value of x in that range there is one and only one value of y in the range $-\frac{1}{2}\pi \leqq y \leqq \frac{1}{2}\pi$. This is called the *principal value* of y and it will be denoted henceforward in this book by $\sin^{-1}x$ [Fig. 102 (ii)], the many-valued function being denoted by $\mathrm{Sin}^{-1}x$.

The graphs of $y = \mathrm{Cos}^{-1}x$ and $y = \cos^{-1}x$ are shown in Fig. 102 (iii) and Fig. 102 (iv). The principal value of $\mathrm{Cos}^{-1}x$ is the value of y in the range $0 \leqq y \leqq \pi$ for any given value of x.

The graphs of $y = \mathrm{Tan}^{-1}x$ and $y = \tan^{-1}x$ are shown in Fig. 102 (v) and Fig. 102 (vi). The principal value of $y = \mathrm{Tan}^{-1}x$ is the value of y in the range $-\frac{1}{2}\pi \leqq y \leqq \frac{1}{2}\pi$ for any given value of x. The graphs show that $\mathrm{Sin}^{-1}x$ and $\mathrm{Cos}^{-1}x$ are defined for $-1 \leqq x \leqq 1$ only but that $\mathrm{Tan}^{-1}x$ is defined for all real values of x.

12.2 The derivatives of the inverse circular functions

(i) If $y = \mathrm{Sin}^{-1}x$, then $x = \sin y$ and

$$\frac{\mathrm{d}x}{\mathrm{d}y} = \cos y;$$

$$\therefore \frac{\mathrm{d}y}{\mathrm{d}x} = \frac{1}{\cos y} = \frac{1}{\pm \sqrt{(1 - x^2)}}.$$

Fig. 102 (i) shows the two-valued nature of the gradient function. Since $\dfrac{1}{\cos y} > 0$ for $-\frac{1}{2}\pi < y < \frac{1}{2}\pi$,

$$\therefore \frac{\mathrm{d}(\sin^{-1}x)}{\mathrm{d}x} = \frac{1}{+\sqrt{(1 - x^2)}}. \tag{12.1}$$

(ii) If $y = \mathrm{Cos}^{-1}x$, then $x = \cos y$ and

$$\frac{\mathrm{d}x}{\mathrm{d}y} = -\sin y;$$

$$\therefore \frac{\mathrm{d}y}{\mathrm{d}x} = -\frac{1}{\sin y} = \frac{-1}{\pm \sqrt{(1 - x^2)}}$$

and, since $\dfrac{1}{\sin y} > 0$ for $0 < y < \pi$,

$$\cdot \frac{\mathrm{d}(\cos^{-1}x)}{\mathrm{d}x} = \frac{1}{-\sqrt{(1 - x^2)}}. \tag{12.2}$$

(iii) If $y = \text{Tan}^{-1}x$, then $x = \tan y$ and

$$\frac{dx}{dy} = \sec^2 y ;$$

$$\therefore \frac{dy}{dx} = \frac{1}{\sec^2 y} = \frac{1}{1 + x^2} .$$

Fig. 102 (v) shows the gradient function to be a single-valued positive function.

$$\therefore \frac{d(\tan^{-1}x)}{dx} = \frac{1}{1 + x^2} . \tag{12.3}$$

Application of the function-of-a-function rule gives the results $(a > 0)$

$$\frac{d}{dx} \sin^{-1}\left(\frac{x}{a}\right) = \frac{1}{\sqrt{(a^2 - x^2)}} , \tag{12.4}$$

$$\frac{d}{dx} \cos^{-1}\left(\frac{x}{a}\right) = - \frac{1}{\sqrt{(a^2 - x^2)}} , \tag{12.5}$$

$$\frac{d}{dx} \tan^{-1}\left(\frac{x}{a}\right) = \frac{a}{x^2 + a^2} . \tag{12.6}$$

The inverse functions $\text{Cot}^{-1}x$, $\text{Sec}^{-1}x$, $\text{Cosec}^{-1}x$ have their principal values in the same ranges as $\tan^{-1}x$, $\cos^{-1}x$, $\sin^{-1}x$ respectively. The derivatives of these functions are best obtained from first principles as they arise.

Examples. (i) Solve the equation $\tan^{-1} 2x - \tan^{-1}x = \tan^{-1} \dfrac{1}{3}$, where $0 \leq 2x \leq \frac{1}{2}\pi$.

If $\tan^{-1} 2x = \alpha$, $\tan^{-1}x = \beta$ and $\tan^{-1} \dfrac{1}{3} = \gamma$, the given equation is equivalent to the equation *concerning angles*:

$$\alpha - \beta = \gamma \text{ where } \tan\alpha = 2x, \quad \tan\beta = x, \quad \text{and} \quad \tan\gamma = \frac{1}{3} .$$

$$\therefore \tan(\alpha - \beta) = \frac{1}{3} .$$

$$\therefore \frac{\tan\alpha - \tan\beta}{1 + \tan\alpha\tan\beta} = \frac{1}{3} ,$$

$$\text{i.e., } \frac{x}{1 + 2x^2} = \frac{1}{3} .$$

$$\therefore 2x^2 - 3x + 1 = 0.$$

$$\therefore x = \tfrac{1}{2} \quad \text{or} \quad x = 1.$$

(ii) If $y = \mathrm{Sin}^{-1}(\tfrac{1}{2}x)$, calculate the values of $\mathrm{d}y/\mathrm{d}x$ in each of the cases (a) $y = -4\pi/3$, (b) $y = \dfrac{1}{3}\,\pi$, (c) $y = 11\pi/4$.

Since $y = \mathrm{Sin}^{-1}(\tfrac{1}{2}x)$ so that $x = 2\sin y$,

$$\therefore \frac{\mathrm{d}x}{\mathrm{d}y} = 2\cos y, \quad \text{i.e.,} \quad \frac{\mathrm{d}y}{\mathrm{d}x} = \frac{1}{2\cos y}.$$

$$\text{(a) When } y = -4\pi/3, \quad \frac{\mathrm{d}y}{\mathrm{d}x} = -1.$$

$$\text{(b) When } y = \frac{1}{3}\,\pi, \quad \frac{\mathrm{d}y}{\mathrm{d}x} = 1.$$

$$\text{(c) When } y = 11\pi/4, \quad \frac{\mathrm{d}y}{\mathrm{d}x} = -\sqrt{2}/2.$$

(iii) Find the expressions for the lengths of the subtangents to the curve $y = x\,\mathrm{Sin}^{-1}x$ at the points where $x = \tfrac{1}{2}$ and $y > 0$.

In the usual notation the length of the subtangent is $|y\cot\psi|$, where $\tan\psi = \mathrm{d}y/\mathrm{d}x$.

But $\dfrac{\mathrm{d}y}{\mathrm{d}x} = \pm\,\dfrac{x}{\sqrt{(1-x^2)}} + \mathrm{Sin}^{-1}x$.

Hence, for $\mathrm{Sin}^{-1}x$ in the first quadrant,

$$\frac{\mathrm{d}y}{\mathrm{d}x} = \frac{1}{\sqrt{3}} + 2n\pi + \frac{\pi}{6} \quad \text{for} \quad n = 0, 1, 2 \cdots.$$

Hence the length of the subtangent is $\dfrac{(12n+1)\pi}{12}\Big/\left\{\dfrac{1}{\sqrt{3}} + \dfrac{(12n+1)\pi}{6}\right\}$ for $n = 0, 1, 2, \ldots$.

For $\mathrm{Sin}^{-1}x$ in the second quadrant,

$$\frac{\mathrm{d}y}{\mathrm{d}x} = -\frac{1}{\sqrt{3}} + \frac{(12n+5)\pi}{6}$$

and the length of the subtangent is

$$\frac{(12n+5)\pi}{12}\Big/\left\{\frac{(12n+5)\pi}{6} - \frac{1}{\sqrt{3}}\right\} \quad \text{for} \quad n = 0, 1, 2 \ldots.$$

$$\text{(iv) } \int_0^{a/2} \sin^{-1}\left(\frac{x}{a}\right)\mathrm{d}x = \left[x\sin^{-1}\left(\frac{x}{a}\right)\right]_0^{a/2} - \int_0^{a/2} \frac{x\,\mathrm{d}x}{\sqrt{(a^2-x^2)}}$$

$$= \left[x\sin^{-1}\left(\frac{x}{a}\right) + \sqrt{(a^2-x^2)}\right]_0^{a/2} = a\left(\frac{\pi}{12} + \frac{\sqrt{3}}{2} - 1\right).$$

12.3 Standard integrals

$$\int \frac{dx}{\sqrt{(a^2 - x^2)}} = \sin^{-1}\left(\frac{x}{a}\right) + C \qquad (a > 0), \qquad (12.7)$$

$$\int \frac{dx}{a^2 + x^2} = \frac{1}{a} \tan^{-1}\left(\frac{x}{a}\right) + C \quad (a > 0). \qquad (12.8)$$

For acute angles $\sin^{-1}\left(\dfrac{x}{a}\right) = \tfrac{1}{2}\pi - \cos^{-1}\left(\dfrac{x}{a}\right)$ and for these angles it is therefore unnecessary to have a standard integral in terms of the inverse cosine, but care must be exercised in the use of these integrals since they refer only to the principal values of the inverse functions.

Examples. (i) $\displaystyle\int \frac{dx}{\sqrt{(1 - 3x^2)}} = \frac{1}{\sqrt{3}} \int \frac{dx}{\sqrt{\left(\frac{1}{3} - x^2\right)}}$

$$= \frac{1}{\sqrt{3}} \sin^{-1}(x\sqrt{3}) + C.$$

(ii) $\displaystyle\int \frac{dx}{\sqrt{(1 - x - x^2)}} = \int \frac{dx}{\sqrt{\left\{\frac{5}{4} - (x + \frac{1}{2})^2\right\}}}$

$$= \sin^{-1}\left\{(x + \tfrac{1}{2})\Big/\frac{\sqrt{5}}{2}\right\} + C = \sin^{-1}\left\{(2x + 1)/\sqrt{5}\right\} + C.$$

(iii) $\displaystyle\int \frac{dx}{2x^2 + 4x + 5} = \frac{1}{2} \int \frac{dx}{(x + 1)^2 + \dfrac{3}{2}}$

$$= \frac{1}{2}\frac{\sqrt{2}}{\sqrt{3}} \tan^{-1}\left\{\frac{(x + 1)}{\sqrt{3}/\sqrt{2}}\right\} + C = \frac{1}{\sqrt{6}} \tan^{-1}\left\{\frac{(x + 1)\sqrt{2}}{\sqrt{3}}\right\} + C.$$

Integrals of the forms $\displaystyle\int \frac{dx}{\sqrt{(ax^2 + bx + c)}}$ and $\displaystyle\int \frac{dx}{ax^2 + bx + c}$ are discussed in § 12.9. The method of completing the square used above should not be used for $\displaystyle\int \frac{dx}{ax^2 + bx + c}$ if the denominator factorizes into rational factors. In such a case the method of partial fractions (Vol. I, § 10.4) should be used.

Exercises 12.3

1. Sketch the graphs of $\operatorname{Sec}^{-1}x$, $\operatorname{Cosec}^{-1}x$ and $\operatorname{Cot}^{-1}x$.

2. Write down the principal values of each of the following
(i) $\operatorname{Sin}^{-1}(\frac{1}{2})$, (ii) $\operatorname{Cos}^{-1}(-\sqrt{3}/2)$, (iii) $\operatorname{Tan}^{-1}(-1)$, (iv) $\operatorname{Cot}^{-1}(0)$, (v) $\operatorname{Cosec}^{-1}(-\sqrt{2})$, (vi) $\operatorname{Sec}^{-1}(-\sqrt{2})$.

3. In each of the following cases obtain the value of θ in the form of an inverse circular function and state the quadrant in which it lies.

(i) $\theta = \tan^{-1}2 + \tan^{-1}1$, (ii) $\theta = \sin^{-1}\left(\frac{5}{13}\right) + \cos^{-1}\left(\frac{3}{5}\right)$,

(iii) $\theta = \tan^{-1}\left(-\frac{3}{4}\right) + \sin^{-1}\left(\frac{4}{5}\right)$, (iv) $\theta = \cos^{-1}(-\frac{1}{2}) + \sin^{-1}(-\frac{1}{2})$,

(v) $\theta = \tan^{-1}\frac{1}{2} - \tan^{-1}\left(-\frac{1}{3}\right)$.

4. If $\operatorname{Tan}^{-1}x = n\pi + \alpha$ write down an expression for $\operatorname{Cot}^{-1}x$ and write down the values of $\sin\alpha$ and $\cos\alpha$.

5. Solve the equation $\tan^{-1}\left(\frac{4x}{3}\right) + \tan^{-1}x = \tan^{-1}7$ where both $\tan^{-1}\left(\frac{4x}{3}\right)$ and $\tan^{-1}x$ are acute.

6. If $\sin^{-1}x + \cos^{-1}2x = \frac{2}{3}\pi$ where both $\sin^{-1}x$ and $\cos^{-1}(2x)$ are acute, prove that $4x\sqrt{(1-x^2)} - 2x\sqrt{(1-4x^2)} + 1 = 0$.

7. If $\sin^{-1}x = -\alpha$ for $\alpha > 0$, write down the value of $\cos^{-1}x$.

8. Prove that $\tan^{-1}\left(\frac{a+x}{a-x}\right) - \tan^{-1}\left(\frac{x}{a}\right) = \frac{\pi}{4}$.

9. If $\operatorname{Sin}^{-1}x = n\pi + (-1)^n\alpha$ and $\operatorname{Cos}^{-1}x = 2n\pi \pm \beta$, prove that

$$\tan(\alpha + 2\beta) = -x/\sqrt{(1-x^2)}.$$

10. Differentiate each of the following with respect to x:
(i) $\sin^{-1}(2x)$, (ii) $\cos^{-1}(x/2)$, (iii) $\tan^{-1}\{(1+x)/(1-x)\}$,

(iv) $\tan^{-1}\{(a+x)/(a-x)\}$, (v) $\sin^{-1}\left(\frac{1}{x}\right)$, (vi) $\tan^{-1}(\sin x)$, (vii) $\tan^{-1}(2\tan x)$,

(viii) $\sin^{-1}\left\{\cos\left(x+\frac{1}{3}\pi\right)\right\}$, (ix) $\cos^{-1}(\sin 2x)$, (x) $\tan^{-1}(e^{2x}+1)$.

11. Write down the value of dy/dx at the point named for each of the following curves; (i) $y = \operatorname{Sin}^{-1}x$, $(-\frac{1}{2}, 7\pi/6)$, (ii) $y = \operatorname{Tan}^{-1}(x/4)$, $(4, 5\pi/4)$, (iii) $y = \operatorname{Sin}^{-1}(5x)$, $(-1/5, 5\pi/2)$, (iv) $y = \operatorname{Cos}^{-1}x$, $(-\frac{1}{2}, 2\pi/3)$.

12. Obtain each of the following derivatives
(i) $\frac{d}{dx}(\sec^{-1}x)$, (ii) $\frac{d}{dx}(\operatorname{cosec}^{-1}x)$, (iii) $\frac{d}{dx}(\cot^{-1}x)$.

13. Verify by using the tables that the tangent to the curve $y = \operatorname{Tan}^{-1}x$ at the origin meets the curve again where $y \doteq 2.03^c$.

14. Find the equations of the tangents to the curve $y = \operatorname{Sin}^{-1}x$ at the points where $x = \frac{1}{2}$ for $0 < y < \pi$. If these tangents intersect at P and meet the x-axis at T and S respectively, calculate the area of the triangle PTS.

15. Calculate an approximate value for the small change in y which results from an increase in the value of x from $x = 1$ to 1.05 when $y = x \tan^{-1}x$.

16. If $y = \sin^{-1}\left(\dfrac{x}{a}\right)$, prove that $\dfrac{d^2y}{dx^2} - x\left(\dfrac{dy}{dx}\right)^3 = 0$.

17. An aeroplane flying at 360 m.p.h. in level flight and in a straight line at a height of 1 mile above the ground is followed by a searchlight beam sited on the ground. The aeroplane passes directly over the searchlight. Calculate the angular velocity and the angular acceleration of the beam at the instant when the aeroplane is directly overhead.

18. A bead P is threaded on to a straight wire and attached to a string which is threaded through a fixed ring O distant 1 ft. from the wire. The bead starts from rest at a point of the wire such that OP is perpendicular to the wire and it is caused to move along the wire with uniform acceleration 2 ft. per sec². the string OP lengthening throughout the motion so as to remain taut. Calculate the angular velocity (i.e., $d\theta/dt$ where $\angle AOP = \theta$) of the string (i) at time 1 sec, (ii) when the bead has moved a distance 4 ft.
Prove that the angular acceleration of the string is zero when $t^4 = \dfrac{1}{3}$.

19. Integrate $1/\sqrt{(x^2 - 1)}$ with respect to x by means of the substitutions $x = \sec\theta$, $\tan\frac{1}{2}\theta = t$.

20. Use the method of integration by parts to obtain each of the following integrals

(i) $\int \cos^{-1}x\,dx$, (ii) $\int \tan^{-1}x\,dx$, (iii) $\int \sec^{-1}x\,dx$, (iv) $\int \cot^{-1}x\,dx$.

21. Sketch the curve $y = a^3/(a^2 + x^2)$ for $a > 0$.
Calculate (i) the area enclosed by the curve, the x-axis and the ordinates at $x = -a$ and $x = +a$.

(ii) the possible values of b in terms of a if the area enclosed by the curve, the x-axis and the ordinates at $x = b$ and $x = 6\,b$ is $\dfrac{1}{4}\pi a^2$ units².

22. Find the volume of rotation about the x-axis of the area defined in question 21 (i).

23. Evaluate each of the following definite integrals

$$\text{(i)} \int_0^1 \frac{dx}{\sqrt{(4 - x^2)}}, \quad \text{(ii)} \int_0^3 \frac{dx}{2x^2 + 6},$$

$$\text{(iii)} \int_{1/6}^{1/3} \frac{dx}{\sqrt{(1 - 9x^2)}}, \quad \text{(iv)} \int_0^2 \frac{dx}{x^2 + 2x + 2},$$

$$\text{(v)} \int_1^{1.5} \frac{dx}{\sqrt{(2x - x^2)}}.$$

24. Obtain each of the following integrals

$$\text{(i)} \int \frac{\mathrm{d}x}{x^2 + 4x + 5}, \quad \text{(ii)} \int \frac{\mathrm{d}x}{2x^2 - 3x + 3},$$

$$\text{(iii)} \int \frac{\mathrm{d}x}{\surd(1 - 2x - x^2)}, \quad \text{(iv)} \int \frac{\mathrm{d}x}{\surd(1 - 6x - 3x^2)},$$

$$\text{(v)} \int \frac{\mathrm{d}x}{5x^2 + 10x + 8}.$$

25. Show that the area enclosed by the ellipse $x^2/a^2 + y^2/b^2 = 1$ is $\pi a b$.

26. Find the area enclosed by the curve $y = \tan x$, the axis of y and the line $y = 1$.

27. Obtain $\int x \tan^{-1} x \, \mathrm{d}x$.

28. Use partial fractions to integrate

$$\frac{5x^2 + 12}{x^4 - 16}$$

with respect to x.

29. Use the method of substitution to evaluate

$$\int\limits_{1}^{4} \frac{x \, \mathrm{d}x}{1 + x^4}.$$

30. In the usual notation, an equation of motion for a particle moving in a straight line is

$$v = 4 \, \surd(a^2 - s^2)$$

where a is a positive constant. Calculate the time taken by the particle to move from $s = \frac{1}{2}a$ to $s = a$.

31. The curve $y = \sin^{-1} x$ intersects the curve $y = \cos^{-1} x$ at P and the latter curve intersects the x-axis at Q. Find the area enclosed by the arcs OP and PQ and the x-axis. (O is the origin.)

12.4 The hyperbolic functions

At this stage we define certain combinations of exponential functions as new functions which are called *hyperbolic functions*.

Hyperbolic sine of $x = \sinh x = \dfrac{\mathrm{e}^x - \mathrm{e}^{-x}}{2} = x + \dfrac{x^3}{3!} + \dfrac{x^5}{5!} + \cdots;$

hyperbolic cosine of $x = \cosh x = \dfrac{\mathrm{e}^x + \mathrm{e}^{-x}}{2} = 1 + \dfrac{x^2}{2!} + \dfrac{x^4}{4!} + \cdots;$

$\tanh x = \dfrac{\sinh x}{\cosh x}; \quad \coth x = \dfrac{1}{\tanh x} = \dfrac{\cosh x}{\sinh x};$

$\operatorname{cosech} x = \dfrac{1}{\sinh x}; \quad \operatorname{sech} x = \dfrac{1}{\cosh x}.$

The names of the new functions suggest an analogy with the circular functions which will be finally explained in Chapter XVI of this book. In this Chapter we show that there is a close analogy between the properties of the hyperbolic functions and those of the circular functions. We discuss some of the properties of the hyperbolic functions below.

(a) $$\cosh^2 x - \sinh^2 x \equiv 1. \qquad (12.9)$$

This result follows from the identity

$$\left(\frac{e^x + e^{-x}}{2}\right)^2 - \left(\frac{e^x - e^{-x}}{2}\right)^2 \equiv e^x \times e^{-x} \equiv 1$$

and is analogous to the identity $\cos^2 x + \sin^2 x \equiv 1$.

The point $(a \cosh t, b \sinh t)$ is a point *on one branch of the hyperbola* $x^2/a^2 - y^2/b^2 = 1$ for all values of t. Thus, in some respects, the hyperbolic functions $\sinh x$ and $\cosh x$ bear the same relation to the rectangular hyperbola $x^2 - y^2 = a^2$ as the circular functions bear to the circle $x^2 + y^2 = a^2$ and it is for this reason that they are so named.

(b) $$\sinh x + \cosh x = e^x. \qquad (12.10)$$

$$\cosh x - \sinh x = e^{-x}. \qquad (12.11)$$

These results follow directly from the definitions; they are very useful in the processes of manipulating hyperbolic functions.

(c) $$\frac{d}{dx}(\sinh x) = \cosh x. \qquad (12.12)$$

$$\frac{d}{dx}(\cosh x) = \sinh x. \qquad (12.13)$$

These results follow directly from the definitions.

(d) $y = sinh x$ is an odd function of x.

    (i) There are no stationary values of y.
    (ii) When $x = 0$, $y = 0$.
    (iii) As $x \to +\infty$, $y \to +\infty$.
        As $x \to -\infty$, $y \to -\infty$.

The graph of $y = \sinh x$ is shown in Fig. 103.

(e) $y = cosh x$ is an even function of x.

    (i) $y \geq 1$ for all values of x.
    (ii) y has a stationary value, which is a minimum value, at $(0, 1)$.
    (iii) As $x \to +\infty$, $y \to +\infty$.
        As $x \to -\infty$, $y \to +\infty$.

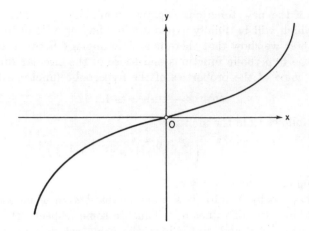

FIG. 103. The graph of $y = \sinh x$.

The graph of $y = \cosh x$ is shown in Fig. 104.

(f) $\dfrac{\mathrm{d}}{\mathrm{d}x} (\tanh x) = \operatorname{sech}^2 x.$ (12.14)

For if $y = \dfrac{\sinh x}{\cosh x}$,

$$\frac{\mathrm{d}y}{\mathrm{d}x} = \frac{\cosh^2 x - \sinh^2 x}{\cosh^2 x} = \frac{1}{\cosh^2 x} = \operatorname{sech}^2 x.$$

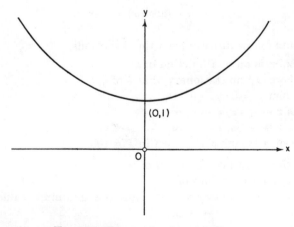

FIG. 104. The graph of $y = \cosh x$.

Similarly $\dfrac{d}{dx}(\operatorname{sech} x) = -\operatorname{sech} x \tanh x$, $\qquad$ (12.15)

$\dfrac{d}{dx}(\operatorname{cosech} x) = -\operatorname{cosech} x \coth x$, $\qquad$ (12.16)

$\dfrac{d}{dx}(\coth x) = -\operatorname{cosech}^2 x$. $\qquad$ (12.17)

(g) $\qquad \sinh(x+y) \equiv \sinh x \cosh y + \cosh x \sinh y$. $\qquad$ (12.18)

$\qquad\qquad \sinh(x-y) \equiv \sinh x \cosh y - \cosh x \sinh y$. $\qquad$ (12.19)

$\qquad\qquad \cosh(x+y) \equiv \cosh x \cosh y + \sinh x \sinh y$. $\qquad$ (12.20)

$\qquad\qquad \cosh(x-y) \equiv \cosh x \cosh y - \sinh x \sinh y$. $\qquad$ (12.21)

These results, which are in most respects analogous to the corresponding results for circular functions, differ in sign in the expanded form in (12.20) and (12.21).

The proof of (12.21) follows here: proofs for the other results follow similar lines.

$$
\begin{aligned}
\cosh(x-y) &\equiv \tfrac{1}{2}(e^{x-y} + e^{-x+y}) \\
&\equiv \tfrac{1}{2}(e^x e^{-y} + e^{-x} e^{y}) \\
&\equiv \tfrac{1}{2}\{(\sinh x + \cosh x)(\cosh y - \sinh y) \\
&\qquad + (\cosh x - \sinh x)(\sinh y + \cosh y)\} \\
&\equiv \cosh x \cosh y - \sinh x \sinh y.
\end{aligned}
$$

This kind of similarity persists for each of the formulae involving hyperbolic functions when compared with the corresponding formula for circular functions. The general rule for changing a formula involving an algebraic relationship concerning circular functions (for *all* values of the variable) into the corresponding formula involving hyperbolic functions is as follows:

Replace each circular function by the corresponding hyperbolic function and change the sign of every product (or implied product) of two sines.

Thus from $\sin x \sin y \equiv \tfrac{1}{2}\{\cos(x-y) - \cos(x+y)\}$ we have

$$\sinh x \sinh y \equiv \tfrac{1}{2}\{\cosh(x+y) - \cosh(x-y)\}.$$

Also from

$$\tan 2x \equiv \frac{2\tan x}{1 - \tan^2 x}$$

we have

$$\tanh 2x \equiv \frac{2\tanh x}{1 + \tanh^2 x},$$

because $\tan^2 x = \sin^2 x / \cos^2 x$ and so implies a product of two sines.

Proofs in particular cases of identities involving hyperbolic functions can be obtained by direct use of the definitions as illustrated in the proof of (12.21). The simpler results such as equations (12.18)–(12.21) can be used to prove the more complicated relations. An alternative method of proof will be given in Chapter XVI.

(h) $\int \sinh x \, dx = \cosh x + C.$ (12.22)

$\int \cosh x \, dx = \sinh x + C.$ (12.23)

Methods of integration, when the integrand involves hyperbolic functions, are similar, in most cases, to those used for integrands involving circular functions. Sometimes the substitution $e^x = u$ can be used to advantage as in example (v) below.

Examples. (i) Solve the equation $\tanh x + 4 \operatorname{sech} x = 4$.
The given equation can be written

$$\frac{e^x - e^{-x}}{e^x + e^{-x}} + \frac{8}{e^x + e^{-x}} = 4.$$

Multiplication by $e^x + e^{-x}$ ($\neq 0$) gives

$$3e^x - 8 + 5e^{-x} = 0.$$

Now multiplication throughout by $e^x (\neq 0)$ gives

$$3e^{2x} - 8e^x + 5 = 0,$$

i.e., $(3e^x - 5)(e^x - 1) = 0.$

$$\therefore 3e^x = 5 \quad \text{or} \quad e^x = 1.$$

i.e., $x = \log_e (5/3) \quad \text{or} \quad 0.$

(ii) Prove that $\dfrac{1 + \sinh x + \cosh x}{1 - \sinh x - \cosh x} \equiv -\coth\left(\dfrac{x}{2}\right)$.

The left hand side $= \dfrac{1 + e^x}{1 - e^x}$.

Dividing numerator and denominator by $e^{x/2}$ gives

left hand side $\equiv \dfrac{e^{-x/2} + e^{x/2}}{e^{-x/2} - e^{x/2}} \equiv -\coth\left(\dfrac{x}{2}\right)$.

(iii) Prove that $\cosh^2 x \cos^2 x - \sinh^2 x \sin^2 x \equiv \frac{1}{2}(1 + \cosh 2x \cos 2x)$.
From equations (12.9) and (12.20)

$$\cosh^2 x \equiv \tfrac{1}{2}(\cosh 2x + 1),$$

$$\sinh^2 x \equiv \tfrac{1}{2}(\cosh 2x - 1).$$

But $\qquad \cos^2 x \equiv \frac{1}{2}(1 + \cos 2x), \ \sin^2 x = \frac{1}{2}(1 - \cos 2x).$

$\therefore \cosh^2 x \cos^2 x - \sinh^2 x \sin^2 x$

$$\equiv \frac{1}{4}(\cosh 2x + 1)(1 + \cos 2x) - \frac{1}{4}(\cosh 2x - 1)(1 - \cos 2x)$$

$$\equiv \frac{1}{2}(1 + \cosh 2x \cos 2x).$$

(iv) Show that the area bounded by the x-axis, the curve $x^2/a^2 - y^2/b^2 = 1$ and the line OP where P is the point $(a \cosh t_1, b \sinh t_1)$ on the x-positive branch of the curve is $|\frac{1}{2} ab t_1|$.

P_2 is a point $(a \cosh t_2, b \sinh t_2)$ where $t_2 < t_1$, on the x-positive, y-positive part of the curve (Fig. 105), N is the foot of the ordinate at P, A is the vertex of the hyperbola and the ordinate at P_2 meets Ox at M and OP at R.

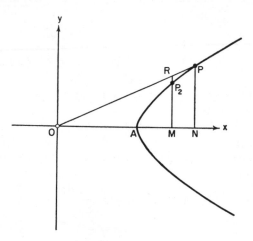

Fig. 105.

Then $P_2 M = b \sinh t_2 = b \cosh t_2 \tanh t_2$ and $RM = b \cosh t_2 \tanh t_1$.

But $\qquad\qquad \tanh t_1 > \tanh t_2$ for $t_1 > t_2 > 0$.

$\therefore P_2 M < RM$ for all positions of P_2 between A and P.

Hence the required area $OPA =$ area of triangle $OPN -$ area bounded by arc AP, PN, and the x-axis.

Area of $\Delta OPN = \frac{1}{2}ab\cosh t_1 \sinh t_1 = \frac{1}{4}ab\sinh 2t_1.$

$$\text{Area of } A\,PN = \int\limits_{a}^{a\cosh t_1} \frac{b}{a}\sqrt{(x^2 - a^2)}\,dx$$

Put $x = a\cosh t$; then $dx/dt = a\sinh t$.

When $x = a\cosh t_1$, $t = t_1$;   when $x = a$, $t = 0$.

$$\therefore \text{ area of } A\,PN = \int\limits_{0}^{t_1} \frac{b}{a}\,a\sinh t\,a\sinh t\,dt$$

$$= ab\int\limits_{0}^{t_1} \frac{1}{2}(\cosh 2t - 1)\,dt$$

$$= ab\left[\frac{1}{4}\sinh 2t - \frac{1}{2}t\right]_{0}^{t_1}$$

$$= \frac{1}{4}\,ab\sinh 2t_1 - \frac{1}{2}\,ab\,t_1.$$

$\therefore$ the required area $OPA = \dfrac{1}{4}\,ab\sinh 2t_1 - \left(\dfrac{1}{4}\,ab\sinh 2t_1 - \dfrac{1}{2}abt_1\right)$
$= \frac{1}{2}abt_1.$

A similar result is obtained for the y-negative part of this branch of the curve where $t_1 < 0$.

Hence the required area $= \frac{1}{2}\,|abt_1|$.

In the special case in which the curve is a rectangular hyperbola, [i.e., $a = b$], the area is $|\frac{1}{2}a^2t_1|$ which is analogous to the sectorial area $\frac{1}{2}a^2\theta$ for a circle.

(v) $$I = \int \operatorname{sech} x\,dx = \int \frac{2\,dx}{e^x + e^{-x}}\,.$$

Putting $e^x = u$, so that $du/dx = e^x$, gives

$$I = \int \frac{2\,du}{u^2 + 1} = 2\tan^{-1}u + C = 2\tan^{-1}(e^x) + C\,.$$

Exercises 12.4

Prove each of the following results for hyperbolic functions.

1. $\sinh 2x \equiv 2\sinh x\cosh x.$

2. $\cosh 2x \equiv \cosh^2 x + \sinh^2 x = 2\cosh^2 x - 1 = 1 + 2\sinh^2 x.$

3. $\tanh 2x \equiv 2\tanh x/(1 + \tanh^2 x).$

4. $\sinh 3x \equiv 3 \sinh x + 4 \sinh^3 x$.

5. $\cosh 3x \equiv 4 \cosh^3 x - 3 \cosh x$.

6. $\sinh x + \sinh y \equiv 2 \sinh \dfrac{x + y}{2} \cosh \dfrac{x - y}{2}$.

7. $\cosh x \cosh y \equiv \frac{1}{2} \{\cosh (x + y) + \cosh (x - y)\}$.

8. $\tanh^2 x + \operatorname{sech}^2 x \equiv 1$.

9. $\tanh x \equiv \sqrt{\left(\dfrac{\cosh 2x - 1}{\cosh 2x + 1} \right)}$ for $x > 0$.

10. $\cosh^2 (x + y) - \cosh^2 (x - y) \equiv \sinh 2x \sinh 2y$.

11. If $\sinh x = 5/12$, calculate $\cosh x$ and $\tanh x$.

12. If $\tanh x = \dfrac{3}{4}$, calculate $\operatorname{sech} x$ and $\sinh x$.

13. Solve the equation $\sinh x + 3 \cosh x = 4.5$.

14. Solve the equation $4 \tanh x = \coth x$.

15. Solve the equation $7 \operatorname{cosech} x + 2 \coth x = 6$.

16. Draw sketch graphs of
(i) $\tanh x$, (ii) $\operatorname{cosech} x$, (iii) $\operatorname{sech} x$.

17. Differentiate each of the following:
(i) $\cosh (2x + 5)$, (ii) $x \sinh x$, (iii) $\sinh 2x \cosh 3x$, (iv) $e^x \tanh x$, (v) $\cosh (\log x)$,
(vi) $(\operatorname{sech} x)/x$, (vii) $\sinh (\sin x)$, (viii) $\tan^{-1} (\sinh x)$, (ix) $\log \tanh 2x$,
(x) $\exp (\sinh x + \cosh x)$, (xi) $\operatorname{cosec}^{-1} (\cosh x)$, (xii) $\cosh^2 x - \sinh^2 x$.

18. Find the equation of the tangent and the equation of the normal to the curve $x = a \cosh t$, $y = b \sinh t$ at the point t_1 on the curve.

19. P is a point on the curve $y = c \cosh (x/c)$. [This curve is called the catenary.] PN is the ordinate at P and the tangent and normal at P meet the x-axis at T, G respectively. NZ is the perpendicular from N to PT and ZK is the perpendicular from Z to Ox. Find expressions in terms of the abscissa of P for PT, PG, TN and NG and prove that
(i) NZ is a constant length,
(ii) $ZK . PN = c^2$.

20. If $x = A \cosh nt + B \sinh nt$, where A, B, n are constants, show that

$$\frac{\mathrm{d}^2 x}{\mathrm{d} t^2} = n^2 x$$

and find A and B given that $\mathrm{d}x/\mathrm{d}t = 4n$ when $t = 0$, and $x = 2$ when $t = 0$.

21. With the usual notation for the motion of a particle moving in a straight line,

$$x = 2 \cosh 4t + \sinh 4t.$$

Find the numerical values of the velocity and the acceleration at time $t = 1$ each correct to 3 significant figures.

22. Find the minimum value of $13 \cosh x + 12 \sinh x$.

C

23. Find the stationary value of the function

$$y = (n^2 + 1)\cosh x + 2n \sinh x \text{ where } |n| \neq 1$$

and discuss the different cases $|n| > 1$ and $|n| < 1$. Sketch the curve in the cases
(i) $n = 2$, (ii) $n = -2$, (iii) $n = 1$, (iv) $n = -1$.

24. Obtain each of the following integrals:

(i) $\int \cosh 2x \, dx$,　(ii) $\int \tanh x \, dx$,　(iii) $\int \coth x \, dx$,　(iv) $\int \operatorname{sech}^2 x \, dx$,

(v) $\int \operatorname{sech} x \tanh x \, dx$,　(vi) $\int \operatorname{cosech} x \coth x \, dx$,　(vii) $\int \cosh^2 x \, dx$,

(viii) $\int \cosh 2x \sinh 4x \, dx$,　(ix) $\int x \sinh x \, dx$,　(x) $\int e^x \cosh x \, dx$.

25. Evaluate each of the following integrals:

(i) $\int_0^1 x \cosh x \, dx$,　(ii) $\int_0^1 \cosh^3 x \, dx$,　(iii) $\int_0^{1/2} \cosh 2x \cosh 4x \, dx$,

(iv) $\int_0^2 \operatorname{sech} x \, dx$,　(v) $\int_0^1 x \operatorname{sech}^2 x \, dx$.

26. Evaluate $\int_0^1 e^x \tanh x \, dx$ in terms of e.

27. Find the area enclosed by the curves $y = \sinh x$, $y = \cosh x$, the y-axis and the ordinate $x = 1$.

28. In the usual notation for a particle moving in a straight line the velocity of such a particle at time t is $5 \tanh 2t$. Calculate, correct to 3 significant figures,

(i) the distance moved by the particle in the first second,

(ii) the acceleration of the particle at time 1 sec.

29. Calculate the volume formed when the area bounded by $y = \tanh x$, the x-axis and the ordinates $x = -1$ and $x = 1$ makes a complete revolution about the x-axis.

12.5 Inverse hyperbolic functions

By analogy with the inverse circular functions, $y = \sinh^{-1} x$ implies $x = \sinh y$ or "y is the quantity whose hyperbolic sine is x".

Graphs of $\sinh^{-1} x$ [Fig. 106 (i)] and $\cosh^{-1} x$ [Fig. 106 (ii)] show that $\cosh^{-1} x$ is a two-valued function of x, but that $\sinh^{-1} x$ is a single-valued function of x. $\tanh^{-1} x$, $\coth^{-1} x$, $\operatorname{sech}^{-1} x$ and $\operatorname{cosech}^{-1} x$ are defined in the same way. $\operatorname{Sech}^{-1} x$ is a two-valued function and all the other are single-valued functions of x.

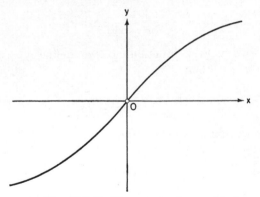

FIG. 106 (i). The graph of $y = \sinh^{-1}x$.

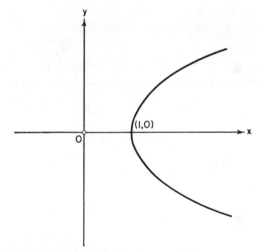

FIG. 106 (ii). The graph of $y = \cosh^{-1}x$.

12.6 Derivatives of the inverse hyperbolic functions

If $y = \sinh^{-1}x$ so that $x = \sinh y$, then

$$\frac{\mathrm{d}y}{\mathrm{d}x} = \frac{1}{\cosh y} = \frac{1}{\sqrt{(\sinh^2 y + 1)}} = \frac{1}{\sqrt{(x^2 + 1)}}$$

the positive sign being taken for the radical because $\cosh y$ is positive for all values of y.

$$\therefore \quad \frac{\mathrm{d}}{\mathrm{d}x}(\sinh^{-1}x) = \frac{1}{\sqrt{(x^2 + 1)}}. \tag{12.24}$$

Similarly $\quad \dfrac{d}{dx}(\text{Cosh}^{-1}x) = \dfrac{1}{\pm \sqrt{(x^2 - 1)}} \quad$ where

$\text{Cosh}^{-1}x$ is the two valued function. If we define $y = \cosh^{-1}x$ as $y = \text{Cosh}^{-1}x$ for $y \geqq 0$, then

$$\frac{d}{dx}(\cosh^{-1}x) = \frac{1}{\sqrt{(x^2 - 1)}}. \tag{12.25}$$

If $y = \tanh^{-1}x$ so that $x = \tanh y$, then

$$\frac{dy}{dx} = \frac{1}{\text{sech}^2 y} = \frac{1}{1 - \tanh^2 y} = \frac{1}{1 - x^2}.$$

$$\therefore \quad \frac{d}{dx}(\tanh^{-1}x) = \frac{1}{1 - x^2}. \tag{12.26}$$

It follows from eqns. (12.24), (12.25), (12.26) that, for $a > 0$

$$\frac{d}{dx}\sinh^{-1}\left(\frac{x}{a}\right) = \frac{1}{\sqrt{(x^2 + a^2)}}, \tag{12.27}$$

$$\frac{d}{dx}\cosh^{-1}\left(\frac{x}{a}\right) = \frac{1}{\sqrt{(x^2 - a^2)}}, \tag{12.28}$$

$$\frac{d}{dx}\tanh^{-1}\left(\frac{x}{a}\right) = \frac{a}{a^2 - x^2}. \tag{12.29}$$

Standard Integrals.

$$\int \frac{dx}{\sqrt{(x^2 + a^2)}} = \sinh^{-1}\left(\frac{x}{a}\right) + C. \tag{12.30}$$

$$\int \frac{dx}{\sqrt{(x^2 - a^2)}} = \cosh^{-1}\left(\frac{x}{a}\right) + C. \tag{12.31}$$

The integral $\displaystyle\int \frac{dx}{a^2 - x^2}$ has already been considered [Vol. I, eqn. (10.2)] and it is more conveniently expressed in the form $\dfrac{1}{2a}\log\left(\dfrac{a + x}{a - x}\right)$. Each of the other integrals (12.30) and (12.31) expressed here in terms of "inverse hyperbolic functions" can be expressed in terms of natural logarithms, i.e., in terms of *inverse exponential functions*. Since the hyperbolic functions are very simple combinations of exponential functions this is a result which might well be expected. We examine this in detail in the next section.

12.7 Logarithmic forms of the inverse hyperbolic functions

If $y = \sinh^{-1}x$ so that $x = \sinh y$, then

$$x = \tfrac{1}{2}(e^y - e^{-y}).$$

$$\therefore \ e^y - 2x - e^{-y} = 0.$$

$$\therefore \ e^{2y} - 2xe^y - 1 = 0.$$

$$\therefore \ e^y = \frac{2x + \sqrt{(4x^2 + 4)}}{2},$$

since

$$e^y > 0 \quad \text{and} \quad \sqrt{(4x^2 + 4)} > 2x.$$

$$\therefore e^y = x + \sqrt{(x^2 + 1)}.$$

$$\therefore \sinh^{-1}x = \log_e \{x + \sqrt{(x^2 + 1)}\}. \tag{12.32}$$

If $y = \cosh^{-1}x$ so that $x = \cosh y$, then

$$x = \tfrac{1}{2}(e^y + e^{-y}).$$

$$\therefore e^{2y} - 2xe^y + 1 = 0.$$

$$\therefore e^y = x \pm \sqrt{(x^2 - 1)}.$$

$$\therefore \ e^y = x + \sqrt{(x^2 - 1)} \quad \text{or} \quad e^y = x - \sqrt{(x^2 - 1)}$$

$$= \frac{\{x - \sqrt{(x^2 - 1)}\}\{x + \sqrt{(x^2 - 1)}\}}{x + \sqrt{(x^2 - 1)}}$$

$$= 1/\{x + \sqrt{(x^2 - 1)}\}.$$

$$\therefore y = \pm \log \{x + \sqrt{(x^2 - 1)}\}.$$

$$\therefore \cosh^{-1}x = \log_e \{x + \sqrt{(x^2 - 1)}\}. \tag{12.33}$$

The working above confirms the two-valued nature of the function $\cosh^{-1}x$ deduced from the graph in Fig. 106 (ii).

Examples. (i)
$$I = \int_0^a \sqrt{(x^2 + a^2)}\,\mathrm{d}x.$$

Put $x = a \sinh t$ so that $\mathrm{d}x/\mathrm{d}t = a \cosh t$.

When $x = a$, $\sinh t = 1$; when $x = 0$, $t = 0$.

Then
$$I = a^2 \int_0^{\sinh^{-1}1} \cosh^2 t \,\mathrm{d}t = a^2 \int_0^{\sinh^{-1}1} \tfrac{1}{2}(\cosh 2t + 1)\,\mathrm{d}t$$

$$= a^2 \left[\frac{1}{4}\sinh 2t + \frac{1}{2}t\right]_0^{\sinh^{-1}1} = a^2 \left[\frac{1}{2}\sinh t \cosh t + \frac{1}{2}t\right]_0^{\sinh^{-1}1}$$

When $\sinh t = 1$, $\cosh t = \sqrt{2}$ and $t = \log_e(1 + \sqrt{2})$.

$$\therefore I = \tfrac{1}{2}a^2\{\sqrt{2} + \log_e(1 + \sqrt{2})\}.$$

It is usual in numerical examples to use the logarithmic forms of the inverse functions, because these are the forms in which they are easily calculated with the help of tables. Most books of Mathematical tables contain tables of natural logarithms but not of inverse hyperbolic functions.

(ii) Solve the equation $3\sinh^2 x - 2\cosh x - 2 = 0$.

The given equation can be written

$$3(\cosh^2 x - 1) - 2\cosh x - 2 = 0,$$

i.e., $$3\cosh^2 x - 2\cosh x - 5 = 0$$

or $$(3\cosh x - 5)(\cosh x + 1) = 0.$$

$$\therefore \cosh x = 5/3 \text{ since } \cosh x \geqq 1.$$

$$\therefore x = \pm\log_e\left\{\frac{5}{3} + \sqrt{\left(\frac{25}{9} - 1\right)}\right\} = \pm\log_e 3.$$

12.8 Methods of integration

Integrals of the types

$$\int\frac{dx}{ax^2 + bx + c}, \quad \int\frac{dx}{\sqrt{(ax^2 + bx + c)}},$$

where a, b, c are constants, can be integrated with the help of one of the standard forms

$$\int\frac{dx}{x^2 + a^2} = \frac{1}{a}\tan^{-1}\left(\frac{x}{a}\right) + C, \tag{12.8}$$

$$\int\frac{dx}{x^2 - a^2} = \frac{1}{2a}\log_e\left(\frac{x - a}{x + a}\right) + C \quad \text{when} \quad |x| > a, \tag{10.1}$$

$$\int\frac{dx}{a^2 - x^2} = \frac{1}{2a}\log_e\left(\frac{a + x}{a - x}\right) + C \quad \text{when} \quad |x| < a, \tag{10.2}$$

$$\int\frac{dx}{\sqrt{(a^2 - x^2)}} = \sin^{-1}\left(\frac{x}{a}\right) + C, \tag{12.7}$$

$$\int\frac{dx}{\sqrt{(a^2 + x^2)}} = \sinh^{-1}\left(\frac{x}{a}\right) + C, \tag{12.30}$$

$$\int\frac{dx}{\sqrt{(x^2 - a^2)}} = \cosh^{-1}\left(\frac{x}{a}\right) + C. \tag{12.31}$$

Examples. (i) $\int \dfrac{\mathrm{d}x}{\sqrt{(3x^2 - 6x + 7)}} = \dfrac{1}{\sqrt{3}} \int \dfrac{\mathrm{d}x}{\sqrt{\left(x^2 - 2x + \dfrac{7}{3}\right)}}$

$$= \dfrac{1}{\sqrt{3}} \int \dfrac{\mathrm{d}x}{\sqrt{\left\{(x - 1)^2 + \dfrac{4}{3}\right\}}} = \dfrac{1}{\sqrt{3}} \sinh^{-1}\left\{\dfrac{(x - 1)\sqrt{3}}{2}\right\} + C.$$

(ii) $\int \dfrac{\mathrm{d}x}{5x^2 + 2x + 6} = \dfrac{1}{5} \int \dfrac{\mathrm{d}x}{\left(x + \dfrac{1}{5}\right)^2 + \dfrac{29}{25}} = \dfrac{1}{\sqrt{29}} \tan^{-1}\left(\dfrac{5x + 1}{\sqrt{29}}\right) + C.$

It is important to notice that integrals of the form $\int \dfrac{\mathrm{d}x}{a x^2 + b x + c}$ *where $a x^2 + b x + c$ can be resolved into rational factors* are best obtained by the method of partial fractions.

Exercises 12.8

1. Show that $\tanh^{-1}\left(\dfrac{x^2 - 1}{x^2 + 1}\right) = \log x$.

2. Solve the equation $\sinh^2 x - 4\cosh x + 5 = 0$.

3. Solve the equation $4\tanh^2 x - \operatorname{sech} x - 1 = 0$.

4. Solve the equations $\sinh x + \sinh y = 0.3$,
$\cosh x + \cosh y = 2.7$.

(Hint: Add and then subtract the equations.)

5. Express each of the following as logarithms

(i) $\tanh^{-1}x$, (ii) $\operatorname{Sech}^{-1}x$, (iii) $\operatorname{cosech}^{-1}x$, (iv) $\sinh^{-1}\left(\dfrac{1}{x}\right)$, (v) $\operatorname{Cosh}^{-1}(1 + x^2)$.

6. Differentiate with respect to x (i) $\coth^{-1}x$, (ii) $\operatorname{sech}^{-1}x$, (iii) $\operatorname{cosech}^{-1}x$.

7. Differentiate each of the following with respect to x:

(i) $x\cosh^{-1}x$, (ii) $\tanh^{-1}\left(\dfrac{x - a}{x + a}\right)$, (iii) $\sinh^{-1}(\tan x)$, (iv) $\cosh^{-1}\left(\dfrac{1}{x}\right)$, (v) $\cosh^{-1}(\sinh 2x)$.

8. Use the method of substitution to evaluate each of the following correct to 3 significant figures:

(i) $\int\limits_{0}^{3} \sqrt{(9 + x^2)}\,\mathrm{d}x$, (ii) $\int\limits_{4}^{8} \sqrt{(x^2 - 16)}\,\mathrm{d}x$.

9. Use the method of substitution to obtain the integrals

(i) $\int \dfrac{x^2\,\mathrm{d}x}{\sqrt{(1 + x^2)}}$, (ii) $\int \dfrac{\sqrt{(x^2 - 1)}\,\mathrm{d}x}{x}$.

10. Obtain each of the following integrals:

(i) $\int \sinh^{-1} x \, dx$,   (ii) $\int \cosh^{-1} dx$,   (iii) $\int \tanh^{-1} x \, dx$,   (iv) $\int x \cosh^{-1} x \, dx$.

11. Find the area enclosed by the curve $x^2/a^2 - y^2/b^2 = 1$ and the line $x = 2a$.

12. Evaluate the following integrals

(i) $\displaystyle \int_{2}^{3} \frac{dx}{\sqrt{(9 + x^2)}}$,   (ii) $\displaystyle \int_{2}^{4} \frac{dx}{\sqrt{(4x^2 - 1)}}$,   (iii) $\displaystyle \int_{1}^{2} \frac{dx}{\sqrt{(x^2 + 2x + 2)}}$,

(iv) $\displaystyle \int_{1}^{2} \frac{dx}{\sqrt{(2x^2 + 4x - 1)}}$,   (v) $\displaystyle \int_{a}^{b} \frac{dx}{\sqrt{\{(x + a)^2 + b^2\}}}$.

13. Find the area between the x-axis, the lines $x = -a$, $x = +a$ and the curve $y\sqrt{(a^2 + x^2)} = a^2$.

14. Prove that, for the curve $y = \log \cosh x$,

$$x = \tfrac{1}{2} \log_e \left\{ \left(1 + \frac{dy}{dx} \right) \middle/ \left(1 - \frac{dy}{dx} \right) \right\}.$$

Obtain each of the following integrals:

15. $\displaystyle \int \frac{dx}{\sqrt{(x^2 - 2x + 3)}}$,   16. $\displaystyle \int \frac{dx}{\sqrt{(1 - 4x - x^2)}}$,   17. $\displaystyle \int \frac{dx}{\sqrt{(x^2 - 2x - 5)}}$,

18. $\displaystyle \int \frac{dx}{2x^2 + 5x + 1}$,   19. $\displaystyle \int \frac{dx}{3 - x - x^2}$,   20. $\displaystyle \int \frac{dx}{x^2 + 2x + 5}$,

21. $\displaystyle \int \frac{dx}{\sqrt{(1 - 4x - 2x^2)}}$,   22. $\displaystyle \int \frac{dx}{\sqrt{(3x^2 - 6x - 5)}}$,

23. $\displaystyle \int \frac{dx}{5 - 4x - 2x^2}$,   24. $\displaystyle \int \frac{dx}{\sqrt{(1 + x + x^2)}}$.

25. Sketch the graph of $y = \tanh^{-1} x$.

12.9   The Integrals $\displaystyle \int \frac{p x + q}{a x^2 + b x + c} \, dx$ and $\displaystyle \int \frac{p x + q}{\sqrt{(a x^2 + b x + c)}} \, dx$.

The first of these integral types can be changed to a sum of two integrals, one of which is of the form $k_1 \displaystyle \int \frac{f'(x)}{f(x)} \, dx$ and the other of the form $k_2 \displaystyle \int \frac{dx}{a x^2 + b x + c}$ where k_1 and k_2 are constants. The second can be changed to a sum of two integrals one of which is of the form $k_3 \displaystyle \int \frac{f'(x)}{\sqrt{\{f(x)\}}} \, dx$ and the other of the form $k_4 \displaystyle \int \frac{dx}{\sqrt{(a x^2 + b x + c)}}$,

Examples. (i) $\displaystyle \int \frac{(3x+7)\,dx}{2x^2+4x+1} = \frac{3}{4}\int \frac{(4x+4)\,dx}{2x^2+4x+1} + \int \frac{4\,dx}{2x^2+4x+1}$

$\displaystyle = \frac{3}{4}\int \frac{(4x+4)\,dx}{2x^2+4x+1} + 2\int \frac{dx}{x^2+2x+\frac{1}{2}}$

$\displaystyle = \frac{3}{4}\int \frac{(4x+4)\,dx}{2x^2+4x+1} + 2\int \frac{dx}{(x+1)^2-\frac{1}{2}}$

$\displaystyle = \frac{3}{4}\log_e(2x^2+4x+1) + \sqrt{2}\log_e\left\{\frac{x+1-1/\sqrt{2}}{x+1+1/\sqrt{2}}\right\} + C$

$\displaystyle = \frac{3}{4}\log_e(2x^2+4x+1) + \sqrt{2}\log_e\left\{\frac{2x+2-\sqrt{2}}{2x+2+\sqrt{2}}\right\} + C.$

(ii) $\displaystyle \int \frac{(x+1)\,dx}{\sqrt{(3x^2-x-4)}} = \frac{1}{6}\int \frac{(6x-1)\,dx}{\sqrt{(3x^2-x-4)}} + \frac{7}{6}\int \frac{dx}{\sqrt{(3x^2-x-4)}}$

$\displaystyle = \frac{1}{6}\int \frac{(6x-1)\,dx}{\sqrt{(3x^2-x-3)}} + \frac{7}{6\sqrt{3}}\int \frac{dx}{\sqrt{\left(x^2-\frac{x}{3}-\frac{4}{3}\right)}}$

$\displaystyle = \frac{1}{6}\int \frac{(6x-1)\,dx}{\sqrt{(3x^2-x-4)}} + \frac{7}{6\sqrt{3}}\int \frac{dx}{\sqrt{\left\{\left(x-\frac{1}{6}\right)^2-\frac{47}{36}\right\}}}$

$\displaystyle = \frac{1}{3}\sqrt{(3x^2-x-4)} + \frac{7}{6\sqrt{3}}\cosh^{-1}\left(\frac{6x-1}{\sqrt{47}}\right) + C.$

Exercises 12.9

Obtain each of the following integrals

1. $\displaystyle \int \frac{(x+2)\,dx}{x^2+5x+1}$,

2. $\displaystyle \int \frac{(2x+3)\,dx}{2x^2+x+3}$,

3. $\displaystyle \int \frac{(3x+2)\,dx}{\sqrt{(x^2+x+1)}}$,

4. $\displaystyle \int \frac{5x\,dx}{\sqrt{(x^2-3x-1)}}$,

5. $\displaystyle \int \frac{(3x+4)\,dx}{2x^2+x+5}$,

6. $\displaystyle \int \frac{x\,dx}{4x^2+2x-3}$,

7. $\displaystyle \int \frac{(x+1)\,dx}{\sqrt{(1-3x-x^2)}}$,

8. $\displaystyle \int \frac{(5x+2)\,dx}{1+4x-x^2}$,

9. $\displaystyle \int \frac{(x+2)\,dx}{\sqrt{(x^2+6x+4)}}$,

10. $\displaystyle \int \frac{(3x^2+5x+6)\,dx}{x^2+x+1}$.

12.10 Summary of standard integrals and methods of integration so far considered

References are to the numbers of the *sections* of Volume I and Volume II in which the integrals first appear or in which the appropriate method of integration was first considered. Constants of integration have been omitted in the following list; in all cases where the logarithm of a function occurs it is implied that reference is to the positive values of the function only; it is also implied that reference is made only to those values of x for which the integrand exists.

1. $\displaystyle\int x^n \, dx = \frac{x^{n+1}}{n+1}, \qquad (n \neq -1).$ (§ 3.1)

2. $\displaystyle\int \frac{1}{x} \, dx = \log x.$ (§ 9.7)

3. $\displaystyle\int \sin x \, dx = -\cos x.$ (§ 5.2)

4. $\displaystyle\int \cos x \, dx = \sin x.$ (§ 5.2)

5. $\displaystyle\int \sec^2 x \, dx = \tan x.$ (§ 5.3)

6. $\displaystyle\int \operatorname{cosec}^2 x \, dx = -\cot x.$ (§ 5.3)

7. $\displaystyle\int \tan x \, dx = -\log \cos x = \log \sec x.$ (§ 9.7)

8. $\displaystyle\int \cot x \, dx = \log \sin x.$ (§ 9.7)

9. $\displaystyle\int \sec x \, dx = \log(\sec x + \tan x) = \log \tan\left(\frac{\pi}{4} + \frac{x}{2}\right).$ (§ 10.5)

10. $\displaystyle\int \operatorname{cosec} x \, dx = -\log(\operatorname{cosec} x + \cot x) = \log \tan\left(\frac{x}{2}\right).$ (§ 10.5)

11. $\int \sin^2 x \, dx$ and $\int \cos^2 x \, dx$ are obtained by using the transformations $\sin^2 x = \frac{1}{2}(1 - \cos 2x)$ and $\cos^2 x = \frac{1}{2}(1 + \cos 2x)$ respectively.

12. Integrals of the type $\int \sin mx \cos nx \, dx$ are obtained by using the product-sum transformations (§ 4.11).

13. $\displaystyle\int \frac{dx}{\sqrt{(a^2 - x^2)}} = \sin^{-1}\left(\frac{x}{a}\right), \qquad (a > 0).$ (§ 12.3)

14. $\displaystyle\int \frac{dx}{x^2 + a^2} = \frac{1}{a} \tan^{-1}\left(\frac{x}{a}\right).$ (§ 12.3)

15. $\int e^{kx}\,dx = \dfrac{1}{k}\,e^{kx}.$ ($\S 9.4$)

16. $\int \dfrac{dx}{ax+b} = \dfrac{1}{a}\,\log(ax+b).$ ($\S 9.7$)

17. $\int \dfrac{dx}{(ax+b)^n} = \dfrac{-1}{a(n-1)(ax+b)^{n-1}},\quad (n \neq 1).$ ($\S 10.4$)

18. $\int \dfrac{f'(x)}{f(x)}\,dx = \log f(x).$ ($\S 9.7$)

19. $\int \sinh x\,dx = \cosh x.$ ($\S 12.4$)

20. $\int \cosh x\,dx = \sinh x.$ ($\S 12.4$)

Integrals $21-24$ follow directly by integration as the reverse of differentiation.

21. $\int \tanh x\,dx = \log \cosh x.$

22. $\int \coth x\,dx = \log \sinh x.$

23. $\int \operatorname{sech}^2 x\,dx = \tanh x.$

24. $\int \operatorname{cosech}^2 x\,dx = -\coth x.$

25. $\int \dfrac{dx}{\sqrt{(x^2+a^2)}} = \sinh^{-1}\left(\dfrac{x}{a}\right) \quad \text{or} \quad \log\{x + \sqrt{(x^2+a^2)}\}$

($\S 12.6$ and $\S 12.7$)

Note $\sinh^{-1}\left(\dfrac{x}{a}\right) = \log\left\{\dfrac{x + \sqrt{(x^2+a^2)}}{a}\right\} = \log\{x + \sqrt{(x^2+a^2)}\} + \log\left(\dfrac{1}{a}\right)$

and the last term forms part of the constant of integration.

26. $\int \dfrac{dx}{\sqrt{(x^2-a^2)}} = \cosh^{-1}\left(\dfrac{x}{a}\right) \quad \text{or} \quad \log\{x + \sqrt{(x^2-a^2)}\}.$

($\S 12.6$ and $\S 12.7$)

27. $\int \dfrac{dx}{x^2-a^2} = \dfrac{1}{2a}\,\log\left(\dfrac{x-a}{x+a}\right).$ ($\S 10.4$)

28. $\int \dfrac{dx}{a^2-x^2} = \dfrac{1}{2a}\,\log\left(\dfrac{a+x}{a-x}\right).$ ($\S 10.4$)

Systematic Integration

(i) Whenever possible an integral is obtained directly from one of the standard forms quoted above.

(ii) If the integrand is a quotient, it should be examined to find out whether

(a) it can be integrated directly as a logarithm using (18), or

(b) the denominator factorizes so that the method of partial fractions (§ 10.4) can be used, or,

(c) it is one of the forms discussed in § 12.9 of this chapter.

(iii) The possibility of simplifying the integration by the method of substitution (§ 10.5) should be considered.

(iv) The possibility of using the method of integration by parts (§ 10.6) should be considered.

The student will find that, with experience, he can proceed directly to the most favourable method of integration in particular cases, but in the early stages a systematic search for the method is desirable

Miscellaneous Exercises XII

1. Find a pair of real values (x, y) satisfying the equations
$$12(\cosh x - \cosh y) = 5,$$
$$12(\sinh x - \sinh y) = 7. \qquad \text{(N.)}$$

2. If x is real, prove that
$$1 - x^2 \leqq \frac{1}{1 + x^2} \leqq 1 - x^2 + x^4,$$
and hence show that, if $x > 0$,
$$x - \frac{1}{3} x^3 < \tan^{-1} x < x - \frac{1}{3} x^3 + \frac{1}{5} x^5.$$

A chord of length $2c$ divides a circle into two segments. The height of one of them is h. Prove that the length S of its arc is given by
$$S = 2 \left(\frac{c^2}{h} + h \right) \tan^{-1} \left(\frac{h}{c} \right).$$

Deduce that, if $a = h/c$,
$$1 + \frac{2 a^2}{3} - \frac{a^4}{3} < \frac{S}{2 c} < 1 + \frac{2 a^2}{3} - \frac{2 a^4}{15} + \frac{a^6}{5} . \qquad \text{(N.)}$$

3. Find the area A enclosed between the two curves $y = c \cosh(x/c)$, $y = c \sinh(x/c)$ and the ordinates $x = 0$, $x = h > 0$.

Find the volume V_1 swept out when the area A revolves about the axis of x. Show that the volume V_2 swept out when the area A revolves about the axis of y is given by $V_2 = 2\pi c^3 \left\{ 1 - \left(\dfrac{h}{c} + 1 \right) e^{-h/c} \right\}$. (N.)

4. If $y = \sqrt{(1 - x^2)} \sin^{-1}x$, find dy/dx. Express $\left(1 - \dfrac{dy}{dx} \right)\left(\dfrac{1}{x} - x \right)$ in terms of y. (N.)

5. Differentiate the functions

(i) $2x \tan^{-1}x - \log_e(1 + x^2)$,   (ii) $\sin^{-1}\sqrt{(x - 1)}$.

State the restrictions on the value of x for the second of these functions. (N.)

6. Prove that $\displaystyle\int_0^1 \dfrac{x^2 + 6}{(x^2 + 4)(x^2 + 9)}\, dx = \dfrac{\pi}{20}$. (N.)

7. If $y = 2/(\sinh x)$, prove that

$$1 + (dy/dx)^2 = \dfrac{1}{4}(y^2 + 2)^2.$$ (N.)

8. Find the real solutions of the simultaneous equations

$$\cosh x \cosh y = 2, \quad \sinh x \sinh y = 1.$$ (N.)

9. Prove that $\displaystyle\int_0^1 \dfrac{x\,dx}{1 + x^4} = \dfrac{\pi}{8}$. (N.)

10. Make in one diagram rough sketches of the curves

$$y = 12e^{-x} \quad\text{and}\quad y = 25\sinh x.$$

Determine the abscissa of their point of intersection. Show that the area of the closed region bounded by the curves and the y-axis is 2 square units. (N.)

11. Find the value of dy/dx in terms of t at a point of the curve whose parametric equations are

$$x = a(t - \tanh t), \quad y = a\,\mathrm{sech}\,t.$$

Prove that $a\dfrac{d^2y}{dx^2} = \dfrac{\cosh^3 t}{\sinh^4 t}$. (N.)

12. Prove that (i) $\displaystyle\int_1^2 \dfrac{dx}{\sqrt{(1 + 6x - 3x^2)}} = \dfrac{\pi}{3\sqrt{3}}$,

(ii) $\displaystyle\int_0^{\pi/4} \cos^3\theta\, d\theta = \dfrac{5\sqrt{2}}{12}$. (N.)

13. (a) Prove that $\sinh 3x = 3 \sinh x + 4 \sinh^3 x$.

If $\sinh 3x = 2$, prove that $\cosh x = \frac{1}{2}\sqrt{5}$.

(b) Prove that $\displaystyle\int_0^{2/3} \frac{dx}{\sqrt{(4x^2+1)}} = \frac{1}{2}\log_e 3$.   (N.)

14. If $6\cosh x + 2\sinh y = 5$ and $3\sinh x = \cosh y$, find the values of x and y as logarithms.   (N.)

15. Prove that

(i) $\displaystyle\int_0^{\pi/4} \frac{\cos x - \sin x}{\cos x + \sin x}\,dx = \frac{1}{2}\log_e 2$,

(ii) $\displaystyle\int_0^1 \frac{5x^2 - 1}{3x^4 + 10x^2 + 3}\,dx = 0$.   (N.)

16. Evaluate

(i) $\displaystyle\int \frac{d\theta}{2\sin\theta + \cos\theta}$,   (ii) $\displaystyle\int \frac{x^2\,dx}{(x-1)(x-2)}$,   (iii) $\displaystyle\int \cos x \cosh x\,dx$.   (L.)

17. Express $\dfrac{x^2+x+1}{x^2(x^2+1)}$ in partial fractions, and hence evaluate

$$\int_1^n \frac{x^2+x+1}{x^2(x^2+1)}\,dx, \qquad n > 1.$$

Find the limit of the integral as $n \to \infty$.   (L.)

18. Use Maclaurin's theorem to obtain the first four terms of the expansion in ascending powers of x which gives an approximation, when x is small, for the function $\log\{x + \sqrt{(1+x^2)}\}$ and show that up to and including the term in x^4 this expansion is the same as the corresponding expansion for $\sin x$.   (C.)

19. (i) Prove that $\tan^{-1}\left(\dfrac{4}{3}\right) + \tan^{-1}\left(\dfrac{12}{5}\right) = \pi - \sin^{-1}\left(\dfrac{56}{65}\right)$.

  (ii) Define $\tanh x$ and prove from your definition that
$$\tanh 2x = 2\tanh x/(1 + \tanh^2 x).$$

  (iii) Find the number of real roots of the equation

$$5\sinh x - 3\cosh x = 3. \tag{O.C.}$$

20. Defining $\sinh\theta$, $\cosh\theta$, $\tanh\theta$ by the formulae $\sinh\theta = \frac{1}{2}(e^\theta - e^{-\theta})$, $\cosh\theta = \frac{1}{2}(e^\theta + e^{-\theta})$, $\tanh\theta = \dfrac{\sinh\theta}{\cosh\theta}$, prove that, for any given θ, there is just one value of φ between $-\frac{1}{2}\pi$ and $\frac{1}{2}\pi$ such that

$$\sin\varphi = \tanh\theta.$$

Prove that
$$\tan\varphi = \sinh\theta, \qquad \sec\varphi = \cosh\theta. \tag{O.C.}$$

21. Evaluate

    (i) $\int \sqrt{(x^2 + 4x - 5)} \, dx,$ (ii) $\int x \cos^2 2x \, dx,$

    (iii) $\int \dfrac{x}{\sqrt{(5 - 2x + x^2)}} \, dx,$ (iv) $\int \dfrac{1}{x(1 - x)^2} \, dx.$

    (v) $\int\limits_0^\infty x^2 \, e^{-x} \, dx.$ (L.)

22. Evaluate the integral $\int\limits_0^1 \tan^{-1}x \, dx.$ (N.)

23. Show that the value of the determinant

$$\Delta = \begin{vmatrix} \cosh(3x + a) & \cosh(2x + a) & \cosh(x + a) \\ \sinh(3x + a) & \sinh(2x + a) & \sinh(x + a) \\ 1 & 1 & 1 \end{vmatrix}$$

is independent of a.

Find the coefficient of x^{2n+1} in the expansion of Δ in ascending powers of x. Show that Δe^{-2x} approaches the limit $\frac{1}{2}$ when $x \to +\infty$, and find the approximate percentage error made in putting $\Delta = \frac{1}{2} e^{2x}$ when x is large and positive.   (N.)

24. Evaluate   (i) $\int\limits_0^1 \dfrac{dy}{(1 + y)\sqrt{(1 + y)}},$

        (ii) $\int\limits_0^1 \dfrac{\tan^{-1}x \, dx}{(1 + x^2)^{3/2}}.$ (N.)

25. If $y = \sin(m \sin^{-1}x)$, show that

$$(1 - x^2)\frac{d^2 y}{dx^2} - x\frac{dy}{dx} + m^2 y = 0. \qquad \text{(N.)}$$

26. (a) Differentiate with respect to x, $\cos^{-1}\left(\dfrac{2 - x}{3 + x}\right).$

    (b) State the number of real roots of the equations

        (i) $\sin^{-1}x = 2x,$ (ii) $2 \cos^{-1}x = x.$

27. Prove that $2 \sinh^2 x = \cosh 2x - 1$ and use this relation to integrate $\sinh^4 x$ with respect to x.   (L.)

28. (i) Solve the equation

$$\sin^{-1}\left\{\frac{x}{\sqrt{(1 + x^2)}}\right\} - \sin^{-1}\left\{\frac{1}{\sqrt{(1 + x^2)}}\right\} = \sin^{-1}\left(\frac{1 + x}{1 + x^2}\right).$$

    (ii) If $\tan^{-1} x + \tan^{-1} y + \tan^{-1} z = \frac{1}{2}\pi$, find the value of $\Sigma yz.$ (L.)

29. If $\tan \frac{1}{2}x = \tan \alpha \tanh \beta$, prove that

$$\tan x = \frac{\sin 2\alpha \sinh 2\beta}{1 + \cos 2\alpha \cosh 2\beta},$$

and that

$$\sin x = \frac{\sin 2\alpha \sinh 2\beta}{\cos 2\alpha + \cosh 2\beta}. \tag{L.}$$

30. (a) Evaluate $\displaystyle\int_0^\pi \frac{1 + \cos x}{5 - 3\cos x}\, \mathrm{d}x$.

(b) Transform $\displaystyle\int_1^2 \frac{\log x}{1 + x^2}\, \mathrm{d}x$ by the substitution

$x = \dfrac{1}{t}$. Deduce the value of $\displaystyle\int_{1/2}^2 \frac{\log x}{1 + x^2}\, \mathrm{d}x$.   (N.)

31. Denoting by $\tan^{-1}x$ the principal value of the inverse tangent (i.e., a function whose values lie between $-\frac{1}{2}\pi$ and $\frac{1}{2}\pi$), prove that $\tan^{-1}x > x - \dfrac{1}{3}x^3$ for all values of $x > 0$.   (N.)

32. Find the integral $\displaystyle\int \frac{\mathrm{d}x}{(x - 3)\,\sqrt{(2x^2 - 12x + 19)}}$.   (N.)

33. Differentiate

$$\tan^{-1}\left(\frac{1}{1 - x}\right) - \tan^{-1}\left(\frac{1}{1 + x}\right), \quad \sin^{-1}\left\{\frac{2x}{\sqrt{(4 + x^4)}}\right\}.$$

Explain why the two answers are numerically equal.

If $y = A + B\sin^{-1}x + (\sin^{-1}x)^2$, where A and B are constants, verify that

$$(1 - x^2)\frac{\mathrm{d}^2 y}{\mathrm{d}x^2} - x\frac{\mathrm{d}y}{\mathrm{d}x} = 2. \tag{O. C.}$$

34. Prove that

$$\frac{\mathrm{d}}{\mathrm{d}x}\left\{\tan^{-1}\left(\frac{\sin x}{\cos 2x}\right)\right\} = \frac{3\cos x - \cos 3x}{2(1 - \sin x \sin 3x)}. \tag{O.C.}$$

35. If $y = \sqrt{\dfrac{1 - x^2}{1 + x^2}}$,   prove that $\dfrac{\mathrm{d}y}{\mathrm{d}x} = -\sqrt{\dfrac{1 - y^4}{1 - x^4}}$.

By differentiation, or otherwise, prove that

$$\int \frac{\mathrm{d}x}{\sqrt{(\cos 2x - \cos 4x)}} = -\frac{1}{\sqrt 6}\sinh^{-1}\sqrt{\frac{\sin 3x}{\sin^3 x}}. \tag{O.C.}$$

36. Evaluate the integrals

(i) $\displaystyle\int \frac{\mathrm{d}x}{\sqrt{(x^2 - 4x + 3)}}$;   (ii) $\displaystyle\int \frac{\mathrm{d}x}{x(x^2 - 4x + 3)}$;   (iii) $\displaystyle\int \sin^4 x\, \mathrm{d}x$.

By using the substitution $x + 1 = 1/y$, or otherwise, prove that

$$\int_1^3 \frac{dx}{(x+1)\sqrt{(4x-3-x^2)}} = \frac{\pi}{2\sqrt{2}}.$$ (O.C.)

37. Evaluate the integrals

(i) $\int \frac{x\,dx}{(x+1)^2\,(x^2+4)}$;   (ii) $\int_0^\pi \frac{dx}{4-3\cos x}$;   (iii) $\int_1^2 \frac{dx}{x\sqrt{(x^2+4)}}$.   (L.)

38. Evaluate the integrals

(i) $\int \left(\frac{1+x}{1-x}\right)^{\frac{1}{2}} dx$,   (ii) $\int x\tan^{-1} x\,dx$.   (L.)

39. Evaluate

(i) $\int_0^1 \frac{dx}{(x+1)\,(x^2+2x+2)}$,   (ii) $\int_0^{\pi/4} x^2 \sin 2x\,dx$.

Find $\int \frac{\sin\theta}{1+\sin\theta}\,d\theta$.   (L.)

40. Evaluate (i) $\int \frac{(2x^3+7)\,dx}{(2x+1)\,(x^2+2)}$,   (ii) $\int \frac{dx}{x\sqrt{(1+2x-x^2)}}$.

Show that $\int_0^{\pi/2} \frac{d\theta}{3+5\cos\theta} = \frac{1}{4}\log 3$.   (L.)

41. Evaluate the integrals

(i) $\int e^x \cos x\,dx$,   (ii) $\int \frac{(x+2)\,dx}{\sqrt{(x^2+2x-3)}}$.   (L.)

42. Integrate $\dfrac{1}{\sinh x + 2\cosh x}$.   (L.)

43. Find indefinite integrals of

(i) $\dfrac{x+1}{x\,(x^2+4)}$,   (ii) $\dfrac{x}{\sqrt{(4x-x^2)}}$,   (iii) $x\,(\log_e x)^2$.   (L.)

44. Show that

$$\int_0^{\pi/4} \frac{2\,dx}{3\sin 2x + 4\cos 2x} = \frac{1}{5}\log 6.$$ (L.)

45. Evaluate

(a) $\int_1^4 \frac{3x-1}{\sqrt{(x^2-2x+10)}}\,dx$,   (b) $\int \frac{e^{3x}}{1+e^{6x}}\,dx$.   (L.)

46. (i) Evaluate the indefinite integral $\int x^n \log x\, dx$,

(a) when $n \neq -1$, (b) when $n = -1$.
Hence or otherwise evaluate

$$\int_1^e \frac{(x+1)^3}{x^2} \log x\, dx.$$

(ii) Show that $\int \sqrt{(x^2 + a^2)}\, dx = \tfrac{1}{2} x \sqrt{(x^2 + a^2)} + \frac{a^2}{2} \sinh^{-1}(x/a)$ and evaluate

$$\int_1^4 \sqrt{(x^2 - 2x + 17)}\, dx.$$

(L.)

CHAPTER XIII

DEFINITE INTEGRALS
FURTHER APPLICATIONS OF INTEGRATION

13.1 Properties of definite integrals

In most cases $\int_a^b f(x)\,\mathrm{d}x$ is found by obtaining the indefinite integral, $\int f(x)\,\mathrm{d}x = \varphi(x) + C$ say, and then using the definition $\int_a^b f(x)\,\mathrm{d}x = \varphi(b) - \varphi(a)$. There are cases, however, in which, because the indefinite integral is difficult to obtain or even unobtainable in terms of functions we have so far defined, it is desirable, if possible, to find the value of the definite integral without obtaining the indefinite integral. The properties of definite integrals which follow are means to this end.

1. $$\int_a^b f(x)\,\mathrm{d}x = - \int_b^a f(x)\,\mathrm{d}x. \qquad (13.1)$$

For, if $\int f(x)\,\mathrm{d}x = \varphi(x) + C$, then

$$\int_b^a f(x)\,\mathrm{d}x = \varphi(a) - \varphi(b)$$

and $$\int_a^b f(x)\,\mathrm{d}x = \varphi(b) - \varphi(a).$$

2. $$\int_a^b f(x)\,\mathrm{d}x = \int_a^c f(x)\,\mathrm{d}x + \int_c^b f(x)\,\mathrm{d}x. \qquad (13.2)$$

For $\int_a^b f(x)\,\mathrm{d}x = \varphi(b) - \varphi(a)$

and $\int_a^c f(x)\,\mathrm{d}x + \int_c^b f(x)\,\mathrm{d}x = \varphi(c) - \varphi(a) + \varphi(b) - \varphi(c) = \varphi(b) - \varphi(a).$

3.
$$\int_a^b f(x)\,\mathrm{d}x = \int_a^b f(y)\,\mathrm{d}y.$$
(13.3)

For each integral is equal to $\varphi(b) - \varphi(a)$. This self-evident property of definite integrals is used frequently in the work which follows.

4.
$$\int_0^a f(x)\,\mathrm{d}x = -\int_0^{-a} f(-x)\,\mathrm{d}x.$$
(13.4)

For, if $I = \int_0^a f(x)\,\mathrm{d}x$, then putting $x = -y$, so that $\mathrm{d}x/\mathrm{d}y = -1$, $x = 0$ when $y = 0$, and $x = a$ when $y = -a$,

$$I = \int_0^{-a} f(-y)\,(-\,\mathrm{d}y) = -\int_0^{-a} f(-y)\,\mathrm{d}y = -\int_0^{-a} f(-x)\,\mathrm{d}x.$$

5.
$$\int_{-a}^a f(x)\,\mathrm{d}x = \int_0^a \{f(x) + f(-x)\}\,\mathrm{d}x.$$
(13.5)

For $\int_{-a}^a f(x)\,\mathrm{d}x = \int_{-a}^0 f(x)\,\mathrm{d}x + \int_0^a f(x)\,\mathrm{d}x$

$$= -\int_a^0 f(-x)\,\mathrm{d}x + \int_0^a f(x)\,\mathrm{d}x$$

$$= \int_0^a f(-x)\,\mathrm{d}x + \int_0^a f(x)\,\mathrm{d}x$$

$$= \int_0^a \{f(x) + f(-x)\}\,\mathrm{d}x.$$

Corollaries of eqn. (13.5) are

$\int_{-a}^a f(x)\,\mathrm{d}x = 2\int_0^a f(x)\,\mathrm{d}x$ when $f(x)$ is an even function of x,

$\int_{-a}^a f(x)\,\mathrm{d}x = 0$ when $f(x)$ is an odd function of x.

6.
$$\int_0^a f(x)\,\mathrm{d}x = \int_0^a f(a-x)\,\mathrm{d}x.$$
(13.6)

For, if $I = \int_0^a f(x)\, \mathrm{d}x$, then putting $x = a - y$ so that $\dfrac{\mathrm{d}x}{\mathrm{d}y} = -1$, $y = 0$ when $x = a$, and $y = a$ when $x = 0$,

$$I = \int_a^0 f(a-y)\,(-\mathrm{d}y) = -\int_0^a f(a-y)\,(-\mathrm{d}y) = \int_0^a f(a-y)\,\mathrm{d}y = \int_0^a f(a-x)\,\mathrm{d}x.$$

The relation between definite integrals and area, as defined in Volume I, Chapter III, is frequently helpful in the consideration of problems concerning definite integrals. Equation (13.2), for example, becomes self-evident when examined in the light of this relationship. It must be remembered that area above the x-axis as defined by the definite integral is positive and area below the x-axis is negative.

Examples. (i) Prove that $\displaystyle\int_{ra}^{(r+1)a} f(x)\, \mathrm{d}x = \int_0^a f(x + ra)\, \mathrm{d}x.$

If $f(x + a) = cf(x)$ where a and c are constants, prove that $f(x + ra) = c^r f(x)$ when r is an integer.

Deduce that, if n is a positive integer and c is not equal to unity,

$$(1 - c) \int_0^{na} f(x)\, \mathrm{d}x = (1 - c^n) \int_0^a f(x)\, \mathrm{d}x.$$

State the corresponding result when $c = 1$. (N.)

If $I = \displaystyle\int_{ra}^{(r+1)a} f(x)\, \mathrm{d}x$, then putting $y = x - ra$, so that $\mathrm{d}y/\mathrm{d}x = 1$, $x = ra$ when $y = 0$, and $x = ra + a$ when $y = a$,

$$I = \int_0^a f(y + ra)\, \mathrm{d}y = \int_0^a f(x + ra)\, \mathrm{d}x.$$

If $f(x + a) = cf(x)$ where a and c are constants,

$$f(x + ra) = cf\{x + (r - 1)\,a\}$$
$$= c^2 f\{x + (r - 2)\,a\}$$
$$= c^3 f\{x + (r - 3)\,a\}$$
$$\dots\dots\dots\dots\dots\dots$$
$$= c^{r-1} f\{x + [r - (r - 1)]\,a\}.$$
$$\therefore f(x + ra) = c^r f(x).$$

$$\int\limits_0^{na} f(x)\,\mathrm{d}x = \int\limits_0^{a} f(x)\,\mathrm{d}x + \int\limits_a^{2a} f(x)\,\mathrm{d}x + \cdots + \int\limits_{(r-1)a}^{ra} f(x)\,\mathrm{d}x + \cdots + \int\limits_{(n-1)a}^{na} f(x)\,\mathrm{d}x$$

$$= \int\limits_0^{a} f(x)\,\mathrm{d}x + \int\limits_0^{a} f(x+a)\,\mathrm{d}x + \cdots + \int\limits_0^{a} f\{x+(r-1)\,a\}\,\mathrm{d}x + \cdots +$$

$$+ \int\limits_0^{a} f\{x+(n-1)\,a\}\,\mathrm{d}x$$

$$= \int\limits_0^{a} f(x)\,\mathrm{d}x + \int\limits_0^{a} c f(x)\,\mathrm{d}x + \cdots + \int\limits_0^{a} c^{r-1} f(x)\,\mathrm{d}x + \cdots + \int\limits_0^{a} c^{n-1} f(x)\,\mathrm{d}x$$

$$= (1 + c + c^2 + \cdots c^{r-1} + \cdots c^{n-1}) \int\limits_0^{a} f(x)\,\mathrm{d}x.$$

Hence, if $c \neq 1$,

$$\int\limits_0^{na} f(x)\,\mathrm{d}x = \frac{1-c^n}{1-c} \int\limits_0^{a} f(x)\,\mathrm{d}x$$

and, if $c = 1$,

$$\int\limits_0^{na} f(x)\,\mathrm{d}x = n \int\limits_0^{a} f(x)\,\mathrm{d}x.$$

(ii) Evaluate $I = \int\limits_{-\pi/2}^{\pi/2} (2+x)^2 \sin 2x\,\mathrm{d}x.$ (N.)

$$I = \int\limits_0^{\pi/2} \{(2+x)^2 \sin 2x - (2-x)^2 \sin 2x\}\,\mathrm{d}x$$

$$= \int\limits_0^{\pi/2} 8x \sin 2x\,\mathrm{d}x$$

$$= 8\left[\frac{-x\cos 2x}{2} + \int \frac{\cos 2x}{2}\,\mathrm{d}x\right]_0^{\pi/2}$$

$$= 8\left[\frac{-x\cos 2x}{2} + \frac{\sin 2x}{4}\right]_0^{\pi/2} = 2\pi.$$

(iii) Show that $\int\limits_0^{a} (a x - x^2)^3 \{x^3 + (x-a)^3\}\,\mathrm{d}x = 0.$ (N.)

Use of eqn. (13.6) gives

$$\int_0^a x^3 (a-x)^3 \left\{ x^3 + (x-a)^3 \right\} \mathrm{d}x = \int_0^a (a-x)^3 x^3 \left\{ (a-x)^3 + (-x)^3 \right\} \mathrm{d}x$$

$$= -\int_0^a x^3 (a-x)^3 \left\{ (x-a)^3 + x^3 \right\} \mathrm{d}x.$$

$$\therefore \int_0^a x^3 (a-x)^3 \left\{ x^3 + (x-a)^3 \right\} \mathrm{d}x = 0.$$

(iv) Evaluate $\quad I = \int_0^1 \dfrac{\mathrm{d}x}{x + \sqrt{(1-x^2)}}.$

Putting $x = \sin\theta$ gives

$$I = \int_0^{\pi/2} \frac{\cos\theta\,\mathrm{d}\theta}{\sin\theta + \cos\theta}. \tag{1}$$

Putting $x = \cos\theta$ in the original integral gives

$$I = \int_{\pi/2}^0 \frac{-\sin\theta\,\mathrm{d}\theta}{\sin\theta + \cos\theta} = \int_0^{\pi/2} \frac{\sin\theta\,\mathrm{d}\theta}{\sin\theta + \cos\theta}. \tag{2}$$

Equation (2) can also be obtained by using Equation (13.6) with $a = \pi/2$. Addition of Equations (1) and (2) gives

$$2I = \int_0^{\pi/2} \frac{\sin\theta + \cos\theta}{\sin\theta + \cos\theta}\,\mathrm{d}\theta = \int_0^{\pi/2} \mathrm{d}\theta = \pi/2.$$

$$\therefore \int_0^1 \frac{\mathrm{d}x}{x + \sqrt{(1-x^2)}} = \frac{\pi}{4}.$$

Exercises 13.1

1. Prove that $\displaystyle\int_a^b f(x)\,\mathrm{d}x = \int_a^b f(a+b-x)\,\mathrm{d}x.$

2. Prove that $\displaystyle\int_a^b f(tx)\,\mathrm{d}x = \frac{1}{t}\int_{ta}^{tb} f(x)\,\mathrm{d}x$ where t is a constant.

3. Prove that $\int_0^a x^m(a-x)^n dx = \int_0^a x^n(a-x)^m dx$.

4. Prove that $\int_{-a}^a x^2 \sinh^3 x\, dx = 0$.

5. Prove that, if n is an even integer,

$$\int_{-\pi}^{\pi} x \sin^n x\, dx = 0,$$

and that, if n is an odd integer,

$$\int_{-\pi}^{\pi} x \sin^n x\, dx = \pi \int_0^{\pi} \sin^n x\, dx.$$

Hence evaluate $\int_{-\pi}^{\pi} x \sin^5 x\, dx$.

6. Prove that $\int_0^{\pi/2} (a \cos^2\theta + b \sin^2\theta) d\theta = \int_0^{\pi/2} (a \sin^2\theta + b \cos^2\theta) d\theta$ where a and b are constants.

7. Deduce from question 6 that $\int_0^{\pi/2} (a \cos^2\theta + b \sin^2\theta) d\theta = \frac{1}{4}(a+b)\pi$.

8. Prove that $\int_0^{\pi} xf(\sin x) dx = \frac{\pi}{2} \int_0^{\pi} f(\sin x) dx$ and evaluate $\int_0^{\pi} \frac{x \sin^3 x}{1+\cos^2 x} dx$. (N.)

9. Evaluate $\int_0^{\pi} (1 + 2\cos x)^3 dx$. (N.)

10. Prove that if n is even $\int_0^{\pi} \frac{\cos n\theta}{\cos \theta} d\theta = 0$. State and prove the corresponding result for $\int_0^{\pi} \frac{\sin n\theta}{\sin \theta} d\theta$.

11. Evaluate $\int_0^{\pi} \cos x \sin^2 2x\, dx$ without first obtaining the indefinite integral. (N.)

12. Evaluate each of the following:

$$\text{(i)} \int_0^3 x^2 \sqrt{(3-x)}\, dx, \quad \text{(ii)} \int_{-\pi/2}^{\pi/2} \frac{\sin^3 x}{1+\cos^2 x}\, dx,$$

$$\text{(iii)} \int_{-2}^2 x^2(2-x)^6\, dx, \quad \text{(iv)} \int_{-\pi/4}^{\pi/4} \sin x \sec^6 x\, dx.$$

13. Prove (i) if $f(x) = f(2a - x)$, then $\int_0^{2a} f(x)\,dx = 2\int_0^a f(x)\,dx$, (ii) if $f(x) = -f(2a-x)$, then $\int_0^{2a} f(x)\,dx = 0$.

14. Use the result of question 13 to evaluate $\int_0^\pi \sin^8 x \cos^3 x\,dx$.

15. Prove (i) $\int_0^{\pi/2} \sin^m x\,dx = \int_0^{\pi/2} \cos^m x\,dx$,

    (ii) $\int_0^{\pi/2} \sin^m x \cos^n x\,dx = \int_0^{\pi/2} \sin^n x \cos^m x\,dx$.

16. Find the value of $\int_{-\alpha}^\alpha \dfrac{\sin^5 x}{1 + \cos^2 x}\,dx$.

17. If $I = \int_0^\pi \dfrac{x \sin x\,dx}{1 + \cos^2 x}$, prove that $I = \int_0^\pi \dfrac{(\pi - x) \sin x}{1 + \cos^2 x}\,dx$

and hence deduce that $I = \pi^2/4$.   (L.)

13.2 Infinite integrals

Thus far, in order to obtain $\int_a^b f(x)\,dx$ we have made the assumptions

(i) that a and b are both finite,

(ii) that the function $f(x)$ is finite for all values of x in the range $a \leq x \leq b$.

We now define $\int_a^\infty f(x)\,dx$ as $\lim\limits_{t\to\infty} \int_a^t f(x)\,dx$. The existence of the infinite integral thus defined is dependent upon the existence of the limit. If this limit exists, the integral is said to converge.

Thus $\int_0^\infty \dfrac{dx}{1 + x}$ does not exist because $\lim\limits_{t\to\infty} \log(1 + t)$ is not finite. But

$$\int_0^\infty \frac{dx}{(1 + x)^{3/2}} = \lim_{t\to\infty} \int_0^t \frac{dx}{(1 + x)^{3/2}} = \lim_{t\to\infty} \int_1^{t+1} \frac{dy}{y^{3/2}}$$

$$= \lim_{t\to\infty} [-2y^{-\frac12}]_1^{t+1} = \lim_{t\to\infty} [-2(t + 1)^{-\frac12} + 2] = 2$$

and hence $\int_0^\infty \dfrac{dx}{(1 + x)^{3/2}}$ converges to the value 2.

Area. The graph of $f(x) = (1 + x)^{-3/2}$ is shown in Fig. 107 (i) and the graph of $f(x) = (1 + x)^{-1}$ is shown in Fig. 107 (ii). In each case the shaded area is equal to $\int_0^t f(x)\,\mathrm{d}x$ for the function concerned.

In the first case there is a limiting value to this area as $t \to \infty$ and $\int_0^\infty f(x)\,\mathrm{d}x$ is *defined* as the area of the unbounded portion of the plane between the curve, the x-axis and the line $x = 0$.

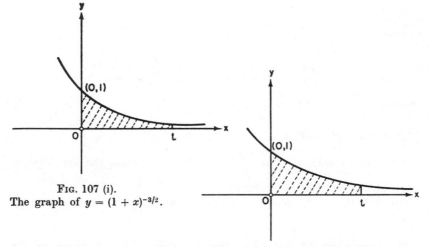

FIG. 107 (i).
The graph of $y = (1 + x)^{-3/2}$.

FIG. 107 (ii). The graph of $y = (1 + x)^{-1}$.

In the second case $\int_0^t f(x)\,\mathrm{d}x$ has no finite limit and the unbounded portion of the plane between the curve, the x-axis and the line $x = 0$ has no finite "area" as defined above.

Now consider $\int_a^b f(x)\,\mathrm{d}x$ where $f(x) \to \infty$ as $x \to b$. If $\varepsilon > 0$ and $\lim\limits_{\varepsilon \to 0} \int_a^{b-\varepsilon} f(x)\,\mathrm{d}x$ is finite and equal to L (say), then we *define* L as the value of $\int_a^b f(x)\,\mathrm{d}x$. Similarly if $\varphi(x) \to \infty$ as $x \to a$ and $\lim\limits_{\varepsilon \to 0} \int_{a+\varepsilon}^b \varphi(x)\,\mathrm{d}x = L'$, then L' is defined as the value of $\int_a^b \varphi(x)\,\mathrm{d}x$. If $f(x) \to \infty$ as $x \to c$ where $a < c < b$ and $f(x)$ is finite and continuous for all other values of x in the given range, then by definition

$$\int\limits_{a}^{b} f(x)\,\mathrm{d}x = \lim_{\varepsilon \to 0} \int\limits_{a}^{c-\varepsilon} f(x)\,\mathrm{d}x + \lim_{\varepsilon' \to 0} \int\limits_{c+\varepsilon'}^{b} f(x)\,\mathrm{d}x$$

if those limits exist, and where ε and ε' tend to zero independently.

These definitions are illustrated geometrically by the graphs of the functions $y = 1/\sqrt{(1 - x^2)}$ [Fig. 108 (i)], $y = \tan x$ [Fig. 108 (ii)] and $y^3 = \dfrac{1}{x}$ [Fig. 108 (iii)].

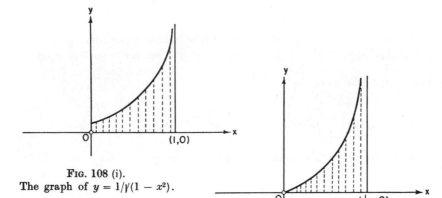

FIG. 108 (i).
The graph of $y = 1/\sqrt{(1 - x^2)}$.

FIG. 108 (ii). The graph of $y = \tan x$.

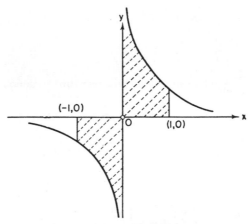

FIG. 108 (iii). The graph of $y^3 = 1/x$.

(i) The unbounded portion of the plane between the line $x = 1$, the x-axis and the curve $y = 1/\sqrt{(1 - x^2)}$ has an "area" defined as $\lim\limits_{\varepsilon \to 0} \int_0^{1-\varepsilon} \dfrac{\mathrm{d}x}{\sqrt{(1 - x^2)}}$. This limit exists and is equal to

$$\lim_{\varepsilon \to 0} \left[\sin^{-1}(1 - \varepsilon) - \sin^{-1}0 \right] = \tfrac{1}{2}\pi.$$

(ii) The unbounded portion of the plane between the curve $y = \tan x$, the x-axis and the line $x = \tfrac{1}{2}\pi$ has no finite "area" as thus defined because $\int \tan x \, \mathrm{d}x = \log \sec x$ and $\lim\limits_{\varepsilon \to 0} \log \sec(\tfrac{1}{2}\pi - \varepsilon)$ is not finite.

(iii) The unbounded portion of the plane between the curve $y^3 = 1/x$, the x-axis and the lines $x = -1$ and $x = +1$ has an "area" defined as

$$\lim_{\varepsilon \to 0} \int_{-1}^{0-\varepsilon} x^{-1/3} \, \mathrm{d}x + \lim_{\varepsilon' \to 0} \int_{0+\varepsilon'}^{1} x^{-1/3} \, \mathrm{d}x$$

$$= \lim_{\varepsilon \to 0} \left[\frac{3}{2} x^{2/3} \right]_{-1}^{0-\varepsilon} + \lim_{\varepsilon' \to 0} \left[\frac{3}{2} x^{2/3} \right]_{0+\varepsilon'}^{1} = -\frac{3}{2} + \frac{3}{2} = 0.$$

The area in the positive quadrant is $3/2$ sq. units.

Examples. (i) $\int_0^{\infty} \cosh x \, \mathrm{d}x$ does not exist because $\lim\limits_{t \to \infty} [\sinh t]$ is not finite.

(ii) $\int_0^{\infty} \dfrac{\mathrm{d}x}{1 + x^2} = \lim\limits_{t \to \infty} \left[\tan^{-1}t - \tan^{-1}0 \right] = \tfrac{1}{2}\pi.$

(iii) $\int_0^{a} \dfrac{\mathrm{d}x}{(x - a)^2}$ does not exist because $\lim\limits_{\varepsilon \to 0} \dfrac{1}{\varepsilon}$ is not finite.

(iv) $\int_0^{2} \dfrac{\mathrm{d}x}{(4x - x^2)^{\frac{1}{2}}} = \lim\limits_{\varepsilon \to 0} \left[\sin^{-1}\left(\dfrac{x - 2}{2} \right) \right]_{0+\varepsilon}^{2} = 0 - (-\tfrac{1}{2}\pi) = \tfrac{1}{2}\pi.$

(v) Evaluate $I = \int_0^{1} \sqrt{\left(\dfrac{1 + x}{1 - x} \right)} \, \mathrm{d}x.$

$$\int \sqrt{\left(\frac{1+x}{1-x}\right)} \, \mathrm{d}x = \int \frac{1+x}{\sqrt{(1-x^2)}} \, \mathrm{d}x$$

$$= -\tfrac{1}{2} \int \frac{-2x \, \mathrm{d}x}{\sqrt{(1-x^2)}} + \int \frac{\mathrm{d}x}{\sqrt{(1-x^2)}} = -\sqrt{(1-x^2)} + \sin^{-1} x.$$

$$\therefore I = \lim_{\varepsilon \to 0} \left[\sin^{-1} x - \sqrt{(1-x^2)}\right]_0^{1-\varepsilon}$$

$$= \tfrac{1}{2}\pi - (-1) = \tfrac{1}{2}\pi + 1.$$

(vi) Evaluate $I = \displaystyle\int_0^\infty \frac{\mathrm{d}x}{(1+x^2)^2}$.

We can use the method of substitution with such an infinite integral provided that the substitution is a valid one.

Let
$$I_1 = \int_0^t \frac{\mathrm{d}x}{(1+x^2)^2}.$$

Put $x = \tan\theta$. Then $\mathrm{d}x/\mathrm{d}\theta = \sec^2\theta$ and θ increases continuously from 0 to $\tan^{-1}t$ as x increases from 0 to t.

Then
$$I_1 = \int_0^{\tan^{-1}t} \cos^2\theta \, \mathrm{d}\theta = \left[\frac{1}{2}\theta + \frac{1}{4}\sin 2\theta\right]_0^{\tan^{-1}t}$$

$$\therefore I = \lim_{\theta \to \pi/2} \left[\frac{1}{2}\theta + \frac{1}{4}\sin 2\theta\right] = \frac{\pi}{4}.$$

(vii) Find the area bounded by the curve $y = x/(1+x^2)$, the co-ordinate axes and the line $x = t$. Find the volume of rotation of this curve about the x-axis. Show that, as $t \to \infty$, the function of t which represents the volume has a finite limit, but the function of t which represents the area has not a finite limit.

The area $= \displaystyle\int_0^t \frac{x \, \mathrm{d}x}{1+x^2} = \left[\tfrac{1}{2}\log(1+x^2)\right]_0^t = \tfrac{1}{2}\log(1+t^2)$,

and this has no finite limit as $t \to \infty$.

$$\text{Volume} = \pi \int_0^t \frac{x^2 \, \mathrm{d}x}{(1+x^2)^2} = \pi \int_0^{\tan^{-1}t} \frac{\tan^2\theta \sec^2\theta \, \mathrm{d}\theta}{\sec^4\theta}$$

$$= \pi \int_0^{\tan^{-1}t} \sin^2\theta \, \mathrm{d}\theta = \pi \left[\frac{\theta}{2} - \frac{\sin 2\theta}{4}\right]_0^{\tan^{-1}t} = \tfrac{1}{2}\pi\left[\tan^{-1}t - \frac{t}{1+t^2}\right].$$

Since $\lim\limits_{t\to\infty} \frac{1}{2}\pi\left[\tan^{-1}t - \dfrac{t}{1+t^2}\right] = \frac{1}{4}\pi^2$, as $t\to\infty$ the volume has the finite limit $\dfrac{1}{4}\pi^2$.

Exercises 13.2

Discuss the existence of each of the integrals in nos. $1-10$ and, where possible, evaluate the integral.

1. $\displaystyle\int_0^1 \frac{\mathrm{d}x}{x}$ 2. $\displaystyle\int_1^\infty \frac{\mathrm{d}x}{x}$ 3. $\displaystyle\int_1^\infty \frac{\mathrm{d}x}{x^2}$ 4. $\displaystyle\int_0^1 \frac{\mathrm{d}x}{x^2}$ 5. $\displaystyle\int_0^\infty \frac{\mathrm{d}x}{4+9x^2}$ 6. $\displaystyle\int_0^4 \sqrt{\left(\frac{x}{4-x}\right)}\,\mathrm{d}x$

7. $\displaystyle\int_1^4 \frac{\mathrm{d}x}{\sqrt{\{(x-1)(4-x)\}}}$.

8. $\displaystyle\int_0^1 x^2 \log x\,\mathrm{d}x$. (Prove and use the result $\lim\limits_{x\to 0+} x^n \log x = 0$ if $n > 0$.)

9. $\displaystyle\int_0^\infty x\,\mathrm{e}^{-x}\,\mathrm{d}x$.   10. $\displaystyle\int_0^\infty \frac{x\,\mathrm{d}x}{(x^2+4)(x^2+1)}$.

11. Evaluate $\displaystyle\int_{-2}^1 \sqrt{\left(\frac{2+x}{1-x}\right)}\,\mathrm{d}x$.   (L.)

12. Evaluate $\displaystyle\int_1^2 \frac{\mathrm{d}x}{x\sqrt{(x-1)}}$.   (L.)

13. Evaluate $\displaystyle\int_a^\infty \frac{\mathrm{d}x}{x^2\sqrt{(a^2+x^2)}}$.

14. Evaluate $\displaystyle\int_0^2 \frac{(x^2+2)\,\mathrm{d}x}{\sqrt{(4-x^2)}}$.   (L.)

15. Evaluate $\displaystyle\int_0^\infty \operatorname{sech} x\,\mathrm{d}x$.   (L.)

16. Show that $\displaystyle\int_2^\infty \frac{(x^2+3)\,\mathrm{d}x}{x^2(x^2+4)} = (\pi + 12)/32$.   (L.)

17. Evaluate $\int_0^\infty e^{-ax} \sin x\, dx \quad (a > 0)$. (L.)

18. Prove that $\int_1^\infty \frac{(x^2 + 2)\, dx}{x^2 (x^2 + 1)} = 2 - \frac{1}{4}\pi$. (L.)

19. Show that $\int_0^1 \frac{dx}{(1 + x)\sqrt{(1 - x^2)}} = 1$. (L.)

20. Evaluate $\int_1^2 \left(\frac{x - 1}{2 - x}\right)^{\frac{1}{2}} dx$. (L.)

In the following questions use the definitions of area and volume given in § 13.2.

21. Sketch the curve $y^2 = \dfrac{x}{4 - x}$ and find the area between the curve and its asymptote. If V_b is the volume formed by the revolution about the x-axis of the area bounded by this curve and the line $x = b$, $(0 < b < 4)$, show that $\lim\limits_{b \to 4} V_b$ is not finite.

22. Find the area of the unbounded region of the coordinate plane between the curves $y = \cosh x$ and $y = \sinh x$ and in the first quadrant.

23. Find the volume of rotation about the x-axis of the area between the curve $y = \dfrac{1}{1 + x^2}$ and the x-axis.

24. Sketch the curve $y = 1/\{x \sqrt{(1 + x)}\}$. Find the unbounded area between the curve for values of $x \geqq 1$, the line $x = 1$ and the x-axis and show that the volume formed by the rotation of this area about the x-axis is $(1 - \log 2)$ cubic units.

25. Sketch the curve $y^2 = \dfrac{1}{(x - 2)(4 - x)}$.
Calculate the unbounded area between the curve and the lines $x = 2$, $x = 4$.

13.3 Reduction formulae

If an integrand involves a constant which is an integer (say n), it may be possible to express the integral in terms of a similar integral which involves $(n - 1)$ or some other integer which is less than n. The formula which expresses the integral in this way is called a *Reduction Formula*, and by a repeated application of the reduction formula the original integral may be reduced to a form in which it can be obtained directly.

Examples. (i) Obtain a reduction formula for $\int x^n e^x \, dx$. Hence evaluate $\int\limits_0^1 x^5 e^x \, dx$.

Let $$u_n = \int x^n e^x \, dx \, .$$

Then $$u_n = x^n e^x - n \int x^{n-1} e^x \, dx \, .$$

$$\therefore u_n = x^n e^x - n u_{n-1}$$

which is the required reduction formula.

$$\therefore u_5 = \int x^5 e^x \, dx = x^5 e^x - 5 u_4$$

$$= x^5 e^x - 5 x^4 e^x + 20 u_3$$

$$= x^5 e^x - 5 x^4 e^x + 20 x^3 e^x - 60 u_2$$

$$= x^5 e^x - 5 x^4 e^x + 20 x^3 e^x - 60 x^2 e^x + 120 u_1$$

$$= x^5 e^x - 5 x^4 e^x + 20 x^3 e^x - 60 x^2 e^x + 120 x e^x - 120 \int e^x \, dx$$

$$= x^5 e^x - 5 x^4 e^x + 20 x^3 e^x - 60 x^2 e^x + 120 x e^x - 120 e^x + C \, .$$

$$\therefore \int\limits_0^1 x^5 e^x \, dx = (1 - 5 + 20 - 60 + 120 - 120) \, e - (-120)$$

$$= 120 - 44 e \, .$$

(ii) Obtain a reduction formula for $\int x^n \sinh x \, dx$ and hence obtain $\int x^5 \sinh x \, dx$.

Let $\quad u_n = \int x^n \sinh x \, dx$.

Then $\quad u_n = x^n \cosh x - n \int x^{n-1} \cosh x \, dx$

$$= x^n \cosh x - n x^{n-1} \sinh x + n(n-1) \int x^{n-2} \sinh x \, dx.$$

$$\therefore u_n = x^n \cosh x - n x^{n-1} \sinh x + n(n-1) u_{n-2}.$$

$$\therefore \int x^5 \sinh x \, dx = x^5 \cosh x - 5 x^4 \sinh x + 20 x^3 \cosh x - 60 x^2 \sinh x$$

$$+ \, 120 x \cosh x - 120 \sinh x + C.$$

(iii) Evaluate $\int\limits_0^{\pi/2} \cos^8 x \, dx$.

Let $\quad u_n = \int\limits_0^{\pi/2} \cos^n x \, dx$ where $n > 1$.

Then $\qquad u_n = \displaystyle\int_{x=0}^{x=\pi/2} \cos^{n-1} x \, \mathrm{d}(\sin x)$

$$= \left[\cos^{n-1} x \sin x\right]_0^{\pi/2} + (n-1)\int_0^{\pi/2} \cos^{n-2} x \sin^2 x \, \mathrm{d}x$$

$$= 0 + (n-1)\int_0^{\pi/2} \cos^{n-2} x \,(1-\cos^2 x)\,\mathrm{d}x$$

$$= (n-1)\,u_{n-2} - (n-1)\,u_n.$$

$$\therefore n u_n = (n-1)\,u_{n-2}.$$

$$\therefore \int_0^{\pi/2} \cos^8 x \, \mathrm{d}x = \frac{7}{8}\cdot\frac{5}{6}\cdot\frac{3}{4}\cdot\frac{1}{2}\int_0^{\pi/2} \mathrm{d}x = \frac{35\pi}{256}.$$

(iv) $\displaystyle\int_0^{\pi/2} \sin^m x \cos^n x \, \mathrm{d}x$, where $m > 1$, $n > 1$.

Note that $\displaystyle\int_0^{\pi/2} \sin^m x \cos^n x \, \mathrm{d}x = \int_0^{\pi/2} \sin^n x \cos^m x \, \mathrm{d}x$. (Exercises 13.1 No. 15)

This integral can be obtained directly by substitution if either of m or n is odd. Thus $\displaystyle\int_0^{\pi/2} \sin^4 x \, \cos^5 x \, \mathrm{d}x = \int_0^1 u^4(1-u^2)^2 \, \mathrm{d}u$ [by the sub-

stitution $\sin x = u$] $= \displaystyle\int_0^1 (u^4 - 2u^6 + u^8) \, \mathrm{d}u = \left(\frac{1}{5} - \frac{2}{7} + \frac{1}{9}\right) = \frac{8}{315}.$

If $u_{m,n} = \displaystyle\int_0^{\pi/2} \sin^m x \cos^n x \, \mathrm{d}x$, then

$$u_{m,n} = \int_{x=0}^{x=\pi/2} \sin^m x \cos^{n-1} x \, \mathrm{d}(\sin x)$$

$$= [\sin^{m+1} x \cos^{n-1} x]_0^{\pi/2}$$

$$- \int_0^{\pi/2} \{m \sin^{m-1} x \cos^n x - (n-1)\sin^{m+1} x \cos^{n-2} x\}\sin x \, \mathrm{d}x$$

$$= 0 - m\,u_{m,n} + (n-1)\int_0^{\pi/2} \sin^{m+2} x \cos^{n-2} x \, \mathrm{d}x$$

$$= -m\,u_{m,n} + (n-1)\int_0^{\pi/2}(\sin^m x \cos^{n-2} x - \sin^m x \cos^n x)\,\mathrm{d}x$$

$$= -m\,u_{m,n} + (n-1)\,u_{m,n-2} - (n-1)\,u_{m,n}.$$

D

$$\therefore u_{m,n} = \frac{n-1}{m+n}\, u_{m,n-2}.$$

This reduction formula reduces the integral $u_{m,n}$ to a multiple of either

(1) $\int_0^{\pi/2} \sin^m x \, \mathrm{d}x$ or (2) $\int_0^{\pi/2} \sin^m x \cos x \, \mathrm{d}x$. Integral (1) can be evaluated

as in Example (iii) above and integral (2) $= \left[\dfrac{\sin^{m+1} x}{m+1}\right]_0^{\pi/2} = \dfrac{1}{m+1}$.

$$\text{Thus} \int_0^{\pi/2} \sin^6 x \cos^4 x \, \mathrm{d}x = \frac{3}{10}\cdot\frac{1}{8}\int_0^{\pi/2}\sin^6 x\, \mathrm{d}x = \frac{3}{10}\cdot\frac{1}{8}\cdot\frac{5}{6}\cdot\frac{3}{4}\cdot\frac{1}{2}\cdot\frac{\pi}{2} = \frac{3\pi}{512}.$$

Wallis' Formulae. The definite integrals $\int_0^{\pi/2} \sin^n \theta \, \mathrm{d}\theta$, $\int_0^{\pi/2} \cos^n \theta \, \mathrm{d}\theta$ and $\int_0^{\pi/2} \sin^m \theta \cos^n \theta \, \mathrm{d}\theta$ are of frequent occurrence and it is useful to remember the reduction formulae associated with them and to quote them where necessary.

$$\text{If } u_n = \int_0^{\pi/2} \sin^n \theta \, \mathrm{d}\theta, \ n > 1, \quad \text{then } u_n = \frac{n-1}{n}\, u_{n-2}. \tag{13.7}$$

$$\text{If } u_n = \int_0^{\pi/2} \cos^n \theta \, \mathrm{d}\theta, \ n > 1, \quad \text{then } u_n = \frac{n-1}{n}\, u_{n-2}. \tag{13.8}$$

$$\text{If } u_{m,n} = \int_0^{\pi/2} \sin^m \theta \cos^n \theta \, \mathrm{d}\theta, \ m > 1, \ n > 1,$$
$$\text{then } u_{m,n} = \frac{m-1}{m+n}\, u_{m-2,n}. \tag{13.9}$$

(v) If n is a constant, prove that

$$\frac{\mathrm{d}}{\mathrm{d}x}\frac{x}{(1+x^2)^n} = \frac{2n}{(1+x^2)^{n+1}} - \frac{2n-1}{(1+x^2)^n}.$$

Hence, or otherwise, obtain a reduction formula for

$$I_n = \int_0^1 \frac{\mathrm{d}x}{(1+x^2)^n}.$$

Prove that $I_3 = (3\pi + 8)/32$ and that $I_{7/2} = 43/60\sqrt{2}$. (L.)

$$\frac{d}{dx}\frac{x}{(1+x^2)^n} = \frac{(1+x^2)^n - 2nx^2(1+x^2)^{n-1}}{(1+x^2)^{2n}}$$

$$= \frac{1}{(1+x^2)^n} - \frac{2nx^2}{(1+x^2)^{n+1}}$$

$$= \frac{1}{(1+x^2)^n} - \frac{2n(1+x^2)}{(1+x^2)^{n+1}} + \frac{2n}{(1+x^2)^{n+1}}$$

$$= \frac{2n}{(1+x^2)^{n+1}} - \frac{2n-1}{(1+x^2)^n}$$

as required. Integrating this relation w.r. to x between 0 and 1, rearranging gives

$$2n I_{n+1} - (2n-1) I_n = \left[\frac{x}{(1+x^2)^n}\right]_0^1 = \frac{1}{2^n} \tag{1}$$

which is a reduction formula for I_{n+1}. Changing n into $n-1$ gives the reduction formula for I_n:

$$2(n-1) I_n - (2n-3) I_{n-1} = \frac{1}{2^{n-1}}.$$

Putting $n = 2$ in Eqn. (1) gives $4I_3 = 3I_2 + \dfrac{1}{4}$.

Putting $n = 1$ in Eqn. (1) gives $2I_2 = I_1 + \frac{1}{2}$.

$$\therefore 4I_3 = \frac{1}{4} + \frac{3}{2}\left(\frac{1}{2} + I_1\right) = 1 + \frac{3}{2}\int_0^1 \frac{dx}{1+x^2} = 1 + \frac{3\pi}{8}.$$

$$\therefore I_3 = (3\pi + 8)/32.$$

The recurrence formula is valid for non-integral values of n.

Putting $n = \dfrac{5}{2}$ in Eqn. (1) gives $5I_{7/2} = 4I_{5/2} + \dfrac{1}{4\sqrt{2}}$.

Putting $n = \dfrac{3}{2}$ in Eqn. (1) gives $3I_{5/2} = 2I_{3/2} + \dfrac{1}{2\sqrt{2}}$.

$$\therefore 5I_{7/2} = \frac{4}{3}\left(2I_{3/2} + \frac{1}{2\sqrt{2}}\right) + \frac{1}{4\sqrt{2}} = \frac{8}{3}I_{3/2} + \frac{11}{12\sqrt{2}}.$$

Putting $n = \frac{1}{2}$ in Eqn. (1) gives $I_{3/2} = 1/\sqrt{2}$.

$$\therefore 5I_{7/2} = 43/12\sqrt{2}, \quad \text{i.e.,} \quad I_{7/2} = 43/60\sqrt{2}.$$

Exercises 13.3

In each of the questions $1-6$ obtain a reduction formula and hence obtain the indefinite integral in the special case quoted.

1. $\int x^n e^{ax} dx;$ $\int x^4 e^{3x} dx.$

2. $\int x^n \sin x \, dx;$ $\int x^3 \sin x \, dx.$

3. $\int \tan^n x \, dx;$ $\int \tan^6 x \, dx.$

4. $\int x (\log x)^n dx;$ $\int x (\log x)^3 dx.$

5. $\int \cosh^n x \, dx;$ $\int \cosh^6 x \, dx.$

6. $\int \sec^n x \, dx;$ $\int \sec^5 x \, dx.$

Evaluate each of the integrals in questions $7-12$.

7. $\int_0^{\pi/2} \sin^8 x \, dx.$ 8. $\int_0^{\pi/2} \cos^7 x \, dx.$

9. $\int_0^{\pi/2} \sin^4 x \cos^3 x \, dx.$ 10. $\int_0^{\infty} x^4 e^{-2x} dx.$

11. $\int_0^a x^2 (a^2 - x^2)^{3/2} \, dx.$ 12. $\int_{-\infty}^{\infty} \frac{x^6}{(a^2 + x^2)^4} \, dx.$

13. Prove that $\int_0^{\pi/2} (2 \sin^5 x - 3 \sin^7 x) \cos^2 x \, dx = 0.$ (O.C.)

14. If $I_n = \int_0^1 x^{n+1/2} (1 - x)^{1/2} dx,$

prove that

$$2(n + 2) I_n = (2n + 1) I_{n-1}, \quad (n \geqq 1).$$

Hence or otherwise show that

$$\int_0^1 x^2 \sqrt{(x - x^2)} \, dx = 5\pi/128.$$ (L.)

15. If $I_n = \int_0^x \frac{t^n \, dt}{\sqrt{(1 + t^2)}}$, prove that, for $n > 1$,

$$n I_n + (n - 1) I_{n-2} = x^{n-1} \sqrt{(1 + x^2)}.$$ (L.)

16. If $I_n = \int_0^{\pi/3} \sin^n \theta \, d\theta$, prove that

$$I_n = \frac{n-1}{n} I_{n-2} - \frac{1}{2n} \left(\frac{\sqrt{3}}{2} \right)^{n-1}.$$

Show by putting $y = a(1 - \cos\theta)$ that

$$\int_0^{a/2} (2ay - y^2)^{5/2}\,\mathrm{d}y = \frac{a^6}{192}(20\pi - 27\sqrt{3}).$$ (L.)

17. If $I_n = \int \cot^n\theta\,\mathrm{d}\theta$ show that

$$(n - 1)[I_n + I_{n-2}] = -\cot^{n-1}\theta.$$

18. If $T_n = \int_0^1 x^n\sqrt{(1 - x^2)}\,\mathrm{d}x$, show that $(n + 2)T_n = (n - 1)T_{n-2}$, when $n \geq 0$, and evaluate T_6.

19. If $I_n = \int_0^\infty x^{2n}e^{-x^2}\,\mathrm{d}x\,(n > -\tfrac{1}{2})$ show that $I_{n+1} = (n + \tfrac{1}{2})I_n$. (L.)

20. If $I_n = \int_0^\infty e^{-x}\sin^n x\,\mathrm{d}x$, $J_n = \int_0^\infty e^{-x}\cos^n x\,\mathrm{d}x$, prove that, for $n > 1$,

$$(n^2 + 1)I_n = n(n - 1)I_{n-2},$$

and

$$(n^2 + 1)J_n = 1 + n(n - 1)J_{n-2}.$$ (L.)

Evaluate $(I_2 + J_2)$ independently and deduce the value of $(I_6 + J_6)$.

13.4 Approximate numerical integration

The use of the formula $\int_a^b y\,\mathrm{d}x$ [Vol. I, § 3.3] for the area under a curve requires a knowledge of the equation of the curve and also requires y to be an integrable function of x. If either of these conditions is not satisfied, an approximate value of the definite integral can be obtained by various methods, two of which are discussed below.

The Trapezium Rule. (Fig. 109.) The area A bounded by $y = f(x)$, the x-axis and the ordinates $x = a$, $x = b$ is divided into n strips of equal width h by ordinates $P_1N_1, P_2N_2, P_3N_3, \ldots, P_{n+1}N_{n+1}$ of lengths $y_1, y_2, y_3, \ldots, y_{n+1}$. Then A is approximately equal to

$$\sum_1^n \text{Area } (P_rN_rN_{r+1}P_{r+1}) = \sum_1^n \left(\frac{y_r + y_{r+1}}{2}\right)h = \tfrac{1}{2}h\left(y_1 + y_{n+1} + 2\sum_2^n y_r\right).$$

i.e., the area is approximately equal to *half the width of a strip* × (*the sum of the first and last ordinates plus twice the sum of the other ordinates*).

This method of evaluation involves the approximation of successive small parts of the curve to straight lines joining their extremities. In general (but not necessarily in every particular case), the result is improved by increasing the number of 'strips' in the range.

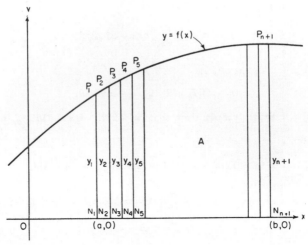

FIG. 109.

Example. Obtain an approximate value for $\int_{0}^{\pi} \sqrt{(\sin x)}\,\mathrm{d}x$ by using the trapezium rule.

A convenient number of strips is 12. The numerical work is best arranged systematically and one such arrangement is suggested here.

x	0	$\pi/12$	$\pi/6$	$\pi/4$	$\pi/3$	$5\pi/12$	$\pi/2$	
$\sin x$	0	0.2588	0.5000	0.7071	0.8660	0.9659	1	The region is symmetrical about $x = \frac{1}{2}\pi$
$\sqrt{(\sin x)}$	0	0.5087	0.7071	0.8409	0.9306	0.9828	1	
	y_1	y_2	y_3	y_4	y_5	y_6	y_7	

$$y_1 + y_{13} = 0.$$

$$2\sum_{2}^{12} y_r = 17.8804$$

$$
\begin{array}{ll}
y_2 & 0.5087 \\
y_3 & 0.7071 \\
y_4 & 0.8409 \\
y_5 & 0.9306 \\
y_6 & \underline{0.9828} \\
 & 3.9701 \\
\times 2 & 7.9402 \\
y_7 & \underline{1} \\
 & 8.9402
\end{array}
$$

$$\therefore \quad \frac{h}{2}(y_1 + y_{13} + 2\sum_{2}^{12} y_r) = 17.88 \times \frac{\pi}{24}$$

$$\doteqdot 2.34 .$$

$$\therefore \quad \int_0^{\pi} \sqrt{(\sin x)}\, \mathrm{d}x \doteqdot 2.34 .$$

N	L
17.88	1.2524
π	0.4971
	1.7495
24	1.3802
	0.3693

Simpson's Rule. If an approximation to $f(x)$ is obtained as a power series in x by the methods of Volume I § 9.2 in the form $f(x) \doteqdot A_0 + A_1 x + A_2 x^2 + \ldots + A_n x^n$ for small values of x, a better approximation to the area, for example, of the region $P_1 N_1 N_3 P_3$ would be obtained by approximating $f(x)$ to $A_0 + A_1 x + A_2 x^2$ (after a change of origin to ensure that the values of x concerned are small) than would be obtained by using the approximation $f(x) \doteqdot A_0 + A_1 x$ which is in fact what we have done in using the trapezium rule.

Figure 110 shows the points $P_1(-h, y_1)$, $P_2(0, y_2)$ and $P_3(h, y_3)$ on the curve $y = ax^2 + bx + c$. $P_1 N_1$, $P_2 O$ and $P_3 N_3$ are the ordinates at P_1, P_2, P_3. The area bounded by the curve, the x-axis, $P_1 N_1$ and $P_3 N_3$, is

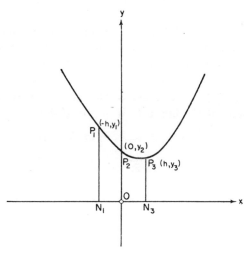

FIG. 110.

$$\int_{-h}^{h}(a\,x^2 + b\,x + c)\,\mathrm{d}\,x = \left[\frac{1}{3}\,a\,x^3 + \frac{1}{2}\,b\,x^2 + c\,x\right]_{-h}^{h}$$

$$= \frac{2}{3}\,a\,h^3 + 2\,c\,h = h\left(\frac{2\,a\,h^2}{3} + 2\,c\right).$$

But $y_1 = ah^2 - bh + c,$

$y_2 = c,$

$y_3 = ah^2 + bh + c.$

$\therefore \; y_1 + y_3 = 2ah^2 + 2y_2.$

$\therefore \;$ Area $P_1\,N_1\,N_3\,P_3 = \dfrac{1}{3}\,h\{(y_1 + y_3 - 2\,y_2) + 6\,y_2\} = \dfrac{1}{3}\,h\,(y_1 + y_3 + 4\,y_2).$

This result is independent of the abscissa of P_1, P_2, P_3 and is dependent only on the ordinates and the width of the strips. Hence (Fig. 109) if n is an *odd* number, i.e., if there is an even number of strips,

$$A \doteqdot \frac{1}{3}\,h\{(y_1 + 4y_2 + y_3) + (y_3 + 4y_4 + y_5) + \ldots + (y_{n-1} + 4y_n + y_{n+1})\}$$

$$\doteqdot \frac{1}{3}\,h\{(y_1 + y_{n+1} + 2(y_3 + y_5 + \ldots + y_{n-1}) + 4(y_2 + y_4 + \ldots + y_n)\}.$$

The area is approximately equal to *one third of the width of a strip* $\times$ *(the sum of the first and last ordinates plus twice the sum of the other odd ordinates plus four times the sum of the even ordinates)*. This is Simpson's Rule.

Precisely this result is obtained in the same way by approximating successive sections of the curve to a curve of the form

$$y = A_0 + A_1 x + A_2 x^2 + A_3 x^3 \text{ (Exercises 13.5 No. 9).}$$

In general the approximation is improved by increasing the number of strips into which the area is divided, and in general the approximation obtained by Simpson's Rule is better than that obtained through a corresponding number of strips with the trapezium rule, but neither of these statements is true for all cases. It is not possible by these methods to be sure of the degree of accuracy to which an approximation is correct, but in general, if an increase in the number of strips does not involve a change in the answer to a certain degree of accuracy, the answer is reliable to that degree of accuracy.

Examples. (i) Use Simpson's Rule to obtain an approximate value of $\int\limits_0^{\pi} \sqrt{(\sin x)}\,\mathrm{d}x$.

We use the table of values prepared for the evaluation of this integral by means of the trapezium rule.

(a) with six strips

Sum of the first and last ordinates	= 0
2 × sum of other odd ordinates	= 3.7224
4 × sum of even ordinates	= 9.6568
	13.3792

$$\therefore \int\limits_0^{\pi} \sqrt{(\sin x)}\,\mathrm{d}x \doteqdot \frac{\pi}{18} \times 13.38 \doteqdot 2.33\,(5).$$

(b) with twelve strips

Sum of first and last ordinates	= 0
2 × sum of other odd ordinates	= 8.5508
4 × sum of even ordinates	= 18.6592
	27.2100

$$\therefore \int\limits_0^{\pi} \sqrt{(\sin x)}\,\mathrm{d}x \doteqdot \frac{\pi}{36} \times 27.21 \doteqdot 2.37.$$

The comparisons between these two answers and between these answers and the one obtained with the trapezium rule are interesting, but *no general conclusion can be drawn from this particular case.*

(ii) $\int\limits_0^1 \dfrac{\mathrm{d}x}{1+x^2} = \left[\tan^{-1}x\right]_0^1 = \dfrac{1}{4}\pi.$

An approximate value of this integral (obtained by the use of Simpson's Rule for example) is therefore an approximate value of $\dfrac{1}{4}\pi$. Ten strips.

x	0	0.1	0.2	0.3	0.4
$1 + x^2$	1	1.01	1.04	1.09	1.16
$1/(1 + x^2)$	1	0.9901	0.9615	0.9174	0.8621
	y_1	y_2	y_3	y_4	y_5

x	0.5	0.6	0.7	0.8	0.9	1.0
$1 + x^2$	1.25	1.36	1.49	1.64	1.81	2
$1/(1 + x^2)$	0.8	0.7353	0.6711	0.6098	0.5525	0.5
	y_6	y_7	y_8	y_9	y_{10}	y_{11}

$$
\begin{array}{ccc}
 & \Sigma y_3 & \Sigma y_2 \\
y_1 + y_{11} = \ 1.5 & 0.9615 & 0.9901 \\
 & 0.8621 & 0.9174 \\
 & 0.7353 & 0.8000 \\
2\Sigma y_3 = \ 6.3374 & 0.6098 & 0.6711 \\
 & \overline{3.1687} & 0.5525 \\
4\Sigma y_2 = \underline{15.7244} & 6.3374 & \overline{3.9311} \\
\overline{23.562} & & 15.7244
\end{array}
$$

$$
\therefore \int_0^1 \frac{\mathrm{d}x}{1 + x^2} \doteqdot \frac{0.1}{3} \times 23.562 \doteqdot 0.7854 .
$$

$$
\therefore \pi \doteqdot 3.14 .
$$

Approximate evaluation of π by means of a power series

If $|x|$ is small, $\dfrac{1}{1 + x^2} \doteqdot 1 - x^2 + x^4 - x^6 + \ldots + (-1)^n x^{2n}$ and therefore if t is small

$$
\int_0^t \frac{\mathrm{d}x}{1 + x^2} = \tan^{-1} t \doteqdot t - \frac{t^3}{3} + \frac{t^5}{5} - \frac{t^7}{7} \cdots + \frac{(-1)^n t^{2n+1}}{(2n + 1)} .
$$

$$
\therefore \tan^{-1} \frac{1}{3} \doteqdot \frac{1}{3} - \frac{1}{81} + \frac{1}{1215} - \frac{1}{15309}
$$

$$
\doteqdot 0.33333 - 0.01235 + 0.00082 - 0.00006
$$

$$
\doteqdot 0.32174 .
$$

Also $\tan^{-1}\dfrac{1}{2} \doteqdot \dfrac{1}{2} - \dfrac{1}{24} + \dfrac{1}{160} - \dfrac{1}{896} + \dfrac{1}{4608}$

$$\doteqdot 0.50000 - 0.04167 + 0.00625 - 0.00112 + 0.00022$$

$$\doteqdot 0.46368.$$

$$\therefore \ \tan^{-1}\frac{1}{3} + \tan^{-1}\frac{1}{2} = \tan^{-1}\left(\frac{\dfrac{1}{3} + \dfrac{1}{2}}{1 - \dfrac{1}{6}}\right) = \tan^{-1}1 = \frac{1}{4}\pi$$

$$\doteqdot 0.32174 + 0.46368$$

$$\doteqdot 0.78542.$$

$$\therefore \ \pi \doteqdot 3.142.$$

This method of obtaining an approximation to the value of a definite integral can be used whenever it is possible to obtain an approximation to the integrand in the form of a power series for values of the variable within the range of integration.

Exercises 13.4

Obtain approximate numerical values for the definite integrals of Nos. 1−9 by the method indicated and with the number of strips indicated. Give answers to 3 significant figures and, in Nos. 1−4, test the accuracy of your answers by direct integration.

1. $\displaystyle\int_{1}^{2} \dfrac{dx}{x}$;   10 strips, Simpson's rule.

2. $\displaystyle\int_{1}^{2} \log_{10}x\, dx$;   10 strips, trapezium rule.

3. $\displaystyle\int_{0}^{\pi/3} \sec x\, dx$;   6 strips, Simpson's rule.

4. $\displaystyle\int_{0}^{4} \sqrt{(1 + x^2)}\, dx$;   8 strips, Simpson's rule.

5. $\displaystyle\int_{0}^{2} e^{-x^2}\, dx$;   4 strips, Simpson's rule.

6. $\displaystyle\int_{0}^{\pi/2} \dfrac{\sin x}{x}\, dx$;   6 strips, trapezium rule.

7. $\int\limits_{0}^{\pi/2} \log_e(1 + \cos x)\, \mathrm{d}x$;   6 strips, trapezium rule.

8. $\int\limits_{0}^{10} \dfrac{\mathrm{d}x}{1 + x^3}$;   10 strips, Simpson's rule.

9. $\int\limits_{0}^{\pi/2} \dfrac{\mathrm{d}\theta}{\sqrt{(3 - 2\sin^2\theta)}}$;   6 strips, Simpson's rule.

10. Evaluate

$$\frac{1}{\pi} \int\limits_{0}^{\pi/2} \sqrt{(4 + \sin^2\theta)}\, \mathrm{d}\theta,$$

correct to three decimal places, by expanding the integrand and integrating term by term.

Also obtain an approximate value using Simpson's rule with five ordinates. (L.)

11. Evaluate approximately $\int\limits_{0}^{1/2} \sqrt{(1 + x^3)}\,\mathrm{d}x$

   (i) by expanding the integrand in powers of x,

   (ii) using Simpson's rule with four intervals.

Give the answers correct to two places of decimals. (L.)

12. (i) By expanding the integrand in powers of x, evaluate

$$\int\limits_{0}^{1/2} \frac{1}{x} \log(1 + x)\, \mathrm{d}x$$

correct to three decimal places.

   (ii) Using Simpson's rule with four intervals, evaluate

$$\int\limits_{0}^{1} \sin\left(\frac{\pi}{2}\, x^2\right) \mathrm{d}x,$$

giving the result to three decimal places. (L.)

13. Evaluate

$$\int\limits_{0}^{0.4} \frac{\mathrm{d}x}{\sqrt{(1 + x^2)}},$$

working to three decimals, (i) using Simpson's rule with 5 ordinates, (ii) by expanding the integrand in powers of x. (The use of logarithms is recommended for the evaluation of the integrand.) (L.)

13.5 Mean values and root mean square

If $y = f(x)$ is a function of x which is continuous over the range $a \leq x \leq b$ (Fig. 111), the *mean value of y with respect to x* over this range is defined as

$$\frac{1}{b-a} \int_a^b f(x)\,\mathrm{d}x.$$

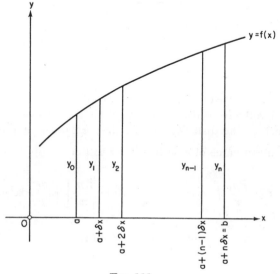

FIG. 111.

If $y = y_0$ when $x = a$, and $y = y_r$ when $x = a + r\delta x$ where $n\delta x = b - a$, the arithmetical average value of the ordinates $y_0, y_1, y_2, \ldots, y_{n-1}$ is

$$\frac{1}{n} \sum_0^{n-1} y_r = \sum_{r=0}^{r=n-1} \frac{y_r\,\delta x}{n\,\delta x} = \frac{1}{b-a} \sum_{r=0}^{r=n-1} y_r\,\delta x$$

and the limiting value of this arithmetical average as $\delta x \to 0$, i.e., as $n \to \infty$, is

$$\frac{1}{b-a} \int_a^b f(x)\,\mathrm{d}x.$$

The mean value as defined above is therefore identified with the limit of the arithmetical average of the ordinates as the interval between the ordinates tends to zero.

If y is expressed in terms of a different variable u so that $y = \varphi(u)$ $u = \alpha$ when $x = a$, $u = \beta$ when $x = b$ and $\varphi(u)$ is a continuous function of u over the range $\alpha \leq u \leq \beta$, the mean value of y with respect to u

is $\dfrac{1}{\beta - \alpha} \displaystyle\int_{\alpha}^{\beta} \varphi(u)\,du$ and this mean value is not necessarily equal to the

mean value of y with respect to x as defined above.

Examples. (i) With the usual notation for a particle moving in a straight line the equation of motion of a particle starting from the origin is

$$v = \omega \sin(pt + \alpha).$$

Calculate (a) the mean value of the velocity with respect to the time for the first $\pi/2p$ sec of the motion, (b) the mean value of the velocity with respect to the distance for the first $\pi/2p$ sec of the motion.

(a) The mean value of v with respect to t is

$$\frac{2p}{\pi} \int_{0}^{\pi/2p} v\,dt = \frac{2p}{\pi} \int_{0}^{\pi/2p} \omega \sin(pt + \alpha)\,dt = \frac{2p}{\pi}\left[-\frac{\omega}{p}\cos(pt+\alpha)\right]_{0}^{\pi/2p}$$

$$= \frac{2\omega}{\pi}(\cos\alpha + \sin\alpha).$$

(b) $s = \displaystyle\int_{0}^{t} v\,dt = \frac{\omega}{p}\{\cos\alpha - \cos(pt+\alpha)\};$ when $t = 0$, $s = 0$,

and when $t = \pi/2p$, $s = \dfrac{\omega}{p}\{\cos\alpha - \cos(\tfrac{1}{2}\pi + \alpha)\} = \dfrac{\omega}{p}(\cos\alpha + \sin\alpha)$

$$= s_1 (\text{say}).$$

Hence the mean value of v with respect to s over the interval $t = 0$ to $t = \pi/2p$ is

$$\frac{1}{s_1} \int_{t=0}^{t=\pi/2p} v\,ds = \frac{1}{s_1} \int_{0}^{\pi/2p} v\frac{ds}{dt}\,dt = \frac{1}{s_1} \int_{0}^{\pi/2p} v^2\,dt$$

$$= \frac{1}{s_1} \int_{0}^{\pi/2p} \omega^2 \sin^2(pt+\alpha)\,dt$$

$$= \frac{\omega^2}{2 s_1} \int_0^{\pi/2 p} \{1 - \cos 2 (p t + \alpha)\}\, \mathrm{d} t$$

$$= \frac{\omega^2}{2 s_1} \left[t - \frac{\sin 2 (p t + \alpha)}{2 p} \right]_0^{\pi/2 p}$$

$$= \frac{\omega^2}{2 s_1} \left(\frac{\pi}{2 p} - \frac{\sin (\pi + 2 \alpha)}{2 p} + \frac{\sin 2 \alpha}{2 p} \right)$$

$$= \frac{\omega^2}{2 s_1} \left(\frac{\pi}{2 p} + \frac{\sin 2 \alpha}{p} \right)$$

$$= \frac{\omega (\pi + 2 \sin 2 \alpha)}{4 (\cos \alpha + \sin \alpha)}.$$

(ii) Calculate the mean distance from a point P on the circumference of a circle of radius r of all the other points on the circumference.

If Q (Fig. 112) is a point on the circumference such that $\angle QOP = \theta$ $QP = 2r \sin \tfrac{1}{2}\theta$. Hence the mean distance of points on the circumference from P is

$$\frac{1}{2 \pi} \int_0^{2\pi} 2r \sin \tfrac{1}{2} \theta\, \mathrm{d} \theta = \frac{4r}{2 \pi} \left[-\cos \tfrac{1}{2} \theta \right]_0^{2\pi} = \frac{4r}{\pi} \doteqdot 1.27 r.$$

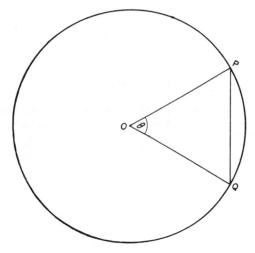

Fig. 112.

The *Root Mean Square* (R.M.S.) value of a function over a given interval is the square root of the mean value of its square over that interval. Thus the R.M.S. of $f(x)$ with respect to x for $a \leq x \leq b$ is

$$\sqrt{\left(\frac{1}{b-a}\int_a^b \{f(x)\}^2\, \mathrm{d}x\right)}.$$

The practical importance of the Root Mean Square lies in the necessity to measure the mean value of the square of such variables as alternating quantities in electricity where, for example, the heating effect of an electric current in a given time is proportional to the square of the current.

Example. Calculate the Root Mean Square value over a period of an electric current given by the formula $I = I_0 \sin(\omega t + \alpha)$.

The *period* of the alternation is $2\pi/\omega$ and it is over this interval of time that the R.M.S. value of a periodic function is expressed.

$$\text{R.M.S. value} = \sqrt{\left\{\frac{\omega}{2\pi}\int_0^{2\pi/\omega} I_0^2 \sin^2(\omega t + \alpha)\, \mathrm{d}t\right\}}$$

$$= \sqrt{\left\{\frac{\omega}{2\pi} I_0^2\left[\frac{t}{\cdot 2} - \frac{\sin(2\omega t + 2\alpha)}{4\omega}\right]_0^{2\pi/\omega}\right\}}$$

$$= \sqrt{\left\{\frac{\omega}{2\pi} I_0^2\left(\frac{\pi}{\omega} - \frac{\sin 2\alpha}{4\omega} + \frac{\sin 2\alpha}{4\omega}\right)\right\}}$$

$$= I_0/\sqrt{2}.$$

Exercises 13.5

In Questions 1–4 calculate the mean value of the function with respect to x over the stated range. Where necessary use Simpson's rule and give mean values to 2 significant figures.

1. $y = \tan x$; $x = 0$ to $x = \dfrac{1}{4}\pi$.

2. $y = \log_{10}\sec x$; $x = 0$ to $x = \dfrac{1}{3}\pi$.

3. $y = x \sinh x$; $x = 0$ to $x = 1$.

4. $y = \sqrt{(a^2 - x^2)}$; $x = 0$ to $x = a$.

5. In the usual notation the equation of motion of a particle moving in a straight line is $x = 4\sin t - 3\cos t$. Calculate (i) the mean value of the velocity with respect

to the time from $t = 0$ to $t = \frac{1}{2}\pi$, (ii) the mean value of the velocity with respect to the distance over this interval of time.

6. The density of a thin rod of length a at a distance x from one end is ϱx^2. Calculate the mean density of the rod.

7. A piece of wire of length $2\,l$ is bent to form a rectangle. Find the mean value of the areas of all the rectangles which can be formed in this way.

8. Calculate the R.M.S. value over a period (π) of an electric current i which is given by $i = 3 \sin\left(2t + \dfrac{1}{3}\pi\right) + 4 \sin\left(4t + \dfrac{1}{4}\pi\right)$.

9. If $f(x) = a + bx + cx^2 + dx^3$, where a, b, c, d are constants, prove that

$$\int\limits_{-h}^{h} f(x)\, \mathrm{d}x = \frac{1}{3}\, h\{f(-h) + 4f(0) + f(h)\},$$

and deduce Simpson's rule for approximate integration.

Find the mean value of $(2 + \sin\theta)^{-1/2}$ in the range $0 \le \theta \le \pi/2$, using Simpson's rule with six strips, giving your answer to three significant figures.

Check that your answer is reasonable from a rough sketch of the graph of the function in the range.   (L.)

13.6 Centre of mass

The position $(\bar{x},\ \bar{y})$ of the centre of mass of a system of coplanar particles of masses m_1, m_2, m_3, $\ldots$, m_n, whose positions in relation to rectangular axes Ox, Oy in the plane are (x_1, y_1), (x_2, y_2), (x_3, y_3), $\ldots$, (x_n, y_n) respectively, is given by

$$\bar{x} = \frac{\displaystyle\sum_{1}^{n} m_r x_r}{\displaystyle\sum_{1}^{n} m_r}, \qquad \bar{y} = \frac{\displaystyle\sum_{1}^{n} m_r y_r}{\displaystyle\sum_{1}^{n} m_r}.$$

The position of the centre of mass of a continuous body, regarded as an aggregate of particles, can be found by using the methods of integration. The limits of the sums involved in the above formulae are replaced by definite integrals.

Examples. (i) Find the position of the centre of mass of a uniform solid hemisphere of radius R.

(Fig. 113.) The centre of mass of the hemisphere will be on the axis of symmetry OP.

Let ϱ be the density of the hemisphere and consider a plane section of the hemisphere parallel to its base and distance x from it. The mass

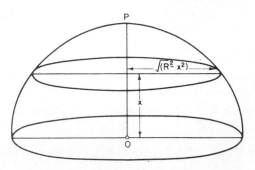

FIG. 113.

of a thin disc, of thickness δx, and with this section as base, is $\pi(R^2 - x^2)\delta x \varrho$ and therefore, if $\bar{x}$ is the distance of the centre of mass of the hemisphere from O,

$$\bar{x} = \frac{\lim\limits_{\delta x \to 0} \sum\limits_{x=0}^{x=R} \pi \varrho \, (R^2 - x^2) \, \delta x \cdot x}{\lim\limits_{\delta x \to 0} \sum\limits_{x=0}^{x=R} \pi \varrho \, (R^2 - x^2) \, \delta x}$$

$$= \frac{\int\limits_0^R \pi \varrho \, (R^2 - x^2) \, x \, \mathrm{d}x}{\int\limits_0^R \pi \varrho \, (R^2 - x^2) \, \mathrm{d}x}$$

$$= \left[\frac{R^2 x^2}{2} - \frac{x^4}{4} \right]_0^R \bigg/ \left[R^2 x - \frac{x^3}{3} \right]_0^R = 3R/8 .$$

The "*centroid of an area*" coincides with the centre of mass of a uniform lamina in the shape of the area, and the *centroid of a volume* of revolution similarly coincides with the centre of mass of a uniform solid in the shape of that volume.

(ii) Find the position of the centroid of the area bounded by the coordinate axes and the curve $y = \cos x$ from $x = 0$ to $x = \frac{1}{2}\pi$.

Consider a thin strip of the area parallel to the y-axis and bounded by the ordinates at (x, y) and $(x + \delta x, y + \delta y)$ on the curve (Fig. 114). The area of the strip is approximately $y \delta x$ and the coordinates of the

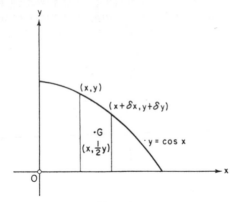

FIG. 114.

centroid, G, of the strip are approximately $(x, \frac{1}{2}y)$. Hence if $(\bar{x}, \bar{y})$ are the coordinates of the centroid of the area,

$$\bar{x} = \int_0^{\pi/2} xy\,\mathrm{d}x \Big/ \int_0^{\pi/2} y\,\mathrm{d}x$$

$$= \int_0^{\pi/2} x\cos x\,\mathrm{d}x \Big/ \int_0^{\pi/2} \cos x\,\mathrm{d}x$$

$$= \left[x\sin x + \cos x \right]_0^{\pi/2} \Big/ \left[\sin x \right]_0^{\pi/2}$$

$$= (\tfrac{1}{2}\pi - 1)/1 = \tfrac{1}{2}\pi - 1,$$

$$\bar{y} = \int_0^{\pi/2} \tfrac{1}{2}yy\,\mathrm{d}x \Big/ \int_0^{\pi/2} y\,\mathrm{d}x$$

$$= \tfrac{1}{2} \int_0^{\pi/2} \cos^2 x\,\mathrm{d}x$$

$$= \tfrac{1}{2}\left[\frac{x}{2} + \frac{\sin 2x}{4} \right]_0^{\pi/2} = \frac{1}{8}\pi.$$

Thus the centroid is at $\left\{ (\tfrac{1}{2}\pi - 1), \dfrac{1}{8}\pi \right\}$.

(iii) Find the position of the centroid of the volume of revolution formed when the x-positive half of the ellipse $\dfrac{x^2}{a^2} + \dfrac{y^2}{b^2} = 1$ makes a complete revolution about the x-axis.

The centroid will be on the x-axis. Consider the thin disc formed by the revolution about the axis of the strip of area bounded by the ordinates at (x, y) and $(x + \delta x, y + \delta y)$, the curve and the x-axis. The volume of the disc is approximately $\pi y^2 \delta x$.

$$\therefore \bar{x} = \int_0^a \pi x y^2 \, \mathrm{d}x \Big/ \int_0^a \pi y^2 \, \mathrm{d}x$$

$$= \int_0^a b^2 \left(x - \frac{x^3}{a^2} \right) \mathrm{d}x \Big/ \int_0^a b^2 \left(1 - \frac{x^2}{a^2} \right) \mathrm{d}x$$

$$= \left(\frac{a^2}{2} - \frac{a^2}{4} \right) \Big/ \left(a - \frac{a}{3} \right) = \frac{3a}{8}.$$

Thus the centroid of the volume is at $(3a/8, 0)$.

Exercises 13.6

Find the position of the centre of mass of each of the solids in questions $1-4$.

1. A uniform solid cone of height h.

2. A thin rod of length $2\,l$ such that the density at a point of the rod varies as the square of the distance from one end.

3. A wire of uniform density in the shape of a semicircle of radius r.

4. A square pyramid of height h.

In each of the questions $5-8$ find the coordinates of the centroid of the area whose boundaries are as described.

5. The parabola $y^2 = 4ax$ and its latus rectum.

6. The curve $y = \sec x$ and the line $y = 2$.

7. The curve $x = a \cos^3 \theta$, $y = a \sin^3 \theta$ and the coordinate axes in the first quadrant.

8. The curves $y^2 = 4x$, $x^2 = 4y$ in the first quadrant.

In each of the questions $9-12$ find the x-coordinate of the centroid of the volume of revolution about the x-axis of the area whose boundaries are as described.

9. The curve $y^2 = x^3$ and the double ordinate $x = 1$.

10. The curve $y = e^x$, the coordinate axes and the line $x = 2$.

11. The curve $y = \log x$, for $x \geqq 1$, the x-axis and the line $x = e$.

12. The curve $y^2 = x^2(a^2 - x^2)$ for $x \geqq 0$.

13.7 The theorem of Pappus concerning volumes

If a plane closed curve makes a complete revolution about a line in its own plane which does not cut the curve, the volume of the solid generated is equal to the area enclosed by the curve multiplied by the length of the path traced out by the centroid of that area.

Suppose that the only tangents to the curve which are parallel to the y-axis are $x = a$ and $x = b$ where $b > a$ (Fig. 115.). Consider a strip $A A' B' B$ of the area enclosed by the curve of width δx and parallel to the y-axis where A is the point (x, y_1) and B is (x, y_2). Then the volume of revolution of this strip about the x-axis is $\pi (y_2^2 - y_1^2)\delta x$. Hence the total volume of revolution about the axis of the area A enclosed by the curve is given by

$$V = \int_a^b \pi\, (y_2 - y_1)\, (y_2 + y_1)\, \mathrm{d}\, x\,.$$

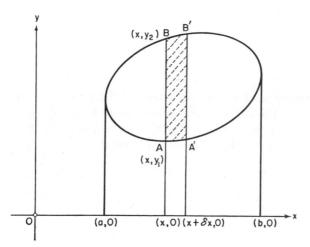

FIG. 115.

But $A = \int_a^b (y_2 - y_1)\mathrm{d}\, x$ and, if $\bar{y}$ is the y-coordinate of the centroid of the area,

$$\bar{y} = \dfrac{\displaystyle\int_a^b (y_2 - y)\, \tfrac{1}{2}\, (y_2 + y_1)\, \mathrm{d}\, x}{\displaystyle\int_a^b (y_2 - y_1)\, \mathrm{d}\, x}\,.$$

$$\therefore\ V = 2\, \pi\, \bar{y}\, A\,.$$

The proof can be extended to any closed area which can be divided into portions each of which has only two tangents perpendicular to the axis of revolution, and it is clearly independent of the position in the plane of the axis of revolution provided that this axis does not cut the area.

Examples. (i) Calculate the volume of the solid generated by the complete revolution of the triangle ABC, where A is $(2, 3)$, B is $(-2, 4)$ and C is $(3, 5)$, about the line $4x - 3y = 12$.

The area of $\triangle ABC$ is

$$\frac{1}{2}\begin{vmatrix} 2 & 3 & 1 \\ -2 & 4 & 1 \\ 3 & 5 & 1 \end{vmatrix} \text{ numerically} = 4\tfrac{1}{2} \text{ square units}.$$

The coordinates of the centroid of the triangle are

$$\left\{\frac{1}{3}(2 - 2 + 3), \frac{1}{3}(3 + 4 + 5)\right\}, \text{ i.e., } (1, 4).$$ The perpendicular distance of $(1, 4)$ from $4x - 3y - 12 = 0$ is

$$\left|\frac{4 - 12 - 12}{5}\right| = 4$$

and the triangle ABC lies completely on the same side of the given line as the origin. Hence the volume of revolution is $2\pi \times 4 \times 4\tfrac{1}{2} = 36\pi$ units^3.

(ii) The area bounded by the curve $y = \cos^2 x$ and the x-axis from $x = 0$ to $x = \tfrac{1}{2}\pi$ makes a complete revolution about the y-axis. Find the volume of revolution and deduce the value of $\int\limits_0^1 (\cos^{-1}\sqrt{x})^2 \, \mathrm{d}x$.

Let $(\bar{x}, \bar{y})$ be the centroid of the area under the curve. The area under the curve is

$$\int\limits_0^{\pi/2} \cos^2 x \, \mathrm{d}x = \left[\frac{1}{2}x + \frac{1}{4}\sin 2x\right]_0^{\pi/2} = \frac{1}{4}\pi.$$

$$\therefore \bar{x} = \frac{4}{\pi}\int\limits_0^{\pi/2} x \cos^2 x \, \mathrm{d}x$$

$$= \frac{4}{\pi}\int\limits_0^{\pi/2} x\left(\frac{1}{2} + \frac{\cos 2x}{2}\right) \mathrm{d}x$$

$$= \frac{4}{\pi}\left[\frac{x^2}{4} + \frac{x\sin 2x}{4} - \int\frac{\sin 2x}{4} \, \mathrm{d}x\right]_0^{\pi/2}$$

$$= \frac{4}{\pi}\left[\frac{x^2}{4} + \frac{x\sin 2x}{4} + \frac{\cos 2x}{8}\right]_0^{\pi/2}$$

$$= \frac{4}{\pi}\left(\frac{\pi^2 - 4}{16}\right) = \frac{\pi^2 - 4}{4\pi}.$$

Hence the volume of revolution about the y-axis is

$$2\pi \left(\frac{\pi^2 - 4}{4\pi} \right) \frac{\pi}{4} = \pi (\pi^2 - 4)/8.$$

$$\therefore \int_0^1 \pi x^2 \, dy = \int_0^1 \pi (\cos^{-1} \sqrt{y})^2 \, dy = \pi (\pi^2 - 4)/8.$$

$$\therefore \int_0^1 (\cos^{-1} \sqrt{x})^2 \, dx = (\pi^2 - 4)/8.$$

Exercises 13.7

1. Use the theorem of Pappus to find the position of the centre of mass of
   (i) a semicircular lamina,
   (ii) a lamina in the shape of a quadrant of a circle of radius r.

2. Calculate the volume of the solid formed when the area bounded by the parabola $y^2 = 4ax$ and the latus rectum makes a complete revolution about the directrix.

3. A square $ABCD$ of side $2a$ makes a complete revolution about a line AP in its plane, where $\angle ABP = \theta$ and AP does not cut the square again. Find the volume of the solid formed and find its greatest and least values as θ varies.

4. Calculate the volume of revolution of the ellipse

$$4x^2 + 9y^2 - 8x - 36y + 4 = 0$$

about the line $4x + 3y + 5 = 0$.

5. Calculate the volumes of revolution of the triangle bounded by the lines $x - y = 0$, $4x - y - 6 = 0$, $5x - 2y - 3 = 0$ about each coordinate axis.

6. Calculate the area and the coordinates of the centroid of the region bounded by $y = \sinh x$, the x-axis and the line $x = 1$. Hence calculate the volume of revolution of this area about the y-axis.

7. Use the theorem of Pappus to calculate the y-coordinate of the centroid of the area in the first quadrant bounded by the hyperbola $xy = c^2$ and the chord joining the points $(2c, \tfrac{1}{2}c)$, $(\tfrac{1}{2}c, 2c)$.

8. Use Simpson's rule and the theorem of Pappus to find an approximate value of the y-coordinate of the centroid of the area bounded by the x-axis, the extreme ordinates, and a curve joining the points

x	0	1	2	3	4
y	0	1.18	3.63	10.02	27.29

13.8 Moments of inertia

The moment of inertia of a system of particles of masses m_1, m_2, m_3, ..., m_n about a given straight line (the axis) is defined as $\sum_1^n m_r x_r^2$, where x_r is the distance of the particle of mass m_r from the axis. If $k^2 \sum_1^n m_r = \sum_1^n m_r x_r^2$, then k is called the *radius of gyration* of the system of particles about the axis.

The concept of moment of inertia is important in theoretical mechanics because the kinetic energy of a body rotating about a fixed axis is equal to $\frac{1}{2} I \omega^2$, where I is the moment of inertia about the axis and ω is the angular velocity of the body.

In the case of a continuous distribution of mass the summation becomes an integral. Examples (i) to (iv), which follow, involve the moments of inertia of bodies of standard shapes and it is useful to remember the results.

Examples. (i) *The moment of inertia of a thin uniform rod of length $2l$ and mass M about an axis through its centre perpendicular to its length.*

Let the line density of the rod be ϱ. (Fig. 116.) Then the M. of I. of a small increment δx of the rod, distance x from the axis, is approximately $\varrho \delta x \cdot x^2$. Hence the total M. of I. of the rod about the axis is

$$I = \int_{-l}^{l} \varrho \, x^2 \, \mathrm{d}x = \frac{2\varrho \, l^3}{3} = \frac{M \, l^2}{3}.$$

(ii) *The moment of inertia of a thin uniform ring of mass M and radius r about an axis through its centre perpendicular to its plane.*

Since every particle of the ring is at a distance r from the axis, $I = M r^2$.

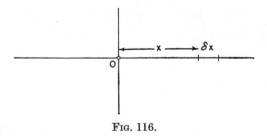

Fig. 116.

(iii) *The moment of inertia of a uniform disc of radius r and mass M about an axis through its centre and perpendicular to its plane.*

Let the surface density of the disc be ϱ. Consider a thin ring of the disc, concentric with the disc, of radius x and width δx (Fig. 117).

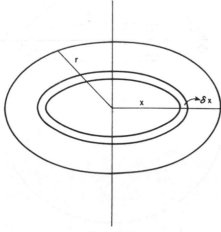

Fɪɢ. 117.

Then the M. of I. of this thin ring about the axis is approximately $2\pi\varrho x\delta x \cdot x^2$

$$\therefore I = \int_0^r 2\pi\varrho x \cdot x^2 \, \mathrm{d}x = \frac{\pi\varrho r^4}{2} = \frac{Mr^2}{2}.$$

(iv) *The moment of inertia of a uniform sphere of radius R and mass M about a diameter.*

Let the density of the sphere be ϱ and its mass M (Fig. 118). Consider a thin disc (of the sphere) whose plane faces are perpendicular to the axis at a distance x from the centre of the sphere. Let the thickness of the disc be δx. Then the mass of the disc is approximately $\pi(R^2 - x^2)\delta x\varrho$ and hence the M. of I. the disc about the axis is approximately $\pi(R^2 - x^2)\delta x\varrho \frac{1}{2}(R^2 - x^2)$. Hence the M. of I. of the sphere about the axis is

$$I = \int_{-R}^{R} \frac{1}{2}\pi\varrho(R^2 - x^2)^2 \, \mathrm{d}x = \frac{1}{2}\pi\varrho \left[R^4 x - \frac{2R^2 x^3}{3} + \frac{x^5}{5} \right]_{-R}^{R} = \frac{8\pi\varrho R^5}{15} = \frac{2MR^2}{5}.$$

(v) Find the moment of inertia about the axis of revolution of the solid formed by the revolution of the ellipse $x^2/a^2 + y^2/b^2 = 1$ about the x-axis.

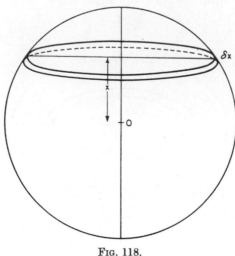

Let the density of the solid be ϱ and its mass be M. Then the M. of I. of the disc formed when the strip of the area of the ellipse bounded by the curve, the x-axis and the ordinates at (x, y), $(x + \delta x, y + \delta y)$ is revolved about the axis is approximately $\pi y^2 \delta x \varrho \cdot \frac{1}{2} y^2$. Hence the M. of I. of the ellipsoid about the axis is

$$\int_{-a}^{a} \frac{\pi \varrho}{2} \left(b^2 - \frac{b^2 x^2}{a^2} \right)^2 \mathrm{d}x = \frac{\pi \varrho \, b^4}{2} \left[x - \frac{2 x^3}{3 a^2} + \frac{x^5}{5 a^4} \right]_{-a}^{a} = \frac{8 \pi \varrho \, a b^4}{15}.$$

Also $\quad M = \int_{-a}^{a} \pi \varrho \left(b^2 - \frac{b^2 x^2}{a^2} \right) \mathrm{d}x = \frac{4}{3} \pi \varrho a b^2.$

Hence the M. of I. about the axis is $\dfrac{2 M b^2}{5}$.

(vi) Find the moment of inertia about each coordinate axis of a lamina in the shape of the area bounded by the curve $y = \mathrm{e}^{-x}$, the coordinate axes and the line $x = 1$.

Consider the thin strip of area bounded by the curve, the x-axis and the ordinates at (x, y) and $(x + \delta x, y + \delta y)$ on the curve (Fig. 119). With the usual notation $M = \varrho \int_0^1 e^{-x}\,dx = \varrho(e - 1)/e$. The M. of I. of this strip of the lamina about the y-axis is approximately $\varrho y \delta x \cdot x^2$.

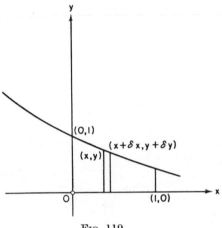

FIG. 119.

Hence the M. of I. of the lamina about the y-axis is

$$\int_0^1 \varrho\, x^2\, e^{-x}\, d x = \varrho \left[- x^2\, e^{-x} - 2 x e^{-x} - 2 e^{-x} \right]_0^1$$

$$= \varrho \left(2 - \frac{5}{e} \right) = M \left(\frac{2e - 5}{e - 1} \right).$$

The moment of inertia of the thin strip about the x-axis is approximately $\varrho y \delta x \cdot \dfrac{y^2}{3}$. $\Big[$The moment of inertia of a uniform rod of length l about an axis through its end perpendicular to its length is $\dfrac{M\, l^2}{3}$.$\Big]$ Hence the M. of I. for the lamina about the x-axis is

$$\varrho \int_0^1 \frac{y^3\, d x}{3} = \varrho \int_0^1 \frac{e^{-3x}\, d x}{3}$$

$$= \varrho \left[\frac{- e^{-3x}}{9} \right]_0^1 = \varrho \left(\frac{1}{9} - \frac{1}{9 e^3} \right) = \varrho \left(\frac{e^3 - 1}{9 e^3} \right) = M \left(\frac{e^2 + e + 1}{9 e^2} \right).$$

Two Theorems

(a) *The theorem of perpendicular axes.* If Ox and Oy are any two rect-angular axes in the plane of a lamina and Oz is an axis at right angles to the plane, then the moment of inertia of the lamina about Oz is equal to the sum of the moments of inertia of the lamina about Ox and Oy. (Fig. 120.)

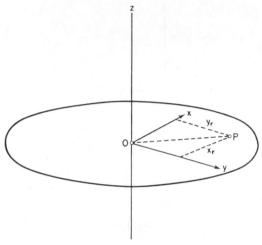

FIG. 120. Illustrating the theorem of perpendicular axes.

For, if m_r is the mass of a particle P of the lamina whose coordinates referred to Ox, Oy are (x_r, y_r), the M. of I. of the particle about Ox is $m_r y_r^2$ and the M. of I. of the particle about Oy is $m_r x_r^2$. But $m_r y_r^2 + m_r x_r^2 = m_r(y_r^2 + x_r^2) = m_r OP^2$ and this is equal to the M. of I. of P about Oz. Hence for the whole lamina, the sum of the moments of inertia about Ox and Oy is equal to the moment of inertia about Oz.

(b) *The theorem of parallel axes.* The moment of inertia of a body about any axis is equal to the moment of inertia about a parallel axis through the centre of mass together with the mass of the body multplied by the square of the distance between the axes.

Fig. 121 shows a *section* of the body through the centre of mass G. This section is at right angles to the axis and meets the axis at T where $GT = h$. Take G as the origin and TG produced as the x-axis. A linear element of the solid of mass m_p which is parallel to the axis meets the section at $P(x_p, y_p)$. Then

$$m_p PT^2 = m_p \{y_p^2 + (h + x_p)^2\}.$$

Hence the M. of I. of the body about the axis through T is

$$\Sigma m_p \, PT^2 = \Sigma \, m_p \, y_p^2 + \Sigma \, m_p \, h^2 + 2 \, \Sigma \, m_p \, h \, x_p + \Sigma \, m_p \, x_p^2$$
$$= \Sigma \, m_p \, (x_p^2 + y_p^2) + \Sigma \, m_p \, h^2 + 2 \, \Sigma \, m_p \, h \, x_p \, .$$

But $\Sigma m_p (x_p^2 + y_p^2) =$ M. of I. of the body about a parallel axis through G,
$\Sigma m_p \, h^2 = M h^2$ where M is the mass of the body,
$\Sigma m_p h \, x_p = h \Sigma m_p x_p = 0$ because the x-coordinate of the centre
of mass of the system is given by $\bar{x} = \Sigma m_p x_p / \Sigma m_p$ and in this case
$\bar{x} = 0$ since G is the origin of coordinates.

Hence the M. of I. of the body about the axis through T is equal to
the M. of I. of the body about the parallel axis through $G + M \cdot TG^2$.

Examples. (i) Find the moment of inertia of a uniform disc of mass M
and radius r about a tangent.

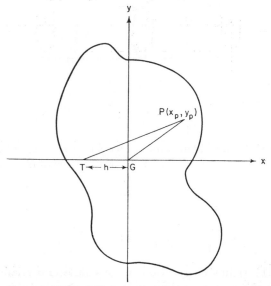

Fig. 121. Illustrating the theorem of parallel axes.

By the perpendicular axes theorem the sum of the moments of inertia
of the disc about each of two diameters at right angles is equal to the
moment of inertia of the disc about an axis through its centre perpendicular to its plane. Hence twice the M. of I. for the disc about a diameter
is $\frac{1}{2} M r^2$.

$\therefore$ the M. of I. for the disc about a diameter is $\dfrac{1}{4} M r^2$.

Hence, by the parallel axes theorem the M. of I. for the disc about the tangent parallel to the chosen diameter is $\dfrac{1}{4} M r^2 + M r^2 = \dfrac{5}{4} M r^2$, i.e., the moment of inertia of a uniform disc about a tangent is $\dfrac{5}{4} M r^2$.

(ii) Find the M. of I. of a uniform cylinder of mass M, radius r and length l, about a diameter of one end.

Consider a thin disc of the cylinder, thickness δx, perpendicular to the axis of the cylinder and distant x from Oy. (Fig. 122.) The M. of I.

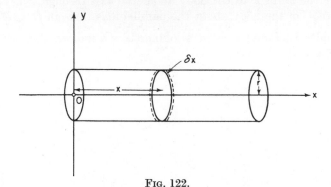

FIG. 122.

of this disc about the given axis is $\varrho \pi r^2 \delta x \cdot \frac{1}{4} r^2 + \varrho \pi r^2 \delta x \cdot x^2$. Hence the total M. of I. of the cylinder about a diameter of one end is

$$\int_0^l \varrho \, \pi \left(\frac{1}{4} r^4 + x^2 r^2 \right) \mathrm{d}x = \varrho \, \pi \left[\frac{r^4 x}{4} + \frac{x^3 r^2}{3} \right]_0^l$$

$$= \varrho \, \pi \left(\frac{r^4 l}{4} + \frac{l^3 r^2}{3} \right) = M \left(\frac{r^2}{4} + \frac{l^2}{3} \right).$$

(iii) Find the radius of gyration of an equilateral triangular lamina of side $2a$ (1) about an axis through one vertex parallel to the opposite side, (2) about an axis through the centroid perpendicular to the lamina. (L.)

(1) Consider a thin strip of the lamina ABC parallel to BC, distance x from A and width δx. (Fig. 123.) Let the length of the side of the lamina be $2a$, its surface density be ϱ and its mass be m. Then the length of the strip is $2x/\sqrt{3}$ and the M. of I. of the strip about an axis through A and parallel to BC is approximately $\dfrac{2 \varrho x \delta x}{\sqrt{3}} \cdot x^2$. Hence the M. of I.

of the lamina about this axis is

$$\int\limits_0^{a\sqrt{3}} \frac{2\,\varrho\,x}{\sqrt{3}}\,x^2\,\mathrm{d}\,x = \frac{2\,\varrho}{\sqrt{3}}\left[\frac{x^4}{4}\right]_0^{a\sqrt{3}} = \frac{3\,\sqrt{3}\,\varrho\,a^4}{2} = \frac{3\,m\,a^2}{2},$$

and the required radius of gyration is $a\,\sqrt{(3/2)}$.

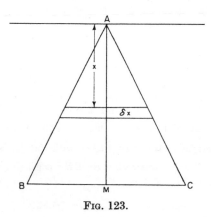

Fig. 123.

(2) By the parallel axes theorem the M. of I. of the lamina about an axis through the centroid parallel to BC is

$$\frac{3\,m\,a^2}{2} - m\left(\frac{2\,\sqrt{3}\,a}{3}\right)^2 = \frac{m\,a^2}{6}.$$

The M. of I. of the strip about the median AM is approximately $\dfrac{2\,\varrho\,x\,\delta\,x}{\sqrt{3}} \cdot \dfrac{x^2}{9}$ and the M. of I. of the lamina about AM is

$$\frac{2\,\varrho}{9\,\sqrt{3}}\int\limits_0^{a\,\sqrt{3}} x^3\,\mathrm{d}\,x = \frac{2\,\varrho}{9\,\sqrt{3}}\,\frac{9\,a^4}{4} = \frac{m\,a^2}{6}.$$

Hence the M. of I. of the lamina about an axis through the centroid perpendicular to the lamina is

$$\frac{m\,a^2}{6} + \frac{m\,a^2}{6} = \frac{m\,a^2}{3},$$

giving the required radius of gyration as $a/\sqrt{3}$.

Exercises 13.8

1. Calculate the M. of I. about its axis of the uniform solid formed when the area bounded by $x^2 = 4ay$ and its latus rectum is revolved completely about the y-axis.

2. Calculate the M. of I. about its axis of the uniform solid formed when the area bounded by the curve $ay^2 = x^3$ and the line $x = a$ revolves completely about the x-axis.

3. Calculate the M. of I. about the x-axis of a uniform lamina in the shape of the area bounded by the curve $y = \sin x$ and the x-axis between $x = 0$ and $x = \pi$.

4. Calculate the M. of I. about each coordinate axis of a lamina in the shape of the area bounded by $y = 1/x$, $x = 1$, $x = 2$ and the x-axis.

5. Calculate the M. of I. of a uniform solid right circular cone of height h and base-radius r (i) about its axis, (ii) about a diameter of its base.

6. Calculate the radius of gyration about a diameter of a hollow sphere formed by removing a concentric sphere of radius b from a sphere of radius a.

7. Prove that the M. of I. of a thin uniform rod, of length $2l$ and mass M, about an axis through its centre at an angle θ with the rod is $\dfrac{1}{3} M l^2 \sin^2\theta$.

Calculate the M. of I. of a framework $ABCDEF$ consisting of a regular hexagon of rods each of length $2l$ and mass m,
   (i) about AD,
   (ii) about an axis through its centre perpendicular to its plane.

8. Calculate the M. of I. about an axis through one end and perpendicular to its length of a thin rod, of length l and of density ϱx at a point distance x from the axis.

9. Calculate the radius of gyration about its axis of revolution of the solid formed by the complete revolution of a circle of radius R about a tangent.

10. Calculate the limiting value of the radius of gyration of the solid formed by the rotation about the x-axis of the area bounded by the curve $y = 1/(1 + x^2)$, the x-axis and the lines $x = -a$, $x = +a$ as a increases indefinitely.

Miscellaneous Exercises XIII

1. (a) By means of the substitution $y = -x$, or otherwise, evaluate $\displaystyle\int_{-1}^{1} x^2 \tan x \, dx$.
   (b) Determine the values of the constants A, B for which

$$\frac{x^4}{x^2 + 1} = A(x^2 - 1) + \frac{B}{x^2 + 1}.$$

Show that this result may be used in the integration of the function $x^3 \tan^{-1} x$, and complete this integration. Deduce the area bounded by the curve $y = x^3 \tan^{-1} x$ and the lines $x = 1$, $y = 0$.   (N.)

2. (a) Show that

$$\int_{0}^{t} \sin \omega x \cos \omega (t - x) \, dx = \tfrac{1}{2} t \sin \omega t.$$

(b) Find the coordinates of the mean centre of the area above the x-axis bounded by the part of the curve $a^2y = x^2(a - x)$ for which $0 \leq x \leq a$. (N.) [The mean centre of an area is the centroid.]

3. (a) Evaluate   (i) $\displaystyle\int_0^2 \frac{1}{\sqrt{(4 - x^2)}}\,\mathrm{d}x$,   (ii) $\displaystyle\int_{\pi/4}^{\pi/2} x \sin 2x\,\mathrm{d}x$.

(b) By means of the substitution $x = \frac{1}{2}\pi - y$, prove that

$$\int_0^{\pi/2} \frac{\cos x}{\cos x + \sin x}\,\mathrm{d}x = \int_0^{\pi/2} \frac{\sin x}{\cos x + \sin x}\,\mathrm{d}x.$$

By considering the sum of these integrals, or otherwise, determine their common value. (N.)

4. Show that the area under the curve $y = 1/x$, from $x = n - 1$ to $x = n + 1$, is $\log_e\{(n + 1)/(n - 1)\}$, provided that $n > 1$.

By applying Simpson's rule to this area, deduce that, approximately,

$$\log_e\left(\frac{n + 1}{n - 1}\right) = \frac{1}{3}\left(\frac{1}{n - 1} + \frac{4}{n} + \frac{1}{n + 1}\right),$$

and show that the error in this approximation is $4/15n^5$, when higher powers of $1/n$ are neglected. (N.)

5. (a) Using Simpson's rule with four intervals calculate $\displaystyle\int_0^4 e^{-x^2/50}\,\mathrm{d}x$ working with four places of decimals throughout.

(b) A particle describes simple harmonic motion in which the displacement x is given in terms of the time t by the equation

$$x = a \sin t.$$

Find, for the interval $t = 0$ to $t = \frac{1}{2}\pi$,

(i) the mean value of the velocity with respect to the time,

(ii) the mean value of the velocity with respect to the distance. (N.)

6. Sketch the curves

$$y = \frac{1}{1 + x}, \qquad y = \frac{1}{1 + x^2}$$

between their points of intersection A $(0, 1)$ and B $(1, \frac{1}{2})$.

If O is the origin and D is the point $(1, 0)$, calculate

(i) the area of the trapezium $OABD$,

(ii) the area between OA, OD, BD and the curve

$$y = \frac{1}{1 + x},$$

(iii) the area between OA, OD, BD and the curve $y = \dfrac{1}{1 + x^2}$.

Give each answer to two decimal places.

Hence find to which of the last two areas the area of the trapezium is the better approximation. (N.)

E

7. The boundary of a uniform lamina consists of a straight part OA of length π and a curved part OPA, the equation of which, referred to rectangular axes OA and OB, is $y = \sin x$. Show that the square of the radius of gyration of the lamina about OB is $\frac{1}{2}\pi^2 - 2$. (N.)

8. By means of Simpson's rule and taking unit intervals of x from $x = 8$ to $x = 12$, find approximately the area enclosed by the curve $y = \log_{10} x$, the lines $x = 8$ and $x = 12$, and the x-axis. Deduce the average value of $\log_{10} x$ between $x = 8$ and $x = 12$. (N.)

9. The distance s fallen by a particle dropped from rest is given in terms of t by the relation $s = 16 t^2$. Find the velocity v both in terms of t and in terms of s. If the final velocity on reaching the ground is V, find in terms of V, the mean value of v both with respect to time and with respect to distance, and show that the former is three-quarters of the latter. (N.)

10. Show that if m and n are positive integers

$$\int_0^\pi \cos m\,x \cos n\,x\,\mathrm{d}x = 0$$

when m and n are unequal, and find the value of the integral when $m = n$.

Prove that

$$2\sin x(\cos x + \cos 3x + \cos 5x + \cos 7x) = \sin 8x.$$

Hence, or otherwise, show that

$$\int_0^\pi \frac{\sin^2 8\,x}{\sin^2 x}\,\mathrm{d}x = 8\,\pi. \tag{N.}$$

11. A sphere of radius a touches a sphere of radius b internally at A and the common diameter of the spheres cuts the inner sphere at B. Calculate the volume bounded by the tangent plane at B and the surfaces of the spheres.

Prove that the centroid of this volume is at a distance $4a/3$ from A, whatever the value of b. (N.)

12. Prove that

$$\int_{-a}^a f(x)\,\mathrm{d}x = \int_0^a \{f(x) + f(-x)\}\,\mathrm{d}x.$$

Hence show that

$$\int_{-\pi/2}^{\pi/2} \frac{\mathrm{e}^x}{1 + \mathrm{e}^x}\sin^4 x\,\mathrm{d}x = \int_0^{\pi/2} \sin^4 x\,\mathrm{d}x$$

and evaluate the latter integral. (N.)

13. If $0 < \alpha < \pi$, show that

$$\int_\alpha^{\pi-\alpha} \frac{x\,\mathrm{d}x}{\sin x} = \int_\alpha^{\pi-\alpha} \frac{\pi - x}{\sin x}\,\mathrm{d}x.$$

Prove that

$$\int_{\pi/3}^{2\pi/3} \frac{x\,\mathrm{d}x}{\sin x} = \tfrac{1}{2}\,\pi \log_{\mathrm{e}} 3. \tag{N.}$$

14. Show that $(\log x)/\sqrt{x}$ has a maximum value when $x = e^2$.

The curve $y = (\log x)/\sqrt{x}$ meets the x-axis at A, and B is the foot of the maximum ordinate PB. Show that the distance from the x-axis of the centroid of the area bounded by AB, BP and the arc AP of the curve is $1/3$. (N.)

15. Using tables where necessary, calculate the value of

$$\int_{0.1}^{0.5} e^{-x} \, dx$$

(i) by direct integration,

(ii) by Simpson's rule, using 5 ordinates spaced at intervals of $\dfrac{1}{10}$ unit.

(Give your answers to four places of decimals.) (N.)

16. Show, by graphical considerations or by using the substitution $u = a - x$, that

$$\int_0^a f(x) \, dx = \int_0^a f(a - x) \, dx.$$

Hence, or otherwise, evaluate $\displaystyle\int_0^2 x(2 - x)^4 \, dx.$ (N.)

17. Find the area, and the coordinates of the centroid, of the triangle whose vertices are the points $(2, 1)$, $(14, 6)$, $(10, 9)$. Deduce the volume swept out by the triangle when it makes one complete revolution about the line $x + y = 0$ as axis of rotation. (N.)

18. Prove that $\displaystyle\int_{-a}^a f(x) \, dx = \int_0^a \{f(x) + f(-x)\} \, dx.$

Hence show that $\displaystyle\int_{-\pi/4}^{\pi/4} \frac{1}{1 + \sin x} \, dx = 2 \int_0^{\pi/4} \sec^2 x \, dx$, and evaluate this integral.

Use the first result to evaluate $\displaystyle\int_{-1}^1 \frac{1}{1 + e^{-x}} \, dx.$ (N.)

19. Show that the area under the curve $y = e^x$ between $x = 0$ and $x = h$ is $e^h - 1$.

By applying Simpson's rule, taking three ordinates at $x = 0$, $x = \frac{1}{2}h$, $x = h$, show that, if h is small, an approximation to the same area is $\dfrac{1}{6} h(1 + 4e^{h/2} + e^h)$.

Expand the correct value and the approximate value each in ascending powers of h and deduce that the difference between them is $h^5/2880$, if h^6 and higher powers of h are neglected. (N.)

20. Evaluate (i) $\displaystyle\int_0^{\pi/2} (\cos \theta - \sin \theta)^3 \, d\theta$, (ii) $\displaystyle\int_0^{\pi/2} \theta (\cos \theta - \sin \theta)^2 \, d\theta.$ (N.)

21. If I_n denotes $\displaystyle\int_0^a (a^2 - x^2)^n \, dx$, prove that, if $n > 0$,

$$I_n = \frac{2 n a^2}{2 n + 1} I_{n-1}.$$ (O. C.)

22. Prove that

$$\int_0^{\pi/2} f(\sin 2x)\sin x\,\mathrm{d}x = \sqrt 2 \int_0^{\pi/4} f(\cos 2x)\cos x\,\mathrm{d}x. \qquad \text{(O.C.)}$$

23. Find the coordinates of the centre of gravity of the area of the half loop of the curve

$$a^2 y^2 = x^2(a^2 - x^2)$$

in the positive quadrant.

Also find the volume obtained by revolving the whole loop about the axis of y. (O.C.)

24. Draw a rough sketch of the curve

$$y^2 = x^4 - 14ax^3 + 45a^2 x^2.$$

The loop of the curve is rotated about the x-axis. Find the distance from the origin of the centre of gravity of the solid generated. (O.C.)

25. The portion of the curve $y^2 = 4ax$ from $(a, 2a)$ to $(4a, 4a)$ revolves round the tangent at the origin. Prove that the volume bounded by the curved surface so formed and plane ends perpendicular to the axis of revolution is $\dfrac{62}{5}\pi a^3$ and find the square of the radius of gyration of this volume about its axis of revolution. (O.C.)

26. Sketch the curve

$$y^2(a + x) = x^2(3a - x).$$

By means of the substitution $a + x = 4a\sin^2\theta$, or otherwise, find the area in the second and third quadrants between the curve and its asymptote. (O.C.)

27. By the methods of the integral calculus find

(i) the area and also the coordinates of the centre of gravity of a uniform plane lamina bounded by the axes of coordinates and that arc of the ellipse

$$\frac{x^2}{a^2} + \frac{y^2}{b^2} = 1$$

which lies in the positive quadrant;

(ii) the volume and also the position of the centre of gravity of a uniform hemisphere whose base is a circle of radius a. (O.C.)

28. (i) Prove that, if

$$I_n = \int \sec^n x\,\mathrm{d}x,$$

then

$$(n - 1)I_n = \tan x\sec^{n-2}x + (n - 2)I_{n-2}.$$

Use this formula to evaluate

$$\int_0^{\pi/4} \sec^5 x\,\mathrm{d}x.$$

(ii) Find the positive value of x for which the definite integral

$$\int_0^x \frac{1-t}{\sqrt{(1+t)}}\,\mathrm{d}t.$$

is greatest.

Evaluate the integral for this value of x. (O.C.)

29.   If $u_n = \int_0^{\pi/2} x^n \sin x \, dx$, and $n > 1$, show that

$$u_n = n(\pi/2)^{n-1} - n(n-1)u_{n-2},$$

and evaluate u_4.   (O.C.)

30. Show that

$$\int_0^a f(x) \, dx = \int_0^a f(a-x) \, dx,$$

Deduce that

$$\int_0^{\pi/4} \left(\frac{1 - \sin 2x}{1 + \sin 2x} \right) dx = \int_0^{\pi/4} \tan^2 x \, dx,$$

and evaluate the integral.   (O.C.)

31. The area enclosed between the curve $y = 3x - x^2$ and the straight line $y = x$ is rotated through four right angles about the axis of x. Prove that the volume of the solid generated is $56\pi/15$.

Calculate the distance from the origin of the centre of gravity of this solid.   (C.)

32. Find the volume of the solid generated when the ellipse $\frac{x^2}{a^2} + \frac{y^2}{b^2} = 1$ is rotated completely about the axis of x.

The solid is cut into two portions by the plane formed by rotation of the line $x = a/2$ about the axis of x. Show that the centre of gravity of the smaller of the two portions is distant $27a/40$ from the origin.   (C.)

33. A sphere of radius r is cut into two portions by a plane which is distant c from the centre of the sphere. Show that the volume of the smaller of the two portions is

$$\frac{1}{3}\pi(r-c)^2(2r+c).$$

Show also that the distance of the centre of gravity of this portion from the centre of the sphere is $3(r+c)^2/4(2r+c)$.

Deduce the position of the centre of gravity of a uniform solid hemisphere.   (C.)

34. The origin O is joined to the point $P(h, k)$ on the parabola $y^2 = 4ax$, and N is the foot of the perpendicular from P to the axis of x. Prove that the area enclosed between the straight line OP and the parabola is one-third of the area of the triangle OPN.

Show also that the x-coordinate of the centre of gravity of the area between OP and the parabola is $2h/5$.   (C.)

35. Two points A, B are taken on a parabola, both points being on the same side of the axis and B being farther from the vertex than A. Lines AC, BC are drawn parallel to the axis and the directrix respectively. If the area ABC is revolved through four right angles about the axis of the parabola, show that the volume of the solid generated is equal to that of a right circular cylinder whose radius is AC and whose length is the semi-latus rectum of the parabola.

Show that the centroid of the solid divides its axis in the ratio $2:1$.   (L.)

36.   Prove that $\int \sec^n \theta \, d\theta = \dfrac{\sec^{n-2} \theta \tan \theta}{n-1} + \dfrac{n-2}{n-1} \int \sec^{n-2} \theta \, d\theta$.

By using the parameter θ given by $y = \tan^3 \theta$, or otherwise, find the area between the curve

$$x^{2/3} = y^{2/3} + 1$$

and the ordinate $x = 2 \sqrt{2}$.   (N.)

37.   Prove that $\int_{-a}^{a} f(x) \, dx = \int_{0}^{a} \{f(x) + f(-x)\} \, dx$ and evaluate

$$\int_{-\pi/2}^{\pi/2} (2 + x)^2 \sin 2 x \, dx.$$ (N.)

38. By means of the substitution $x = 1/t$ and considering the intervals -1 to 0 and 0 to 1 separately, prove that

$$\int_{-1}^{1} \frac{dx}{1 - 2 x \cos \theta + x^2} = \tfrac{1}{2} \int_{-\infty}^{\infty} \frac{dx}{1 - 2 x \cos \theta + x^2}$$

provided that θ is *not* an integral multiple of π.

Evaluate $\int_{-1}^{1} \dfrac{dx}{1 - 2 x \cos \theta + x^2}$ when θ lies between 0 and π.

What is the value of the integral (i) when $\theta = -\dfrac{1}{4} \pi$, (ii) when $\theta = 5\pi/4$?   (N.)

39. Given that $C = \displaystyle\int_{0}^{\pi/2} \frac{\cos^2 x}{a^2 \sin^2 x + b^2 \cos^2 x} \, dx$

and

$$S = \int_{0}^{\pi/2} \frac{\sin^2 x}{a^2 \sin^2 x + b^2 \cos^2 x} \, dx,$$

prove that $C + S = \pi/2ab$ when $ab > 0$. Obtain another such relation between C and S and hence determine them when $a \neq b$.

Show that $\displaystyle\int_{0}^{\pi} \frac{\cos^2 x}{a^2 \cos^2 x + b^2 \sin^2 x} \, dx = \frac{\pi (b - a)}{a b (a + b)}$.   (N.)

40. If $u_n = \displaystyle\int_{0}^{\pi/2} \frac{\sin (2 n + 1) \theta}{\sin \theta}$, evaluate $u_n - u_{n-1}$, and hence show that

$u_n = \tfrac{1}{2}\pi$ for all positive integers n. Using a similar method, prove that

$$\int_{0}^{\pi/2} \frac{\sin^2 n \theta}{\sin^2 \theta} \, d\theta = \tfrac{1}{2} n \pi,$$

and find the value of

$$\int_0^{\pi/2} \frac{(2n+1)\sin\theta - \sin(2n+1)\theta}{\sin^3\theta}\,d\theta,$$

where n is a positive integer in each case.   (N.)

41. A particle moves along the x-axis, its speed dx/dt at time t being given by

$$dx/dt = \sqrt{(9t - t^3)},\ 0 \le t \le 3.$$

Use Simpson's rule to calculate approximately the distance travelled by the particle during the interval from $t = 0$ to $t = 3$. (Use steps of 0.5 in t.)

Show that the average speed is approximately three-quarters of the greatest speed of the particle in this interval.   (N.)

42. If $u_n = \int_0^{\alpha} \dfrac{\sin(2n-1)x}{\sin x}\,dx$ and $v_n = \int_0^{\alpha} \dfrac{\sin^2 nx}{\sin^2 x}\,dx$

prove that (i) $n(u_{n+1} - u_n) = \sin 2n\alpha$, (ii) $v_{n+1} - v_n = u_{n+1}$.
Deduce that, if n is a positive integer,

$$\int_0^{\pi/2} \frac{\sin^2 nx}{\sin^2 x} = \frac{n\pi}{2}.$$ (N.)

43. The horizontal velocity v of a point P is given by

$$v = v_0\{\sin\omega t + |\sin\omega t|\}$$

where v_0 and ω are positive constants. Sketch the graphs of v and dv/dt from $t = 0$ to $t = 6\pi/\omega$.

If P is a point on one of the feet of a pedestrian whose average speed is V, find the ratio of v_0 to V.   (N.)

44. If A denotes the area bounded by the coordinate axes, the ordinate $x = 1$ and the arc of the curve $y(1 + x^2) = 1$ from $x = 0$ to $x = 1$, calculate the volumes obtained by rotating A about (i) the x-axis (ii) the y-axis. Hence, by using Pappus' theorem, find the coordinates of the centroid of the area A.   (N.)

45. (a) A number a can take any value between 0 and n, all values being equally likely. Prove that the mean value of $a(n - a)$ is $\dfrac{1}{6}n^2$.

(b) In the quadratic equation $x^2 - px + q = 0$ both p and q are positive and p is fixed. If all positive values of q are equally likely, subject to the condition that the roots of the equation are real, prove that the mean value of the larger root is $5p/6$.   (N.)

46. Evaluate $\displaystyle\int_0^{\infty} x^2 e^{-x/2}\,dx$. It may be assumed that $x^n e^{-x/2} \to 0$ as $x \to \infty$.   (L.)

47. Show that

$$\int_0^\infty \frac{x^2\,(x+1)}{(x^2+1)^3}\,\mathrm{d}x = (\pi+4)/16.$$ (L.)

48. Evaluate $(1/\pi)\int_0^{\pi/2} (9+\cos^2\theta)^{1/2}\mathrm{d}\theta$, correct to three decimal places, by expanding the integrand, using the binomial theorem and integrating term by term. Also obtain an approximate value of this integral using Simpson's rule with seven ordinates.   (L.)

49. A uniform solid body consists of a cylinder with rounded ends. The cylinder is of length $2a$, radius a, and density ϱ; the ends are hemispheres of radius a, and density 3ϱ. Find the mass of the body and its moment of inertia about a line through its centre perpendicular to the axis of the cylinder.   (L.)

50. The area bounded by the curve $y^2 = 4a(b-x)$ and the y-axis is rotated through two right angles about the x-axis. Show that the volume of the solid of revolution is $2\pi a b^2$, and find the radius of gyration of this solid about the y-axis.   (L.)

51. Show that the centre of mass of a uniform semi-circular disc of mass M and radius a is at a distance $4a/3\pi$ from its bounding diameter, AB, and find, by integration, the moment of inertia of the disc about AB.

Hence determine the moment of inertia of the disc about the tangent which is parallel to AB.   (L.)

52. An area in the x, y plane is bounded by the arc of the curve $y = a\cos(x/a)$, between the points $(0, a)$, $(\frac{1}{2}\pi a, 0)$, and the parts of the coordinate axes between the origin and these points. Show that the radii of gyration of the area about the axes of x and y are $\dfrac{1}{3}a\sqrt{2}$ and $\frac{1}{2}a\sqrt{(\pi^2-8)}$.   (L.)

53. The equation of a curve is

$$y^2 = \frac{x-1}{x-2}.$$

Show that no part of the curve lies between $x = 1$ and $x = 2$, and sketch the curve for the remaining values of x.

The portion of the curve between $x = 0$ and $x = 1$ is rotated through $360°$ about the axis of x. Find the volume generated and the distance of its centroid from the origin.   (L.)

54. A solid of revolution is generated by rotation about the axis of x of the area bounded by the curve $y = \sec x$ and the lines $x = 0$, $x = \pi/4$ and $y = 0$. Find the volume of the solid, the position of its centroid and its radius of gyration about its axis.   (L.)

55. Find the values of (i) $\int_0^\pi \cos pt \cos qt\ \mathrm{d}t$, (ii) $\int_0^\pi \cos^2 pt\ \mathrm{d}t$ where p, q are integers, $p \neq q$.

If an alternating current i is given by the formula $i = a_1 \cos pt + a_2 \sin pt$ where a_1, a_2 are constants, show that the mean value of i^2 taken over the half-period from $t = 0$ to $t = \pi/p$ is $\frac{1}{2}(a_1^2 + a_2^2)$.   (L.)

CHAPTER XIV

SOME PROPERTIES OF CURVES

14.1 Points of inflexion

In Volume I we considered points of inflexion on a curve $y = f(x)$ at which the tangents were parallel to the x-axis and also at which $f'(x)$ had either maximum or minimum values. We now *define* a point of inflexion on a curve as a point at which the gradient has a maximum or minimum value. Thus, $y = f(x)$ has a point of inflexion at the point (x_1, y_1) if $f''(x_1) = 0$ and if $f''(x_1)$ increases or decreases continuously as x passes through the value x_1. This is equivalent to the statement that, if $f''(x)$ is continuous at $x = x_1$, necessary and sufficient conditions for an inflexion at $x = x_1$ are $f''(x_1) = 0$ and $f''(x)$ *changes sign* as x passes through the value $x = x_1$. Sufficient conditions for an inflexion at $x = x_1$ are therefore $f''(x_1) = 0$, $f'''(x_1) \neq 0$. But, if $f''(x_1) = 0$ and $f'''(x_1) = 0$, it is necessary to investigate further the behaviour of the function $f''(x)$ as x passes through the value $x = x_1$.

In Fig. 124 (i) the tangent at the point P (x_1, y_1) lies *above* the curve $y = f(x)$ for points on the curve near P. [The tangent at P is *above* the

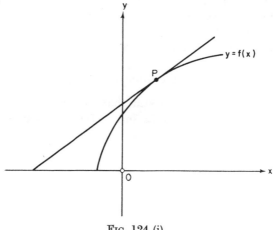

FIG. 124 (i).

curve if $f(x_1 + \varepsilon) < f(x_1) + \varepsilon f'(x_1)$ however small $|\varepsilon|$ may be.] In such a case the curve is said to be concave downwards. In this case $f'(x)$ decreases continuously as x passes through the value $x = x_1$ and $f''(x_1)$ is therefore negative.

In Fig. 124 (ii) the tangent at P lies below the curve $y = f(x)$ for points of the curve near P. The curve is said to be *concave upwards* at P, $f'(x)$ increases continuously as x passes through the value x_1, and $f''(x_1)$ is positive.

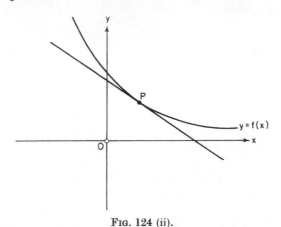

FIG. 124 (ii).

In Fig. 124 (iii) the tangent is above the curve for values of x less than x_1 and however near to x_1, the tangent is below the curve for values of x greater than x_1 and however near to x_1, i.e., $f''(x_1) < 0$ for $x = x_1 - \varepsilon$ and $f''(x_1) > 0$ for $x = x_1 + \varepsilon$ where $\varepsilon > 0$. Hence $f'(x)$ has a minimum at P, *the curve crosses the tangent at P* and P is a point of inflexion.

In Fig. 124 (iv) $f'(x)$ has a minimum at P which is a point of inflexion.

Examples. (i) Investigate the stationary points and the point(s) of inflexion on the curve $y = x^5 - 5x$ and sketch the curve.

Let $f(x) = x^5 - 5x$. Then

$$f'(x) = 5x^4 - 5.$$

Hence the stationary points occur where $x = \pm 1$. Also

$$f''(x) = 20x^3$$

so that $f(x)$ has a minimum at $(1, -4)$ and a maximum at $(-1, 4)$.

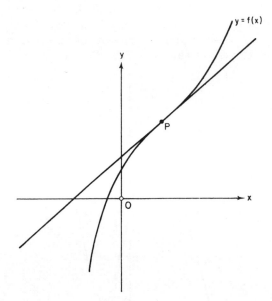

FIG. 124 (iii).

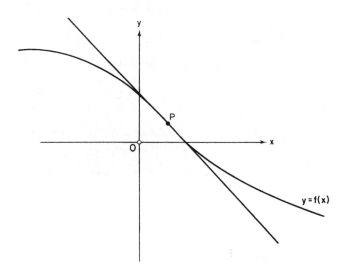

FIG. 124 (iv).

But $f''(x) = 0$ at $(0, 0)$ and $f'''(x) = 0$ at $(0, 0)$. However $f''(0 - \varepsilon) < 0$ and $f''(0 + \varepsilon) > 0$. Hence there is an inflexion at the origin.

At $(0, 0)$, $f'(x) = -5$ and the equation of the tangent at $(0, 0)$ is $y = -5x$. The curve is as shown in Fig. 125.

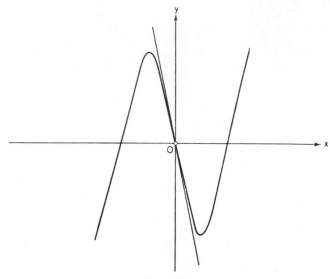

Fig. 125.

(ii) Show that the three points of inflexion on the curve $y = \dfrac{k - x}{k^2 + x^2}$, where $k > 0$, are collinear. Find the equation of the line joining the three points in the case $k = 1$.

$$\frac{dy}{dx} = \frac{x^2 - 2kx - k^2}{(k^2 + x^2)^2},$$

$$\frac{d^2y}{dx^2} = \frac{-2x^3 + 6kx^2 + 6k^2x - 2k^3}{(k^2 + x^2)^3}.$$

(a) If there are points of inflexion they occur where

$$x^3 - 3kx^2 - 3k^2x + k^3 = 0,$$

i.e., where

$$(x + k)(x^2 - 4kx + k^2) = 0,$$

i.e., where $x = -k$ or α or β, where α and β are the roots of $x^2 - 4kx + k^2 = 0$ and $\alpha > \beta > -k$. The signs of $(x + k)$ and $(x^2 - 4kx + k^2)$, and therefore the sign of $(x + k)(x^2 - 4kx + k^2)$ as

x passes through the critical values $-k$, α and β, are determined as follows:

x	$x < -k$	$-k < x < \beta$	$\beta < x < \alpha$	$x > \alpha$
$x + k$	$-$	$+$	$+$	$+$
$x^2 - 4kx + k^2$	$+$	$+$	$-$	$+$
$\dfrac{d^2 y}{d x^2}$	$-$	$+$	$-$	$+$

Hence $\dfrac{d^2 y}{d x^2}$ changes sign as x increases through each of the values $-k$, β, α and therefore there is an inflexion at each of these points.

(b) These inflexions are collinear if (§ 11.11)

$$\Delta = \begin{vmatrix} -k & \dfrac{1}{k} & 1 \\[2mm] \alpha & \dfrac{k - \alpha}{k^2 + \alpha^2} & 1 \\[2mm] \beta & \dfrac{k - \beta}{k^2 + \beta^2} & 1 \end{vmatrix} = 0.$$

By taking $\varrho_2 - \varrho_1$, $\varrho_3 - \varrho_1$,

$$\Delta = \frac{(k + \alpha)(k + \beta)}{k} \begin{vmatrix} 1 & \dfrac{-\alpha}{k^2 + \alpha^2} \\[2mm] 1 & \dfrac{-\beta}{k^2 + \beta^2} \end{vmatrix} = \frac{(k + \alpha)(k + \beta)(k^2 - \alpha\beta)(\alpha - \beta)}{k(k^2 + \alpha^2)(k^2 + \beta^2)}.$$

But $\alpha\beta = k^2$ and hence $\Delta = 0$. Thus the three points of inflexion are collinear.

When $k = 1$, the line joining the three points of inflexion is the line joining the points $(-1, 1)$ and $\left(\alpha, \dfrac{1 - \alpha}{1 + \alpha^2}\right)$. The slope of this line is

$$m = \frac{\dfrac{1 - \alpha}{1 + \alpha^2} - 1}{\alpha + 1} = -\frac{\alpha}{\alpha^2 + 1}.$$

But $\alpha^2 - 4\alpha + 1 = 0$ so that $\dfrac{\alpha}{\alpha^2 + 1} = \dfrac{1}{4}$. Hence the required line is

$$(y - 1) = -\frac{1}{4}(x + 1),$$

i.e.,

$$x + 4y - 3 = 0.$$

Exercises 14.1

1. Show that each of the following curves has an inflexion at the origin and in each case find the equation of the tangent at the origin:

(i) $y = x^7 + 7x$,   (ii) $y = \tan x$,   (iii) $y = \sinh x$,

(iv) $y = \sinh^{-1} x$,   (v) $y = \dfrac{x}{1 + x^2}$.

2. Find the coordinates of the stationary points, stating the nature of each, and the points of inflexion where they exist, for each of the following curves and, in each case, sketch the curve and its tangents at the points of inflexion:

(i) $y = x^3 - 12x^2 + 1$,   (ii) $y = x^4 - 2x^3$,

(ii) $y = x/(1 - x^2)$,   (iv) $y = ax^2/(a^2 + x^2)$,

(v) $y = \sin x + \cos x$.

3. Show that the curve $x^3 - y^3 = 1$ cuts the y-axis at a point of inflexion and calculate the angle at which it cuts the axis there.

4. Find the points of inflexion on the curve $y = e^{-x^2}$. Sketch the curve and its tangents at the points of inflexion.

5. Find the stationary points and the point of inflexion on the curve $y = xe^{-x}$. Find the equation of the tangent at the point of inflexion and sketch the curve. (Consider $\lim\limits_{x \to 0} xe^{-x}$ as $\lim\limits_{x \to 0} x/e^x$ and $\lim\limits_{x \to \infty} xe^{-x}$ as $\lim\limits_{x \to \infty} x/e^x$ in each case expanding the denominator and dividing by the numerator.)

6. For the curve $x = a(1 - \cos\theta)$, $y = a(\theta + \sin\theta)$, find dy/dx and d^2y/dx^2 each in terms of θ and hence find the coordinates of the stationary points and show that the curve has no points of inflexion.

7. Find the points of inflexion on the curve $y = x - \cos x$ and sketch the curve for $-2\pi \leq x \leq 2\pi$.

8. The equations of the curves in the shapes of beams supported and loaded in different ways are given as follows:

(i) resting on two supports at its ends with a load W suspended from its middle point, $y = \dfrac{W}{48\,EI}\, x(3l^2 - 4x^2)$,

(ii) fixed at one end and uniformly loaded, $y = \dfrac{W}{24\,EI}\, x^2(6l^2 - 4lx + x^2)$,

(iii) resting on supports at its extremities and uniformly loaded,

$y = \dfrac{W}{384\,EI}\, (16x^4 - 24x^2l^2 + 5l^4)$,

(iv) clamped horizontally at both ends and uniformly loaded,

$y = \dfrac{W}{384\,EI}\, (4x^2 - l^2)^2$,

where E and I are constants for the beam.

In each case investigate the positions of the inflexions (if any) in the curves of the loaded beams.

9. Find the stationary points and the points of inflexion on the curve $y = \cos 2x - 4 \sin x$. Sketch the curve.

10. Show that for the curve $y = x \cosh x$ there is a point of inflexion at $(a, a \cosh a)$, where $a = \frac{1}{2} \log\left(\dfrac{2 - a}{2 + a}\right)$.

14.2 The length of a curve

Definitions. If $P_1, P_2, P_3, \ldots, P_n$ are vertices of a polygon inscribed in a curve $P_1 P_n$ (Fig. 126), the length of the curve $P_1 P_n$ is defined as $\lim\limits_{n \to \infty} \sum\limits_{1}^{n-1} P_r P_{r+1}$ if that limit exists and if the length of *every one* of $P_r P_{r+1}$ tends to zero as $n \to \infty$.

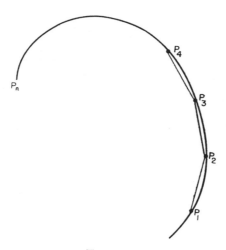

FIG. 126.

If the length of the arc measured from a chosen fixed point on the curve is denoted by s, the direction in which s increases may also be chosen arbitrarily. It is conventional to choose this direction so that s increases with x. When the equation of the curve is given in terms of a parameter t in the form $x = f(t)$, $y = g(t)$, it is conventional to choose as the direction in which s increases the direction in which t increases.

The positive direction of the tangent at a point P on the curve is defined as the direction along the tangent which coincides with the direction of s increasing, and one of the angles which this direction makes with the positive direction of the x-axis is denoted by ψ. If possible, that value of ψ which ensures that ψ changes continuously with s is chosen. Thus, for example, Fig. 127 (i) shows the direction of s increasing from the origin O, the positive direction of the tangent, and the angle ψ in three positions on the curve $y = \sin x$; Fig. 127 (ii) shows these three variables for the curve $x = a \cos t$, $y = b \sin t$.

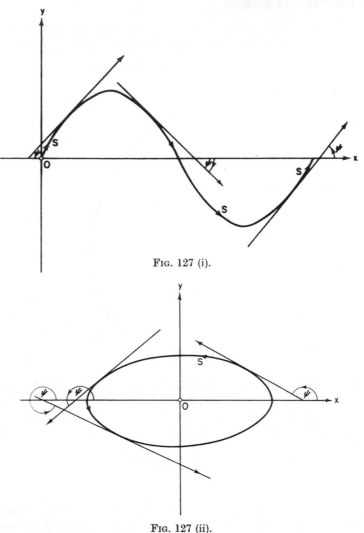

Fig. 127 (i).

Fig. 127 (ii).

Calculation of the length of a curve. (Fig. 128.) If s is the length of an arc of the curve $y = f(x)$ measured from a fixed point P $\{a, f(a)\}$, A is the point (x, y) on the curve and B the point $(x + \delta x, y + \delta y)$, then s is a function of x and the length of the arc AB is δs, the increase in s which corresponds to the increase δx in x. Then

$$\sin BAM = \frac{\delta y}{\text{chord } AB} = \frac{\delta y}{\delta s} \cdot \frac{\delta s}{AB}.$$

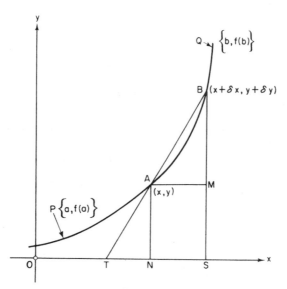

FIG. 128.

Similarly

$$\cos BAM = \frac{\delta x}{\delta s} \cdot \frac{\delta s}{AB}.$$

If now we assume that $\lim\limits_{B \to A} \dfrac{\delta s}{AB} = 1$, which is an extension of the assumption we make in arriving at the result $\lim\limits_{\theta \to 0} \sin\theta/\theta = 1$ in the case of a circle, then

$$\lim_{B \to A} \sin BAM = \frac{dy}{ds},$$

$$\lim_{B \to A} \cos BAM = \frac{dx}{ds}.$$

$$\therefore \sin \psi = \frac{\mathrm{d}y}{\mathrm{d}s}, \qquad\qquad (14.1)$$

$$\cos \psi = \frac{\mathrm{d}x}{\mathrm{d}s}, \qquad\qquad (14.2)$$

where ψ is the angle which the tangent at A makes with the positive Ox direction.

$$\therefore \left(\frac{\mathrm{d}s}{\mathrm{d}x}\right)^2 = \sec^2 \psi = 1 + \tan^2 \psi = 1 + \left(\frac{\mathrm{d}y}{\mathrm{d}x}\right)^2.$$

Similarly
$$\left(\frac{\mathrm{d}s}{\mathrm{d}y}\right)^2 = 1 + \left(\frac{\mathrm{d}x}{\mathrm{d}y}\right)^2.$$

Since s has been defined so that it increases with x,

$$\frac{\mathrm{d}s}{\mathrm{d}x} = + \sqrt{\left\{1 + \left(\frac{\mathrm{d}y}{\mathrm{d}x}\right)^2\right\}}.$$

Hence the length of the arc PQ, where Q is $\{b, f(b)\}$, is given by

$$s = \int_a^b \sqrt{\left\{1 + \left(\frac{\mathrm{d}y}{\mathrm{d}x}\right)^2\right\}}\, \mathrm{d}x. \qquad\qquad (14.3)$$

Similarly
$$s = \int_{f(a)}^{f(b)} \sqrt{\left\{1 + \left(\frac{\mathrm{d}x}{\mathrm{d}y}\right)^2\right\}}\, \mathrm{d}y, \qquad\qquad (14.4)$$

if s is specified so that it increases with y.

If the equation of the curve is given by $x = f(t)$, $y = g(t)$ where t increases steadily from t_1 at A to t_2 at B, then

$$s = \int_{t_1}^{t_2} \sqrt{\left\{1 + \left(\frac{\mathrm{d}y}{\mathrm{d}x}\right)^2\right\}} \frac{\mathrm{d}x}{\mathrm{d}t}\, \mathrm{d}t.$$

($\mathrm{d}x/\mathrm{d}t$ is positive if x and t both increase with s.)

$$\therefore s = \int_{t_1}^{t_2} \sqrt{\left\{\left(\frac{\mathrm{d}x}{\mathrm{d}t}\right)^2 + \left(\frac{\mathrm{d}y}{\mathrm{d}t}\right)^2\right\}}\, \mathrm{d}t. \qquad\qquad (14.5)$$

Care must be taken to ensure that (14.3) is not used over an interval which includes a point at which $\psi = \frac{1}{2}\pi$ since $\mathrm{d}y/\mathrm{d}x$ would not exist there; (14.4) must not be used over an interval which includes a point $\psi = 0$.

Examples. (i) The curve into which a uniform heavy thin rope hangs when freely suspended from its ends is called a *catenary* and its equation, referred to horizontal and vertical axes of x and y chosen so that the coordinates of its lowest point are $(0, c)$, is $y = c \cosh(x/c)$. Show that the length of an arc of this catenary measured from the lowest point to the point (x, y) is given by

$$s = c \sinh(x/c)$$

and that, in the usual notation, $s = c \tan \psi$ and $y^2 = c^2 + s^2$.

$$y = c \cosh(x/c).$$

$$\therefore \frac{dy}{dx} = \sinh(x/c).$$

$$\therefore s = \int_0^x \sqrt{\{1 + \sinh^2(x/c)\}}\, dx = \int_0^x \cosh(x/c)\, dx.$$

$$\therefore s = c \sinh(x/c).$$

$$\tan \psi = \frac{dy}{dx} = \sinh\left(\frac{x}{c}\right).$$

$$\therefore s = c \tan \psi.$$

An equation in the form $s = f(\psi)$, where s and ψ have the meanings defined in this section, is known as *the intrinsic equation of the curve*.

$$y^2 - s^2 = c^2 \{\cosh^2(x/c) - \sinh^2(x/c)\} = c^2.$$

$$\therefore y^2 = c^2 + s^2.$$

(ii) Find the length of the arc of the curve $x = a \sin^3\theta$, $y = a \cos^3\theta$ in the first quadrant.

This curve, Fig. 129, is the *astroid*. The condition that s increases with θ is satisfied if the direction of s increasing is chosen as from $\theta = 0$ to $\theta = \frac{1}{2}\pi$. Then

$$\frac{dx}{d\theta} = 3a \sin^2\theta \cos\theta,$$

$$\frac{dy}{d\theta} = -3a \cos^2\theta \sin\theta.$$

$$\therefore s = \int_0^{\pi/2} 3a \sin\theta \cos\theta \sqrt{(\sin^2\theta + \cos^2\theta)}\, d\theta$$

$$= \int_0^{\pi/2} \frac{3a}{2} \sin 2\theta\, d\theta = \frac{3a}{2}\left[-\frac{\cos 2\theta}{2} \right]_0^{\pi/2} = \frac{3a}{2}$$

so that the length of the arc of the curve in the first quadrant is $3a/2$.

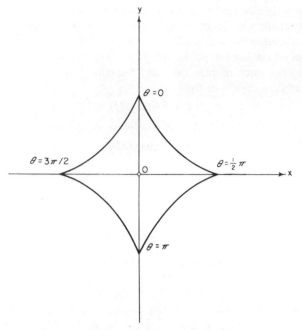

Fig. 129. The astroid.

14.3 The cycloid

If P is a point on the circumference of a circle which rolls without slipping on a straight line, the path of P is called a *cycloid*. In Fig. 130 the initial position of P in contact with the line is taken as the origin

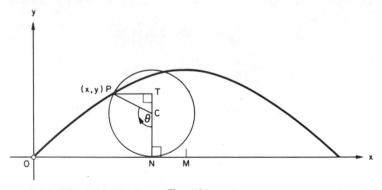

Fig. 130.

and the line itself as the x-axis. The radius of the circle is a and the figure shows one of the positions of the circle along the line. N is the point of contact of the circle with the line, C is the centre of the circle, $\angle NCP = \theta$ and PT is perpendicular to NC. Then, because there is no slipping, $ON = $ arc NP and, if the coordinates of P are (x, y),

$$x = ON - PT = a\theta - a\sin\theta,$$
$$y = CN + CT = a - a\cos\theta.$$

Hence the parametric equations of the cycloid referred to these axes are

$$x = a(\theta - \sin\theta), \quad y = a(1 - \cos\theta).$$

The length of the arc of the cycloid from $O(\theta = 0)$ to A $(\theta = 2\pi)$, i.e., the length of one *arch* of the cycloid, is

$$\int_0^{2\pi} a\sqrt{\{(1 - \cos\theta)^2 + \sin^2\theta\}}\, d\theta = a\int_0^{2\pi} \sqrt{(2 - 2\cos\theta)}\, d\theta$$

$$= a\int_0^{2\pi} 2\sin\left(\frac{\theta}{2}\right) d\theta = \left[-4a\cos\left(\frac{\theta}{2}\right)\right]_0^{2\pi} = 8a.$$

In general, $s = 4a(1 - \cos\frac{1}{2}\theta)$ if s is measured from O. If, however, s is measured from $M(\theta = \pi)$ on the cycloid, and in the reverse direction, $s = 4a\cos\frac{1}{2}\theta$. But

$$\tan\psi = \frac{dy}{d\theta}\Big/\frac{dx}{d\theta} = \frac{\sin\theta}{1 - \cos\theta} = \frac{2\sin\frac{1}{2}\theta\cos\frac{1}{2}\theta}{2\sin^2\frac{1}{2}\theta} = \cot\frac{1}{2}\theta.$$

$$\therefore\ \psi = \tfrac{1}{2}\pi - \tfrac{1}{2}\theta \quad \text{and} \quad \cos\tfrac{1}{2}\theta = \sin\psi.$$

Hence the intrinsic equation of the cycloid with this origin is

$$s = 4a\sin\psi.$$

Exercises 14.3

Calculate the lengths of the arcs of the curves specified in Nos. 1−5.

1. $y = x^2/2$ from $x = 0$ to $x = 1$.
2. $ay^2 = x^3$ from $x = 0$ to $x = a$ in the first quadrant.
3. $x = at^2$, $y = 2at$ from $t = 0$ to $t = 1$
4. $y = \log_e \sec x$ from $x = 0$ to $x = \dfrac{1}{4}\pi$.
5. $x = \tanh t$, $y = \operatorname{sech} t$ from $t = 0$ to $t = 1$.
6. Sketch the curve $y^2 = \dfrac{1}{6}x(2 - x)^2$ and find the length of the whole loop.

14.4 Areas of surfaces of revolution

Definition. If the curve $y = f(x)$ [Fig. 131] from P_1 to P_n describes a complete revolution about the axis of x, each of the chords $P_r P_{r+1}$ will generate the surface of a frustum of a right circular cone. This surface is a developable one and its area S is given by the formula $S = 2\pi l \left(\dfrac{r_1 + r_2}{2} \right)$, where r_1 and r_2 are the radii of the ends of the frustum and l is the slant height. The limit of the sum of the surface areas of these frusta as $n \to \infty$ is defined as the *surface area* generated by this part of the curve in its revolution about the x-axis.

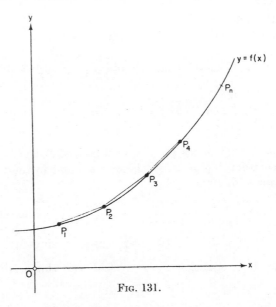

FIG. 131.

In Fig. 132, P is (x, y) and Q is $(x + \delta x, y + \delta y)$ on the curve $y = f(x)$. A is the point $\{a, f(a)\}$ and B is the point $\{b, f(b)\}$. If arc $AP = s$, then s is a function of x and $PQ = \delta s$ where δs is the increment of s resulting from an increase δx in x. Then the surface generated by the arc PQ is approximately $2\pi \left(\dfrac{y + y + \delta y}{2} \right) \delta s$ and hence the total surface area generated by AB is

$$S = \int_{x=a}^{x=b} 2\pi y \, ds \qquad (14.6)$$

and, since
$$\frac{\mathrm{d}s}{\mathrm{d}x} = \sqrt{\left\{1 + \left(\frac{\mathrm{d}y}{\mathrm{d}x}\right)^2\right\}},$$

$$S = 2\pi \int_a^b y \sqrt{\left\{1 + \left(\frac{\mathrm{d}y}{\mathrm{d}x}\right)^2\right\}} \, \mathrm{d}x, \tag{14.7}$$

$$= 2\pi \int_{f(a)}^{f(b)} y \sqrt{\left\{1 + \left(\frac{\mathrm{d}x}{\mathrm{d}y}\right)^2\right\}} \, \mathrm{d}y. \tag{14.8}$$

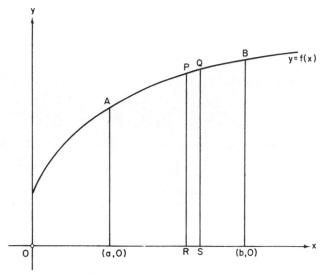

FIG. 132.

If $x = f(t)$, $y = g(t)$, $t = t_1$ when $x = a$ and $t = t_2$ when $x = b$, the total surface generated is

$$S = 2\pi \int_{t_1}^{t_2} g(t) \sqrt{\left\{\left(\frac{\mathrm{d}x}{\mathrm{d}t}\right)^2 + \left(\frac{\mathrm{d}y}{\mathrm{d}t}\right)^2\right\}} \, \mathrm{d}t. \tag{14.9}$$

In all cases we have chosen the sign of S so that it increases with the variable of integration.

Examples. (i) Find the area S of the surface of a sphere of radius R. Consider the sphere generated by the rotation of $x = R\cos\theta$, $y = R\sin\theta$ about the x-axis. Then

$$\frac{\mathrm{d}x}{\mathrm{d}\theta} = -R\sin\theta, \quad \frac{\mathrm{d}y}{\mathrm{d}\theta} = R\cos\theta.$$

$$\therefore S = 2\pi \int_0^\pi R\sin\theta \sqrt{(R^2\sin^2\theta + R^2\cos^2\theta)}\, \mathrm{d}\theta$$

$$= 2\pi \int_0^\pi R^2\sin\theta\, \mathrm{d}\theta = 2\pi R^2\Big[-\cos\theta\Big]_0^\pi = 4\pi R^2.$$

(ii) Find the surface area generated, when the curve

$$y^2 = \frac{1}{16}\, x^2\,(2 - x^2)$$

is rotated completely about the x-axis.

$$\frac{\mathrm{d}y}{\mathrm{d}x} = \pm\, \frac{1 - x^2}{2\sqrt{(2 - x^2)}}.$$

The curve is symmetrical about both axes and $|x| \leq \sqrt{2}$. There are stationary points at $\left(\pm 1, \pm\frac{1}{4}\right)$. Also $\mathrm{d}x/\mathrm{d}y = 0$ at $(\pm\sqrt{2}, 0)$, $\mathrm{d}y/\mathrm{d}x = \pm \sqrt{2}/4$ at $(0, 0)$. The curve is as shown in Fig. 133.

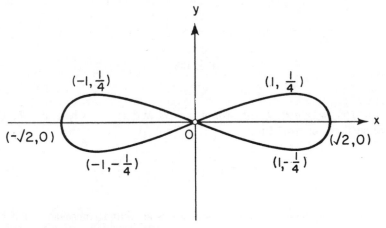

Fɪɢ. 133.

Consider the surface generated by the rotation of the portion of the curve $y = +\dfrac{1}{4}\,x\sqrt{(2-x^2)}$ from $x = 0$ to $x = +\sqrt{2}$.

$$1 + \left(\frac{dy}{dx}\right)^2 = 1 + \frac{(1-x^2)^2}{4(2-x^2)} = \frac{x^4 - 6x^2 + 9}{4(2-x^2)}$$

Then the area of the surface generated is

$$2\int_0^{\sqrt{2}} 2\pi\,\frac{x\sqrt{(2-x^2)}}{4}\,\sqrt{\left\{\frac{x^4-6x^2+9}{4(2-x^2)}\right\}}\,dx = \frac{\pi}{2}\int_0^{\sqrt{2}} x(3-x^2)\,dx.$$

[Note that $\sqrt{(x^4 - 6x^2 + 9)}$ here is $(3-x^2)$ not (x^2-3), since throughout the argument the positive square root is implied.] Hence the surface area generated is

$$\frac{\pi}{2}\left[\frac{3x^2}{2} - \frac{x^4}{4}\right]_0^{\sqrt{2}} = \pi \text{ units}^2.$$

Exercises 14.4

Calculate the areas of the surfaces generated when the following curves, within the stated limits, are rotated completely about the x-axis.

1. $y = \cosh x$; $x = 0$ to $x = 1$.
2. One arch of the cycloid $x = a(\theta - \sin\theta)$, $y = a(1 - \cos\theta)$.
3. $x = a\sin^3 t$, $y = a\cos^3 t$; $x = 0$ to $x = a$.
4. $y^2 = 4x$; $x = 0$ to $x = 1$.
5. $x = t^3$, $y = 3t^2$; $t = 0$ to $t = 2$.
6. $y = e^{-x}$; $x = 0$ to $x = 1$.
7. $x = \operatorname{sech} t$, $y = \tanh t$; $t = 0$ to $t = 1$.
8. $x^2 - y^2 = 1$; $x = 1$ to $x = \sqrt{5}$.

14.5 The theorem of Pappus concerning surfaces of revolution

This is a second theorem of Pappus corresponding to that of § 13.7 and can be stated as follows.

The area of the surface generated by the rotation about an axis of an arc of a curve which does not cut that axis is equal to the length of the arc multiplied by the length of the path described by the centroid of the arc in the rotation.

The y-coordinate of the centroid of the arc AB of Fig. 132 is given by $\bar{y} = \dfrac{1}{s_1} \int\limits_0^{s_1} y\,\mathrm{d}s$, where s_1 is the length of the arc AB. The area of the surface of revolution about the x-axis is given by

$$S_1 = \int\limits_0^{s_1} 2\pi\,y\,\mathrm{d}s \qquad\qquad (14.6)$$

so that the area of the surface of revolution $S_1 = 2\pi\bar{y}s_1$, and in general

$$S = 2\pi\,\bar{y}\,s \qquad\qquad (14.10)$$

which proves the required result.

Examples. (i) Find the area of the surface generated by the complete revolution of a circle of radius r about a tangent.

The centroid of the circumference of the circle is at its centre. Hence the surface S of the solid of revolution is the length of the path traced by the centroid of the arc $\times$ the circumference of the circle, i.e.,

$$S = 2\pi r \cdot 2\pi r = 4\pi^2 r^2.$$

(ii) A uniform heavy chain is suspended from a point A 8 ft above the ground and part of it hangs in the form of a catenary and part lies on the horizontal ground. The chain is horizontal at B, the point of contact with the ground. The length of the chain from A to B is 12 ft. Calculate the height of the centre of mass of the suspended part of the chain above the ground and hence write down the area of the surface that would be generated by the complete revolution of this part of the chain about its projection on the ground.

From example (i) § 14.2, for the catenary $y^2 = c^2 + s^2$. Hence at A

$$(8 + c)^2 = c^2 + 12^2.$$

$$\therefore 64 + 16c + c^2 = c^2 + 12^2.$$

$$\therefore c = 5.$$

$$\therefore \bar{y} = \frac{1}{12} \int\limits_0^{12} y\,\mathrm{d}s = \frac{1}{12} \int\limits_0^{12} \sqrt{(25 + s^2)}\,\mathrm{d}s.$$

Putting $s = 5\sinh t$ gives

$$\bar{y} = \frac{25}{12} \int\limits_{s=0}^{s=12} \cosh^2 t\,\mathrm{d}t = \frac{25}{12}\left[\frac{t}{2} + \frac{\sinh 2t}{4}\right]_{s=0}^{s=12}.$$

When $\qquad\qquad s = 12, \ \sinh t = \dfrac{12}{5}.$

$$\therefore \ \cosh t = \frac{13}{5}, \quad \text{and} \quad \sinh 2t = \frac{2 \cdot 12}{5} \cdot \frac{13}{5} = \frac{312}{25}.$$

Also $\qquad\qquad t = \log_e \left[\frac{12}{5} + \sqrt{\left\{ 1 + \left(\frac{12}{5} \right)^2 \right\}} \right] = \log_e 5.$

$$\therefore \ \bar{y} = \frac{25}{12} \left(\frac{1}{2} \log_e 5 + \frac{78}{25} \right) = \frac{25}{24} \log_e 5 + \frac{13}{2},$$

i.e., the height of the centre of mass of the suspended part of the chain above the ground is

$$\bar{y} - c = \left(\frac{25}{24} \log_e 5 + \frac{3}{2} \right) \text{ft}$$

and the required surface of revolution is

$$\pi (25 \log_e 5 + 36) \text{ ft}^2.$$

Exercises 14.5

1. Calculate the surface of the anchor ring generated by the revolution of a circle of radius r about a line in its plane and distance c from its centre. ($r < c$.)

2. Calculate the area of the surface of revolution of the circle $x^2 + y^2 - 4x + 2y - 4 = 0$ about the line $5x + 12y + 54 = 0$.

3. Use the answer obtained for § 14.2, Example (ii) to calculate the area of the surface of revolution of the astroid $x^{2/3} + y^{2/3} = a^{2/3}$ about the line $y = 2a$.

4. Calculate the area of the surface of revolution of an equilateral triangular framework of thin uniform wires each of length $2a$ about a line in its plane through one vertex at right angles to one of the sides which intersect there.

5. A groove is cut completely round a cylinder of radius a so that the plane of symmetry of the groove is at right angles to the axis of the cylinder and the section of the groove by a plane through the axis of the cylinder is a semicircle of radius $a/8$. Calculate the surface area of the groove. (The centroid of a semicircular arc of radius r is distant $2r/\pi$ from the centre of the circle.)

6. Prove that if the area of the surface of revolution of the ellipse $x = a \cos \theta$, $y = b \sin \theta$ about the tangent at $\theta = \frac{1}{4}\pi$ is one third of the area of the surface of revolution of the ellipse about the line $x = 2a$, the eccentricity of the ellipse is $\sqrt{(5/7)}$.

7. Semicircles are described on each of the two shorter sides of a right-angled triangle in which the hypotenuse is of length a and one of the acute angles is θ. If the semicircular arcs are rotated completely about the hypotenuse, find an expression in terms of a and θ for the surface area generated and verify that it has a maximum value as θ varies when $\theta = 45°$. Calculate this maximum value.

14.6 Curvature

The curvature $(\varkappa)$ at a point on the curve $y = f(x)$ is defined as the rate of change of ψ with respect to s, i.e., as $\dfrac{\mathrm{d}\psi}{\mathrm{d}s}$, for that point.

$$\varkappa = \frac{\mathrm{d}\psi}{\mathrm{d}s}. \tag{14.11}$$

With the conventions we have adopted, s increases with x and therefore $\dfrac{\mathrm{d}\psi}{\mathrm{d}s}$ will be positive if ψ increases with x and negative if ψ decreases with x. If, however, the equations of the curve are given in such a way that these conventions do not apply the best course is to consider each case from first principles. For most purposes it is sufficient to obtain $|\varkappa|$.

The curvature at any point on a circle of radius r

In Fig. 134, s is measured from the point A on the circle, CA is parallel to the y-axis, P is any point on the circle, the arc $AP = s$ and $\angle ACP = \psi$.

$$\therefore s = r\psi,$$

$$\frac{\mathrm{d}\psi}{\mathrm{d}s} = \frac{1}{r}.$$

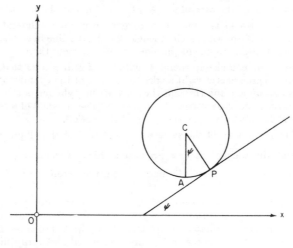

FIG. 134.

The curvature of a circle is inversely proportional to its radius

This result illustrates the definition of curvature given above. Curvature is the measure of the rate at which the tangent to the curve rotates compared with the distance moved along the curve, or *the rate at which the curve curves*. The reciprocal of the modulus of the curvature, $|\varkappa|$, at a point P on a curve is the radius of a circle which has the same curvature as the curve at P. It is called the *radius of curvature* of the curve at P and is denoted by ϱ.

$$\varrho = \frac{1}{|\varkappa|} = \left| \frac{\mathrm{d}s}{\mathrm{d}\psi} \right|. \qquad (14.12)$$

Fig. 135 (i) shows the point $P(x, y)$ on the curve $y = f(x)$ for which ψ increases with s at P. If the normal GP is produced to C so that $PC = \varrho$, C is called the *centre of curvature* for the curve at P and the circle with centre C and radius CP is called the *circle of curvature*. The centre of curvature is at a distance ϱ from P along the normal in the inward direction.

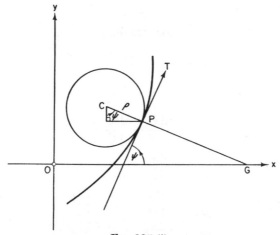

Fig. 135 (i).

The positive direction of the normal is defined as that direction which makes an anticlockwise angle (i.e., a positive angle) of $\frac{1}{2}\pi$ with the positive direction of the tangent.

Fig. 135 (ii) shows the centre and circle of curvature in a case in which the curvature is negative at the point.

If P is (x, y), the coordinates of the centre of curvature are (ξ, η) where $\xi = x \mp \varrho \sin\psi$, $\eta = y \pm \varrho \cos\psi$, the ambiguous sign being determined by the sign of $\varkappa$ at P. The locus of the centre of curvature as P moves on the curve is called the *evolute*.

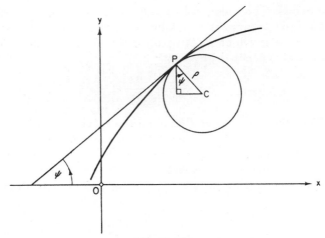

FIG. 135 (ii).

Curvature in cartesian coordinates

$$\tan\psi = \frac{\mathrm{d}y}{\mathrm{d}x}.$$

$$\therefore \sec^2\psi \frac{\mathrm{d}\psi}{\mathrm{d}s} = \frac{\mathrm{d}^2y}{\mathrm{d}x^2} \frac{\mathrm{d}x}{\mathrm{d}s}.$$

But, from § 14.2, $\qquad \dfrac{\mathrm{d}x}{\mathrm{d}s} = \cos\psi.$

$$\therefore \varkappa = \frac{\mathrm{d}\psi}{\mathrm{d}s} = \frac{\dfrac{\mathrm{d}^2y}{\mathrm{d}x^2}}{\sec^3\psi} = \frac{\dfrac{\mathrm{d}^2y}{\mathrm{d}x^2}}{(1 + \tan^2\psi)^{3/2}}.$$

$$\therefore \varkappa = \frac{\mathrm{d}^2y}{\mathrm{d}x^2} \bigg/ \left\{ 1 + \left(\frac{\mathrm{d}y}{\mathrm{d}x}\right)^2 \right\}^{3/2}, \tag{14.13}$$

$$\varrho = \left| \left\{ 1 + \left(\frac{\mathrm{d}y}{\mathrm{d}x}\right)^2 \right\}^{3/2} \bigg/ \frac{\mathrm{d}^2y}{\mathrm{d}x^2} \right|. \tag{14.14}$$

The curvature is positive or negative according as $\mathrm{d}^2y/\mathrm{d}x^2$ is positive or negative [the positive sign of the radical being taken in the formulae

(14.13) and (14.14)], i.e., according to whether the curve is concave upwards, Fig. 124 (ii), or concave downwards, Fig. 124 (i). A necessary condition for an inflexion is $d^2y/dx^2 = 0$; therefore the curvature at a point of inflexion is zero.

Curvature in parametric coordinates

If

$$x = f(t), \ y = g(t)$$

and if $\cdot dx/dt$ is denoted by $\dot{x}$ and d^2x/dt^2 by $\ddot{x}$ etc., then

$$\frac{dy}{dx} = \frac{\dot{y}}{\dot{x}}.$$

$$\therefore \frac{d^2y}{dx^2} = \frac{d\left(\dfrac{dy}{dx}\right)}{dt} \bigg/ \frac{dx}{dt} = \frac{d}{dt}\left(\frac{\dot{y}}{\dot{x}}\right)\bigg/\frac{dx}{dt}$$

$$= \frac{\dot{x}\,\ddot{y} - \dot{y}\,\ddot{x}}{\dot{x}^2}\,\frac{1}{\dot{x}} = \frac{\dot{x}\,\ddot{y} - \dot{y}\,\ddot{x}}{\dot{x}^3}.$$

$$\therefore \varkappa = \frac{\dot{x}\,\ddot{y} - \dot{y}\,\ddot{x}}{\dot{x}^3}\,\frac{1}{\left\{1 + \dfrac{\dot{y}^2}{\dot{x}^2}\right\}^{3/2}}.$$

$$\therefore \varkappa = \frac{\dot{x}\,\ddot{y} - \dot{y}\,\ddot{x}}{(\dot{x}^2 + \dot{y}^2)^{3/2}}, \tag{14.15}$$

$$\therefore \varrho = \left| \frac{(\dot{x}^2 + \dot{y}^2)^{3/2}}{\dot{x}\,\ddot{y} - \dot{y}\,\ddot{x}} \right|. \tag{14.16}$$

Examples. (i) The normal to the curve $4y = x^2$ at the point $P(2\sqrt{2}, 2)$ meets the curve again at Q. Show that PQ is the radius of curvature at P. (N.)

$$4y = x^2.$$

$$\therefore \frac{dy}{dx} = \tfrac{1}{2}x, \qquad \frac{d^2y}{dx^2} = \tfrac{1}{2}.$$

$$\therefore \varrho = 2\left(1 + \frac{1}{4}x^2\right)^{3/2}$$

Hence at P,

$$\varrho = 6\sqrt{3}.$$

The normal at P is

$$y - 2 = -\frac{1}{\sqrt{2}}(x - 2\sqrt{2}),$$

i.e.,

$$x + \sqrt{2}y - 4\sqrt{2} = 0.$$

This meets the curve where

$$\sqrt{2}\,x^2 + 4x - 16\sqrt{2} = 0,$$

i.e., where $\qquad (x - 2\sqrt{2})(\sqrt{2}x + 8) = 0.$

Hence the normal meets the curve again at $(-4\sqrt{2}, 8)$. Thus the length of PQ is $\sqrt{\{(6\sqrt{2})^2 + 6^2\}} = 6\sqrt{3}$ so that PQ is the radius of curvature at P. Since PQ is the positive direction of the normal, Q is the centre of curvature corresponding to the point P.

(ii) Find an expression for $\varkappa$ in terms of θ for the astroid $x = a\sin^3\theta$, $y = a\cos^3\theta$.

$$\frac{dx}{d\theta} = 3a\sin^2\theta\cos\theta, \qquad \frac{dy}{d\theta} = -3a\cos^2\theta\sin\theta.$$

$$\therefore \frac{dy}{dx} = -\cot\theta.$$

$$\frac{d^2y}{dx^2} = \frac{d\left(\dfrac{dy}{dx}\right)}{d\theta} \cdot \frac{d\theta}{dx}$$

$$= \frac{\operatorname{cosec}^2\theta}{3a\sin^2\theta\cos\theta} = \frac{1}{3a\sin^4\theta\cos\theta}.$$

$$\therefore \varkappa = \frac{1}{3a(1 + \cot^2\theta)^{3/2}\sin^4\theta\cos\theta} = \frac{1}{3a\sin\theta\cos\theta}.$$

(iii) Find the radius of curvature for the curve

$$x^2 + y^2 - 5xy + 3y - 1 = 0$$

at the point $(1, 2)$.

Differentiation with respect to x gives

$$2x + 2y\frac{dy}{dx} - 5x\frac{dy}{dx} - 5y + 3\frac{dy}{dx} = 0. \tag{1}$$

Hence at $(1, 2)$, $\qquad\qquad dy/dx = 4.$

Differentiating eqn. (1) with respect to x gives

$$2 + 2\left(\frac{dy}{dx}\right)^2 + 2y\frac{d^2y}{dx^2} - 5\frac{dy}{dx} - 5x\frac{d^2y}{dx^2} - 5\frac{dy}{dx} + 3\frac{d^2y}{dx^2} = 0.$$

Hence at $(1, 2)$, using $dy/dx = 4$ there,

$$2 + 32 + 4\frac{d^2y}{dx^2} - 20 - 5\frac{d^2y}{dx^2} - 20 + 3\frac{d^2y}{dx^2} = 0.$$

Thus at $(1, 2)$, $d^2y/dx^2 = 3$. Hence ϱ at $(1, 2)$ is $\dfrac{(1 + 16)^{3/2}}{3} = \dfrac{1}{3}\,17\sqrt{17}.$

14.7 Newton's formula for radius of curvature at the origin

If the curve $y = f(x)$ touches the x-axis at the origin, $f(0) = 0$ and $f'(0) = 0$ and therefore $\varrho = |1/f''(0)|$. From the Maclaurin series (Vol. I, § 9.2):

$$y = f(0) + x f'(0) + \frac{x^2}{2} f''(0) + \cdots$$

and in our case

$$y = \frac{x^2}{2} f''(0) + \text{terms involving higher powers of } x.$$

$$\therefore \frac{2y}{x^2} = f''(0) + \text{terms containing } x \text{ as a factor.}$$

$$\therefore \varkappa \text{ at the origin} = f''(0) = \lim_{x \to 0} \frac{2y}{x^2}. \tag{14.17}$$

$$\therefore \varrho \text{ at the origin} = \left| \lim_{x \to 0} \frac{x^2}{2y} \right|. \tag{14.18}$$

If the curve touches the y-axis at the origin,

$$\varkappa \text{ at the origin} = \lim_{x \to 0} \frac{2x}{y^2}, \tag{14.19}$$

$$\varrho \text{ at the origin} = \left| \lim_{x \to 0} \frac{y^2}{2x} \right|. \tag{14.20}$$

Newton's formula has a limited application because it applies only to curves which *touch* one axis at the origin. It is sometimes convenient to move the origin in order to make use of the formula.

Examples. (i) Calculate the radius of curvature at the origin of the curve

$$a y^2 = x(a - x)^2.$$

The substitution $x = 0$ gives an equation in y with a repeated root. The curve, therefore, touches the y-axis at the origin. Hence ϱ at the origin is

$$\left| \lim_{x \to 0} \frac{y^2}{2x} \right| = \left| \lim_{x \to 0} \frac{x(a - x)^2}{2xa} \right| = \frac{a}{2}.$$

(ii) Calculate the curvature of $y = \sin^2(x - \pi/6)$ at the point $(\pi/6, 0)$.

F

If we move the origin to $(\pi/6, 0)$, the equation becomes $Y = \sin^2 X$ and this curve touches the X-axis at the new origin. Hence the curvature at the new origin is

$$\lim_{X \to 0} \frac{2Y}{X^2} = \lim_{X \to 0} 2\left(\frac{\sin X}{X}\right)^2 = 2,$$

i.e., the curvature of $y = \sin^2(x - \pi/6)$ at $(\pi/6, 0)$ is 2.

Exercises 14.7

Calculate the curvature for each of the curves in Questions $1-10$ at the points specified. Leave answers, where necessary, in surd form.

1. $y^2 = 4ax$; $(a, 2a)$.

2. $x = a \cos\theta,\ y = b \sin\theta$; $(\theta = \frac{1}{4}\pi)$.

3. $x = a \cosh t,\ y = b \sinh t$; $(t = 1)$.

4. $y = \log x$; $(x = 1)$.

5. $s = a \sec^3\psi$; $(\psi = \frac{1}{4}\pi)$.

6. $ay^2 = x^3$; $(4a, 8a)$.

7. $y = c \cosh(x/c)$; $(0, c)$.

8. $y = \log \sin x$; $(x = \frac{1}{2}\pi)$.

9. $x^3 + y^3 + 3xy + 3x = 0$; $(1, -1)$.

10. $x^2 + 3xy + 3y^2 + 6x + 3y - 6 = 0$; $(0, 1)$.

Use Newton's formula to calculate the radius of curvature at the point specified in each of Nos. $11-15$.

11. $y^2 = a^2 x/(a - x)$; $(0, 0)$.

12. $a^2 y^2 = x(a - x)^3$; $(0, 0)$.

13. $y = x^2/(1 + x^2)$; $(0, 0)$.

14. $y = c \cosh^2(x/c)$; $(0, c)$.

15. $x^2/a^2 + y^2/b^2 = 1$; $(a, 0)$.

16. Show that the radius of curvature at any point of the cycloid

$$x = a(\theta - \sin\theta),\quad y = a(1 - \cos\theta)$$

is of magnitude $4a \sin(\theta/2)$ and that the corresponding centre of curvature is the point $\{a(\theta + \sin\theta), a(\cos\theta - 1)\}$.   (L.)

17. Show that the curves $x^2 - y^2 = 3a^2$ and $xy = 2a^2$ intersect at right angles, and that at the point of intersection their radii of curvature are in the ratio $4 : 3$. (L.)

18. Prove that the circle of curvature at the point $\theta = \pi/6$ on the astroid $x = a \cos^3\theta,\ y = a \sin^3\theta$ touches the y-axis.   (L.)

19. Prove that the centre of curvature at the point $\{\alpha, f(\alpha)\}$ on the curve $y = f(x)$ has coordinates given by the equations

$$x = \alpha - f'(\alpha)[1 + \{f'(\alpha)\}^2]/f''(\alpha),$$
$$y = f(\alpha) + [1 + \{f'(\alpha)\}^2]/f''(\alpha).$$

The normal at the point P of the catenary $y = c \cosh(x/c)$ meets the axis of x in N. Prove that PN is equal in magnitude to the radius of curvature of the curve at P.   (L.)

20. Show that the equation of the normal to the parabola $y^2 = 4ax$ at the point $P(at^2, 2at)$ is

$$y + tx - 2at - at^3 = 0.$$

Find the coordinates of the centre of curvature of the parabola at P. Hence, or otherwise, show that the locus of the centre of curvature of the parabola is the curve

$$27ay^2 = 4(x - 2a)^3.$$ (L.)

Miscellaneous Exercises XIV

1. Find the radius of curvature at the point t on the ellipse $x = \sqrt{2} \cos t$, $y = \sin t$. Show that the circle of curvature at either end of the major axis of the ellipse passes through the centre of the ellipse.   (N.)

2. Prove that the radius of curvature ϱ of the curve $y = x^n$ $(n > 1)$ at the point $(1, 1)$ is given by

$$\varrho = \frac{(1 + n^2)^{3/2}}{n^2 - n}.$$

Show that $\varrho \to \infty$ when $n \to 1$ and when $n \to \infty$, and that ϱ is a minimum when $n = \frac{1}{2}(3 + \sqrt{5})$.   (N.)

3. Show that there is just one turning point on the curve $y = x \log_e x$, and determine the nature of this point. Sketch the curve.

Find the curvature at a general point on this curve, and show that at the turning point the radius of curvature has the same magnitude as the y-coordinate. (N.)

4. If $y = 2/(\sinh x)$, prove that

$$1 + (dy/dx)^2 = \frac{1}{4}(y^2 + 2)^2.$$

Deduce that the radius of curvature at any point (x, y) on the curve $y = 2/(\sinh x)$ is equal to $(y^2 + 2)^2/(4y)$.   (N.)

5. Prove that the graph of the function

$$x^4 - 2x^3 + 8x - 4$$

(i) crosses the axis of x between $x = 0$ and $x = 1$ and also between $x = -2$ and $x = -1$, (ii) has a minimum point between $x = -1$ and $x = 0$. Find the coordinates of the points of inflexion and sketch the graph.   (N.)

6. Calculate the area of the surface of revolution of the curve $y = x^3$ from $x = 0$ to $x = 1$ about the x-axis.

7. Obtain the equations of the tangents at the two points of inflexion on the curve $y = 3x^2 - 1/x^2$ and determine the coordinates of the points where these tangents meet the curve again.   (N.)

8. A fixed point P of a circular disc of radius $2a$ is at a distance a from its centre. Show that the curve traced out by P when the disc rolls without slipping on the outside of a fixed circle of radius $6a$ may be expressed by the parametric equations

$$x = 8a \cos\theta - a \cos 4\theta, \quad y = 8a \sin\theta - a \sin 4\theta.$$

Show that the path of P has zero curvature at six points, situated at the vertices of two equilateral triangles.   (N.)

9. Find the tangents at the origin and at the point $(-a, 0)$ to each of the following curves:

   (i) $y^2(a - x) = x^2(a + x)$,

   (ii) $a^3 y^2 (a - x) = x^2 (a + x)^4$.

Sketch the graphs of the above two curves. Find (by Newton's method or otherwise) the radius of curvature of the curve (i) at the point $(-a, 0)$.   (N.).

10. The line $x = 4a$ cuts the parabola $y^2 = 4ax$ at the points A and B. Find

   (i) the gradient of the curve at A and at B;

   (ii) the area enclosed between the chord AB and the arc AB of the parabola;

   (iii) the length of the arc AB.   (O.C.)

11. The coordinates of a point of a curve are given in terms of a parameter t by the equations $x = te^t$, $y = t^2 e^t$. Find dy/dx in terms of t, and prove that

$$\frac{d^2 y}{dx^2} = \frac{t^2 + 2t + 2}{(t + 1)^3} e^{-t}.$$

Prove also that the radii of the circles of curvature at the two points at which the curve is parallel to the x-axis are in the ratio $e^2 : 1$.   (O.C.)

12. Prove that the curve

$$y = x^2 + \frac{32}{x + 3}$$

has only one turning point, which is at $x = 1$, and state whether this point determines a maximum or a minimum. Prove also that the curve has one and only one point of inflexion.

Draw a rough sketch of the curve.   (N.)

13. Find the equation connecting s and ψ, where s is the length of arc OP and $\tan \psi$ is the slope of the tangent at P, if P is any point on the cycloid

$$x = a(\theta + \sin\theta), \quad y = a(1 - \cos\theta).$$

Hence or otherwise show that the radius of curvature at P is twice the distance PN where N is the point of intersection of the line $y = 2a$ with the normal at P.   (O.C.)

14. Sketch the curve $y = \sin^2 2x$ between $x = -\frac{1}{2}\pi$ and $x = \frac{1}{2}\pi$. Use Newton's formula to find the radius of curvature at the point $(0, 0)$.   (N.)

15. If $f(x) = (2 - x^2)\sin x - 2x\cos x$, show that

(i) the expansion of $f(x)$ in ascending powers of x is of the form

$$f(x) = kx^3 + \text{higher powers of } x,$$

and obtain the value of k,

(ii) the graph of $f(x)$ has a stationary point at $x = 0$ and another at $x = \frac{1}{2}\pi$, and determine in each case whether the point is a maximum or minimum point or a point of inflexion.

(iii) the equation $f(x) = 0$ has a root between $x = \frac{1}{2}\pi$ and $x = \pi$. Sketch the graph of $f(x)$ from $x = 0$ to $x = \pi$. (N.)

16. Determine the equations of the tangents at the points of inflexion on the curve $y = x^2(x - 4)(x - 6)$ and the coordinates of the points where these tangents meet the curve again. (N.)

17. Sketch the arc of the curve $x = a\cos^3 t$, $y = a\sin^3 t$ for $0 \leq t \leq \frac{1}{2}\pi$.
Find the radius of curvature of the curve at the point t.
The normal at the point P whose parameter is t meets the x-axis at G, and C is the centre of curvature at P. Find the non-zero value of t for which P is the mid-point of CG. (N.)

18. Find the equation of the tangent to the curve $x = 3t^2$, $y = 2t^3$ at the point P whose parameter t is $\sqrt{2}$.
Determine the parameter of the point Q where this tangent meets the curve again and prove that PQ is the normal at Q.
Find the radius of curvature at Q. (N.)

19. Find the value of dy/dx in terms of t at a point on the curve whose parametric equations are

$$x = a(t - \tanh t), \quad y = a\,\text{sech}\,t.$$

Prove that

$$a\frac{d^2 y}{dx^2} = \frac{\cosh^3 t}{\sinh^4 t}.$$

Hence find the length of the radius of curvature at the point whose parameter is t. (N.)

20. Given that $y = (x + 1)e^{-x}$ find dy/dx and $d^2 y/dx^2$. Calculate the coordinates of the turning point and of the point of inflexion on the graph of y, and sketch the graph. (N.)

21. A curve is given by the parametric equations

$$x = a(t\sin t + \cos t - 1), \quad y = a(\sin t - t\cos t).$$

Find dy/dx and $d^2 y/dx^2$ in terms of t and prove that, at the point P whose parameter is t, the length of the radius of curvature is $|at|$.
Obtain the equation of the normal at P. Show that all normals touch the circle $x^2 + y^2 + 2ax = 0$. (N.)

22. Find a point of inflexion on the curve

$$x = a(1 + t^3),$$
$$y = a(1 + 9t + 2t^3 + t^6).$$

23. Find the length of the curve $2y = c(e^{x/c} + e^{-x/c})$ between the points $(0, c)$ and $\left(c, \dfrac{c\,e^2 + c}{2\,e} \right)$.

Find also the curved surface and volume of the solid obtained by revolving the curve between these two points about the axis of x.

24. A curve is given in the form

$$x = \cosh t - t,$$
$$y = \cosh t + t.$$

Express t in terms of the length s of the arc of the curve measured from the point $(1, 1)$.

The coordinates of any point of the curve are expressed in terms of s and are then expanded in series of ascending powers of s. Prove that the first few terms of these expansions are

$$x = 1 - \frac{s}{\sqrt{2}} + \frac{s^2}{4} + \cdots,$$
$$y = 1 + \frac{s}{\sqrt{2}} + \frac{s^2}{4} + \cdots. \tag{O.C.}$$

25. Trace the curve $x = at$, $y = a(1 - \cos t)$ for values of t between -2π and 2π.

Determine the inflexions on the curve, and prove that if the tangents at two real points P and Q are at right angles, then both P and Q are inflexions. (O.C.)

26. An arc of a circle of radius c subtending an angle 2α at the centre is rotated about its chord. Prove that the area of the surface of revolution so formed is $4\pi c^2(\sin\alpha - \alpha\cos\alpha)$, and that its volume is

$$\frac{2}{3}\pi c^3(2\sin\alpha + \sin\alpha\cos^2\alpha - 3\alpha\cos\alpha). \tag{O.C.}$$

27. A cycloid, given by the equations

$$x = a(1 - \cos\theta), \ y = a(\theta + \sin\theta),$$

revolves about its base $x = 2a$; find the area of the surface, and the volume of the solid of revolution so formed. (O.C.)

28. Show that the arc of the curve given by

$$x = a\cos t, \ y = \frac{1}{4}a\cos 2t$$

between points for which $t = 0$, $t = \frac{1}{2}\pi$, is equal to

$$\int_0^1 a(1 + z^2)^{\frac{1}{2}}\, dz. \tag{O.C.}$$

29. Sketch the curve $y^2 = (2 - x)/(x + 6)$.

Show that there are points of inflexion at $(0, \pm 3^{-\frac{1}{2}})$, and find the radius of curvature at the point $(-2, 1)$. (O.C.)

30. Show that the radius of curvature of the cycloid $x = a(\theta - \sin\theta)$, $y = a(1 - \cos\theta)$ at the point θ is $4a\sin(\theta/2)$. (L.)

31. Find the coordinates of the point of intersection K, other than the origin, of the parabolas $y^2 = 4ax$ and $2x^2 = ay$.

Show that the parabolas intersect at K at an angle $\tan^{-1}(3/5)$ and find their radii of curvature at this point. (L.)

32. The parametric equations of a curve are

$$x = a(\cos\theta + \theta\sin\theta), \quad y = a(\sin\theta - \theta\cos\theta),$$

where θ is the parameter and a a constant.

Find the radius of curvature, ϱ, in terms of θ, and the coordinates of the centre of curvature. Show that the centre of curvature lies on a circle of radius a. (L.)

33. Sketch the graph of the plane curve

$$x = \sin^3\theta, \quad y = 3\cos^3\theta$$

for values of θ lying in the range 0 to 2π.

P is the point on the curve for which $\theta = \pi/3$ and C is the centre of curvature at P. Show that the equation of the line CP is $\sqrt{3}\,y = x$ and find the length of CP. (L.)

34. A curve is given parametrically by $x = e^t\cos t$, $y = e^t\sin t$.

Show that the tangent at any point t makes an angle $(t + \pi/4)$ with the x-axis, and that the radius of curvature at 't' is $\sqrt{2}e^t$. (L.)

35. Show that, at any point on the curve whose parametric equations are $x = c\log_e\tan(\tfrac{1}{2}\theta)$, $y = c\operatorname{cosec}\theta$ (i) the normal makes an angle θ with the x-axis, (ii) the radius of curvature is given by y^2/c, (iii) the distance of the centre of curvature from the x-axis is $2y$. (L.)

36. Sketch the curve $ay^2 = x^2(x + a)$ where $a > 0$.

Find the equations of the tangents to the curve at the origin. Show that the radius of curvature of the curve at the point $\left(-\dfrac{a}{2}, \dfrac{a}{2\sqrt{2}}\right)$ is $\dfrac{27\,a}{40}$. (L.)

37. Obtain the parametric equations of the cycloid in the form $x = a(\theta - \sin\theta)$, $y = a(1 - \cos\theta)$. If O is the origin and P is any point on the curve, prove that $s = 4a(1 - \sin\psi)$ where s is the length of the arc OP and ψ is the angle between the tangent at P and the x-axis. Deduce the radius of curvature ϱ at the point P in terms of ψ and prove that $\varrho^2 = s(8a - s)$. (L.)

38. Show that the radius of curvature at the point t on the curve $x = 2a\sin t + a\sin 2t$, $y = 2a\cos t + a\cos 2t$ is

$$\frac{8}{3}a\cos\left(\frac{t}{2}\right). \tag{L.}$$

39. Prove that the length of the arc of the parabola $y^2 = 4ax$, between $y = 0$ and $y = 2a$, is

$$a\{\sqrt{2} + \log(1 + \sqrt{2})\}.$$

This arc is rotated through 2π radians about the x-axis. Find the area of the surface of revolution generated. Hence, or otherwise, find the distance of the centroid of the arc from the line $y = 0$. (L.)

40. Prove that the length s of the arc of the catenary $y = c\cosh(x/c)$ from the point where the curve cuts the y-axis to the point where $x = a$ is given by $s = c\sinh(a/c)$.

From any point P on the catenary the ordinate PQ is drawn, meeting the x-axis at Q. From Q a perpendicular QR is drawn to the tangent at P, meeting the latter at R. Prove that the length of QR is constant and equal to c, and that the length of PR is equal to s.   (L.)

41. Sketch the curve given by $x = 4t$, $y = t^2 - 2 \log_e t$ where t is a variable parameter, for values of t from 1 to 3.

If A and B are the points corresponding to $t = 1$ and $t = 3$ respectively, calculate

(i) the length of the arc AB,

(ii) the area bounded by the arc AB, the x-axis and the ordinates through A and B.   (L.)

42. Sketch the curve $3ay^2 = x(x - a)^2$, where $a > 0$.
Show that

$$\frac{dy}{dx} = \pm \frac{(3x - a)}{2\sqrt{(3ax)}}$$

and, hence or otherwise, show that the perimeter of the loop of the curve is $\dfrac{4a}{\sqrt{3}}$.   (L.)

CHAPTER XV

POLAR COORDINATES

15.1 Definitions

Thus far we have fixed the position of a point in a plane by means of cartesian coordinates only. There are many curves, however, for which cartesian equations are algebraically clumsy and in such cases different coordinate systems are used. One of these systems is provided by *Polar Coordinates*. In this system the position of a point is fixed by reference to its distance from a fixed point called the *Pole*, and to the angular displacement of the line which joins it to the pole from a fixed line through the pole, which is called the *Initial Line*. In Fig. 136, O is the pole and Ol the initial line. If $OP = r$ and $\angle lOP = \theta$, the polar coordinates of P are (r, θ).

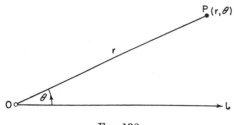

FIG. 136.

With this definition r is a scalar and any point of the plane has a unique pair of coordinates. In two cases, however, variations from the definitions are conventionally accepted.

(i) Negative values of r are defined as values of r measured from the pole in a direction opposite to the direction of r-positive. Thus, the coordinate pairs (r_1, θ_1), $(-r_1, \pi + \theta_1)$ represent the same point. This extension of the definition allows such an equation as $r = 1 + 2\cos\theta$ to be represented by a curve which includes the negative values of r arising from the equation.

(ii) For curves such as $r = k\theta$, values of $\theta > 2\pi$ are considered. A point of the curve, therefore, which, considered merely as a point in the coordinate plane, has coordinates (r_1, θ_1), when referred to the equation, can have coordinates $(r_1, \theta_1 + 2n\pi)$ where n is a positive integer. With this extension of the definition no point of the plane has a unique coordinate pair, but any one of its coordinate pairs defines the point uniquely.

In the light of such variations it is necessary, when working with polar coordinates, to consider many of the problems from first principles.

15.2 Loci in polar coordinates

The straight line. (i) If a straight line goes through the pole and makes an angle α with the initial line, its equation is

$$\theta = \alpha. \tag{15.1}$$

(This equation assumes points in one part of the line to have negative r-coordinates.)

(ii) In Fig. 137 the straight line AB makes an angle α with the initial line and cuts the initial line at $A(c, 0)$; $P(r, \theta)$ is any point on the line and ON is perpendicular to AB. Then from the triangle OPA

$$\frac{r}{\sin\alpha} = \frac{c}{\sin(\alpha - \theta)},$$

so that
$$r = c \sin\alpha \,\operatorname{cosec}(\alpha - \theta) \tag{15.2}$$

is the equation of a straight line which cuts the initial line at $(c, 0)$ and makes an angle α with the initial line.

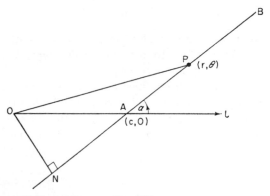

Fig. 137.

If $ON = p$, then

$$r = p \operatorname{cosec}(\alpha - \theta) \tag{15.3}$$

is the equation of the straight line which makes an angle α with the initial line and which is at a perpendicular distance p from the pole.

The circle. (i) The equation of a circle with radius a and centre at the pole is

$$r = a. \tag{15.4}$$

(ii) Fig. 138 shows a circle of radius a which passes through the pole and has its centre on the initial line. If $P(r, \theta)$ is any point on the circle, then

$$r = 2a \cos\theta \tag{15.5}$$

and this is the equation of a circle through the pole with its centre on the initial line, and of radius a.

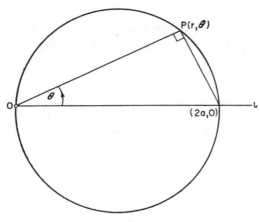

FIG. 138.

(iii) Fig. 139 shows a circle centre $C(\varrho, \alpha)$ and radius a. P is any point (r, θ) on the circle. Then from the triangle POC

$$a^2 = r^2 + \varrho^2 - 2r\varrho \cos(\theta - \alpha).$$

$$\therefore \ r^2 - 2r\varrho \cos(\theta - \alpha) + (\varrho^2 - a^2) = 0 \tag{15.6}$$

is the equation of a circle with centre (ϱ, α) and radius a.

The line $\theta = \theta_1$ cuts this circle where $r^2 - 2r\varrho \cos(\theta_1 - \alpha) + (\varrho^2 - a^2) = 0$ and the roots of this equation in r give the r-coordinates

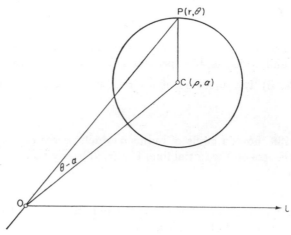

FIG. 139.

of the two points in which the line cuts the circle. If the roots are equal, the line is a tangent to the circle and if they are not real the line does not cut the circle.

Examples. (i) The centre of a circle is $\left(4a, \dfrac{1}{4}\pi\right)$ and its radius is a. The line $\theta = \dfrac{1}{3}\pi$ cuts the circle at P_1, P_2.

(a) Calculate the value of $OP_1 \cdot OP_2$ where O is the pole.

(b) Find the equations of the tangents from O to the circle.

(c) Calculate the angle subtended by the circle at the pole.

The equation of the circle (15.6) is

$$r^2 - 8ra \cos\left(\theta - \frac{1}{4}\pi\right) + 15a^2 = 0. \qquad (1)$$

The line $\theta = \dfrac{1}{3}\pi$ cuts this circle where

$$r^2 - 8ra \cos(\pi/12) + 15a^2 = 0.$$

Then (a) OP_1, OP_2 are the roots of this equation in r.

$$\therefore \; OP_1 \cdot OP_2 = 15a^2.$$

(b) Equation (1) has a repeated root if

$$64a^2 \cos^2\left(\theta - \frac{1}{4}\pi\right) = 60a^2,$$

i.e., if

$$\cos\left(\theta - \frac{1}{4}\pi\right) = \pm\frac{1}{4}\sqrt{15}, \ .$$

whence

$$\theta - \frac{1}{4}\pi = \pm\cos^{-1}\left(\frac{1}{4}\sqrt{15}\right).$$

Hence the equations of the tangents from the pole are

$$\theta = \frac{1}{4}\pi + \cos^{-1}\left(\frac{1}{4}\sqrt{15}\right) \quad \text{and} \quad \theta = \frac{1}{4}\pi - \cos^{-1}\left(\frac{1}{4}\sqrt{15}\right).$$

(c) The angle subtended by the circle at the pole is

$$2\cos^{-1}\left(\frac{1}{4}\sqrt{15}\right) = \cos^{-1}\left(\frac{7}{8}\right).$$

This method of obtaining these results has been used as an illustration of the use of the quadratic equation in r. The results could have been obtained directly from the geometrical properties of the figure.

(ii) A circular disc rolls without slipping round the outside of an equal circular disc. Find a polar equation of the locus of a point on the circumference of the rolling disc, referred to a suitably chosen pole and initial line.

Fig. 140 shows the position of the rolling disc which started with the point P_2 of its circumference in contact with the point P_1 of the circumference of the fixed disc. The line joining the centres A, B of the discs has rotated through an anticlockwise angle P_1AB. If K is the point of contact of the discs, then

$$\text{arc } P_1K = \text{arc } KP_2.$$

$$\therefore \ \angle P_1AB = \angle ABP_2.$$

But

$$\Delta BP_2A \equiv \Delta AP_1B$$

and therefore these triangles are equal in area.

$$\therefore \ AB \text{ is parallel to } P_1P_2.$$

$$\therefore \ \angle BP_2R = \angle AP_1R = \angle ARP_1.$$

$$\therefore \ AR \text{ is parallel to } BP_2.$$

$$\therefore \ ABP_2R \text{ is a parallelogram}.$$

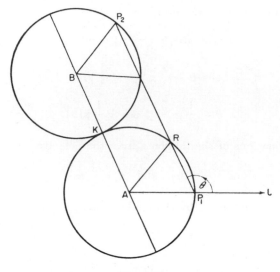

F_{IG}. 140.

$\therefore P_2R = AB = 2a$ where a is the radius of each circle, and, if $\angle RP_1l = \theta$,

$$P_1R = -2a\cos\theta.$$
$$\therefore \ P_1P_2 = 2a - 2a\cos\theta.$$

Hence referred to P_1 as pole and P_1l as initial line, the locus of P_2 is

$$r = 2a(1 - \cos\theta).$$

This curve is called a *Cardioid*. We investigate its shape in example (i) of the next section and we find that it is roughly "heart-shaped". Whence it derives its name.

15.3 Curve sketching in polar coordinates

Investigation of the shape of a polar curve is usually best carried out by plotting several points of the locus, and by consideration of the factors listed below.

(i) symmetry about the initial line, or about $\theta = \frac{1}{2}\pi$,

(ii) the values of θ for which r increases or decreases with θ,

(iii) the angle between the curve and the radius vector, or the angle between the tangent to the curve and the initial line at particular points on the curve (§ 15.9),

(iv) asymptotes,

may help to decide the shape of the polar curve and they should be used whenever it is appropriate to do so.

Special *polar graph paper* is used for accurate drawing of polar curves. The easily drawn grid shown in Fig. 141 is satisfactory for the purpose of obtaining a sketch of most polar curves.

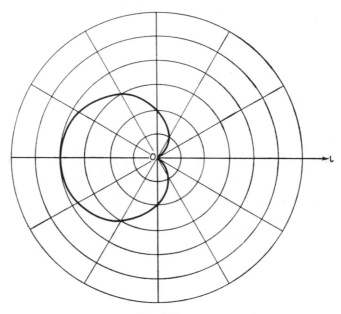

FIG. 141.

Examples. (i) Sketch the cardioid $r = a(1 - \cos\theta)$.

Because the only function of θ involved is $\cos\theta$, the curve is symmetrical about the initial line. (If r is a function of $\sin\theta$ only, the curve is symmetrical about the line $\theta = \frac{1}{2}\pi$.)

It is sufficient, therefore, to obtain values of r for $0 \leqq \theta \leqq \pi$.

θ	0	$\pi/6$	$\pi/4$	$\pi/3$	$\pi/2$	$2\pi/3$	$3\pi/4$	$5\pi/6$	π
r/a	0	0.13	0.29	0.5	1	1.5	1.71	1.87	2

(In § 15.9 we shall consider the shape of the curve at O and at $\theta = \pi$.) The sketch of the curve is therefore as shown in Fig. 141.

(ii) Sketch the curve $r = 2 \sin 3\theta$.

The curve is symmetrical about $\theta = \frac{1}{2}\pi$ because $\sin 3\theta = 3\sin\theta - 4\sin^3\theta$.

(a) r is positive and increases with θ for $0 < \theta < \pi/6$ and r is positive and decreases with θ for $\pi/6 < \theta < \frac{1}{3}\pi$.

(b) r is negative and $|r|$ decreases with θ for $\frac{1}{3}\pi < \theta < \frac{1}{2}\pi$ and r is negative and $|r|$ increases with θ for $\frac{1}{2}\pi < \theta < \frac{2}{3}\pi$.

(c) r is positive and r increases with θ for $\frac{2}{3}\pi < \theta < 5\pi/6$ and r is positive and r decreases with θ for $5\pi/6 < \theta < \pi$.

(d) For $\pi < \theta < 2\pi$, $2\sin 3\theta = -2\sin 3(\theta - \pi)$ and therefore the curve for this range of values of θ retraces the curve for $0 < \theta < \pi$.

θ	0	$\dfrac{\pi}{12}$	$\dfrac{\pi}{6}$	$\dfrac{\pi}{4}$	$\dfrac{\pi}{3}$
r	0	$\sqrt{2}$	2	$\sqrt{2}$	0

The curve is therefore as shown in Fig. 142.

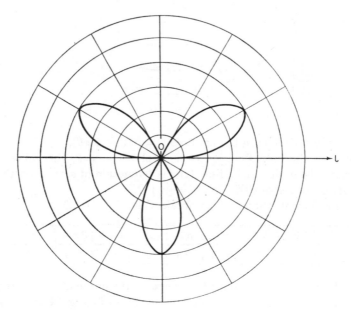

Fig. 142.

(iii) Sketch the curve $r = a \tan \frac{1}{2}\theta$ for $0 \leqq \theta \leqq 2\pi$.

θ	0	$\frac{1}{3}\pi$	$\frac{1}{2}\pi$	$\frac{2}{3}\pi$
r/a	0	$1/\sqrt{3}$	1	$\sqrt{3}$

r increases with θ for $0 \leqq \theta < \pi$.

As $\theta \to \pi$, $r \to \infty$.

$$\tan \tfrac{1}{2}(2\pi - \theta) = -\tan \tfrac{1}{2}\theta.$$

The curve is therefore as shown in Fig. 143. Note that the point P_1 on the curve is the point $\left(a/\sqrt{3}, \dfrac{1}{3}\pi\right)$ but that the point P_2 is the point $\left(-\dfrac{a}{\sqrt{3}}, \dfrac{5}{3}\pi\right)$. (See also Exercises 15.10, Question 5.)

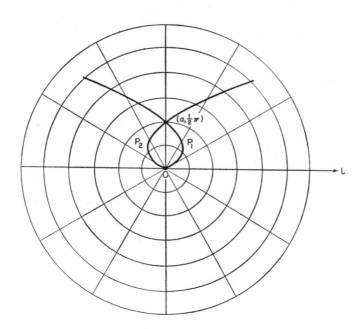

Fig. 143.

15.4 The lengths of chords of polar curves which are drawn through the pole

Example. The chord POP' of the curve $r = a/(1 + \cos\theta)$ passes through the pole O. If the length of the chord is $8a/3$, find the coordinates of P and P'.

If P is (r_1, θ_1) and P' is $(r_2, \pi + \theta_1)$ where r_1 and r_2 are each positive (Fig. 144) then

$$r_1 = a/(1 + \cos\theta_1), \qquad r_2 = a/\{1 + \cos(\pi + \theta_1)\} = a/(1 - \cos\theta_1).$$

$$\therefore r_1 + r_2 = \frac{a}{1 + \cos\theta_1} + \frac{a}{1 - \cos\theta_1},$$

$$\text{i. e.,} \quad \frac{8a}{3} = \frac{2a}{(1 + \cos\theta_1)(1 - \cos\theta_1)}.$$

$$\therefore \sin^2\theta_1 = \frac{3}{4}.$$

$$\therefore \theta_1 = \frac{1}{3}\pi \quad \text{or} \quad \frac{2}{3}\pi.$$

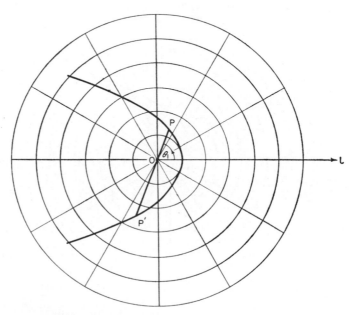

FIG. 144.

Hence P is $\left(\dfrac{2}{3}\,a,\dfrac{1}{3}\,\pi\right)$ or $\left(2\,a,\dfrac{2}{3}\,\pi\right)$

and P' is $\left(2\,a,\dfrac{4}{3}\,\pi\right)$ or $\left(\dfrac{2}{3}\,a,\dfrac{5}{3}\,\pi\right).$

Exercises 15.4

In each of Questions 1—8 find, from first principles, the polar equation of the locus defined there and sketch the locus.

1. The straight line at right angles to the initial line and distant c from the pole.

2. A straight line parallel to the initial line and distant c from it.

3. The straight line joining $(a, 0)$ to $(a, \tfrac{1}{2}\pi)$.

4. The straight line through $(a, \pi/6)$ which makes an angle $\dfrac{1}{3}\pi$ with the initial line.

5. The circle, radius a, centre $(a, 0)$.

6. Two circles, one of radius a centre $(a, 0)$ and the other of radius a centre (a, π).

7. The parabola with focus at the pole and directrix $r = -a \sec\theta$.

8. The ellipse with one focus at the pole, eccentricity $\tfrac{1}{3}$, and the line $r = -2a\sec\theta$ as the corresponding directrix.

9. A is a fixed point on the circumference of a circle of radius a. A chord AP of the circle is produced to Q so that $PQ = 2a$. Find the polar equation of the locus of Q as P varies, with reference to A as pole and a diameter of the circle as initial line. Sketch the locus.

10. A point moves so that the sum of its distances from the pole and from the point $(2ae, 0)$ is constant and equal to $2a$. Prove that the equation of the locus of the point is $r = a(1 - e^2)/(1 - e \cos\theta)$.

11. In Fig. 143, if OP_1 $\left(\theta = \dfrac{1}{3}\pi\right)$ is produced to meet the curve again at Q, write down the length of P_1Q.

12. The curve $r = a/(1 + \cos\theta)$ cuts the line $\theta = \tfrac{1}{2}\pi$ at A and B and POQ is a chord through the pole. Prove that $1/OP + 1/OQ = 1/AB$.

13. Sketch the polar curve $r = 4(1 + \cos\theta)$ and find the angles between the initial line and the chords through the pole which are divided by the pole in the ratio $1 : 3$.

14. The pole O is joined to a point P on the circle $r = a \cos\theta$ and OP is produced to meet the tangent at $(a, 0)$ at R. If Q is the point on OP such that $OQ = PR$, find the polar equation of the locus of Q as P varies and sketch this locus. (This curve is called a *cissoid*.)

15. Sketch the curves (i) $r = a \cos\theta$, (ii) $r = a(1 + \cos\theta)$, (iii) $r = a(2 + 3\cos\theta)$ (a *limaçon*).

16. Sketch the curve $r = k\theta$ (the spiral of Archimedes) for $0 \leqq \theta \leqq 4\pi$. If a straight line through the pole in the positive r direction meets this part of the curve at P and Q, write down the length of PQ. If P is $(k\alpha, \alpha)$, the line $\theta = 5\alpha/6$ meets the curve in its first revolution at A, AO produced meets this part of the curve at B, and $OA \cdot OB = OP \cdot PQ$, prove that $\alpha = 42\pi/25$.

17. Sketch the curve $r = k/\theta$ (the *reciprocal spiral*).

18. Sketch the curve $r = e^{k\theta}$ (the *equiangular* or *logarithmic* spiral). Show that if a radius vector cuts the curve at P, Q, R etc., then OP, OQ, OR etc. are in geometric progression and hence that $\log OP, \log OQ, \log OR$ etc. are in arithmetic progression.

19. Sketch the curve $r^2 = a^2 \cos 2\theta$ (the *lemniscate*).

20. Sketch the curves $r = a \sin 2\theta$ and $r^2 = a^2 \sin 2\theta$.

21. Sketch the curve $r = a \sin 3\theta$.

22. The pole O is joined to a point P on the curve $r = a/(1 - \cos\theta)$ and OP is produced to Q so that $OP \cdot OQ = k^2$. Find the polar equation of the locus of Q as P varies. Sketch both curves in one diagram. (Each curve is said to be the *inverse* of the other with respect to the pole as centre and k as *radius of inversion*.)

23. Using the definition given in Question 22, prove that the inverse of a circle with respect to a point on its circumference is a straight line at right angles to the diameter of the circle through that point.

24. Find the inverse of a circle with respect to a point outside the circle.

15.5 Transformations from polar to cartesian equations and the reverse process

If the pole is taken as the origin and the initial line as x-axis, and if with these axes a point has polar coordinates (r, θ) and cartesian coordinates (x, y), then

$$r \cos\theta = x, \; r \sin\theta = y, \qquad (15.7)$$

and the sign conventions for the cartesian coordinates are preserved in this relationship.

$$\therefore r = \sqrt{(x^2 + y^2)} \quad \text{and} \quad \tan\theta = y/x \qquad (15.8)$$

where θ has a unique value in the first revolution determined by the respective signs of y and x.

If the polar equation of a curve is $r = f(\theta)$, then $x = r \cos\theta = f(\theta) \cos\theta$ and when x is a stationary function of θ the tangent to the curve is parallel to Oy. Similarly $y = r \sin\theta = f(\theta) \sin\theta$ and when y is a stationary function of θ the tangent to the curve is parallel to Ox.

In the Examples and Exercises which follow it will be assumed, unless it is stated otherwise, that the origin of the cartesian coordinates is the pole and that the positive part of the x-axis is the initial line.

Examples. (i) Transform $r = a\,(\sec\theta - \cos\theta)$ into cartesian coordinates. The equation transforms to

$$\sqrt{(x^2 + y^2)} = a\left\{\frac{\sqrt{(x^2 + y^2)}}{x} - \frac{x}{\sqrt{(x^2 + y^2)}}\right\},$$

i.e.,
$$x\,(x^2 + y^2) = a\,(x^2 + y^2 - x^2),$$

i.e.,
$$y^2\,(x - a) + x^3 = 0,$$

i.e.,
$$y^2 = x^3/(a - x).$$

(ii) Transform $x^2 + y^2 = 4xy$ to polar coordinates and hence show that $\tan(\pi/12) = 2 - \sqrt{3}$ and $\tan(5\pi/12) = 2 + \sqrt{3}$.

The equation transforms to

$$r^2 = 4r^2 \sin\theta \cos\theta,$$

i.e.,
$$\sin 2\theta = \tfrac{1}{2}.$$

$$\text{i.e.,}\quad \theta = \frac{\pi}{12} \quad \text{and} \quad \theta = \frac{5\pi}{12}.$$

These equations represent two straight lines through the origin making angles of $\dfrac{\pi}{12}$ and $\dfrac{5\pi}{12}$ with the initial line. But

$$x^2 - 4xy + y^2 = 0$$

represents the lines

$$y = (2 \pm \sqrt{3})\,x.$$

$$\therefore\ \tan\frac{\pi}{12} = 2 - \sqrt{3}, \qquad \tan\frac{5\pi}{12} = 2 + \sqrt{3}.$$

(iii) Find the points at which the tangents to the lemniscate $r^2 = a^2 \cos 2\theta$ are parallel to the initial line.

On the given curve

$$y = r \sin\theta = a \sin\theta \sqrt{(\cos 2\theta)}.$$

The tangent is parallel to the initial line (Ox) when y is stationary. But

$$\frac{dy}{d\theta} = a \cos\theta \sqrt{(\cos 2\theta)} - \frac{a \sin\theta \sin 2\theta}{\sqrt{(\cos 2\theta)}} = \frac{a \cos 3\theta}{\sqrt{(\cos 2\theta)}}.$$

Hence y is stationary when $\cos 3\theta = 0$, i.e., when $\theta = \dfrac{\pi}{6}, \dfrac{5\pi}{6}, \dfrac{7\pi}{6}, \dfrac{11\pi}{6}$. Thus the required points are $(\tfrac{1}{2}a, \pi/6)$, $(\tfrac{1}{2}a, 5\pi/6)$, $(\tfrac{1}{2}a, 7\pi/6)$, $(\tfrac{1}{2}a, 11\pi/6)$.

(iv) Transform the equation $y^2 = x^2(a^2 - x^2)/(a^2 + x^2)$ to polar form and sketch the curve it represents.

The equation transforms to

$$r^2 \sin^2\theta = r^2 \cos^2\theta (a^2 - r^2 \cos^2\theta)/(a^2 + r^2 \cos^2\theta),$$

i.e., $$r^4(\sin^2\theta \cos^2\theta + \cos^4\theta) = r^2 a^2(\cos^2\theta - \sin^2\theta),$$

i.e., $$r = 0 \quad \text{and} \quad r^2 = a^2(2 - \sec^2\theta).$$

The point $(0, \tfrac{1}{4}\pi)$ is included in the second part of the locus. There is no part of the curve for values of θ such that $\sec^2\theta > 2$.

$$x = r \cos\theta = \pm a \cos\theta \sqrt{(2 - \sec^2\theta)} = \pm a \sqrt{(\cos 2\theta)}.$$

$$dx/d\theta = \mp a \sin 2\theta / \sqrt{(\cos 2\theta)}.$$

$$\therefore dx/d\theta = 0 \quad \text{when} \quad \theta = 0, \ r = \pm a.$$

There is no point on the curve corresponding to $\theta = \tfrac{1}{2}\pi$. Hence the curve is parallel to the y-axis at $(a, 0)$ and at (a, π). Similarly it can be shown that $dy/d\theta = 0$ where $\cos^4\theta = \tfrac{1}{2}$. There will therefore be points of the curve between $\theta = 0$ and $\theta = \dfrac{1}{4}\pi$ and between $\theta = \dfrac{3}{4}\pi$ and $\theta = \pi$ at which the curve is parallel to the initial line.

The sketch of the curve is shown in Fig. 145.

Exercises 15.5

1. Transform each of the following equations to cartesian form
   (i) $r^2 = a^2 \cos 2\theta$, (ii) $r = a/(1 + \cos\theta)$, (iii) $r = a(1 + 2\cos\theta)$, (iv) $r^2 = a^2 \tan 2\theta$, (v) $r = a \sec(\theta - \alpha)$.

2. Transform each of the following equations to polar form
   (i) $x^2 + y^2 - 2ax = 0$, (ii) $x^2 + y^2 - 2ay = 0$, (iii) $(x^2 + y^2)^2 - 4xy = 0$, (iv) $x^2 - y^2 = a^2$, (v) $xy = c^2$.

3. Sketch the curve $a^2x^2 = (x^2 + y^2)^2$.

4. Sketch the curve $(x^2 + y^2)^3 = a^2(x^2 - y^2)^2$.

5. Sketch the curve $(x^2 + y^2)^2 = a(x^3 - 3xy^2)$.

15.6 Areas in polar coordinates

In Fig. 146, P_1 is the point (r, θ) and P_2 is the point $(r + \delta r, \theta + \delta\theta)$ on the arc AB of the curve $f(r, \theta) = 0$. It will be assumed that θ increases steadily from α at A to β at B. The circle with centre O and radius OP_1 cuts OP_2 at S, and the circle with centre O and radius OP_2 cuts OP_1 produced at R.

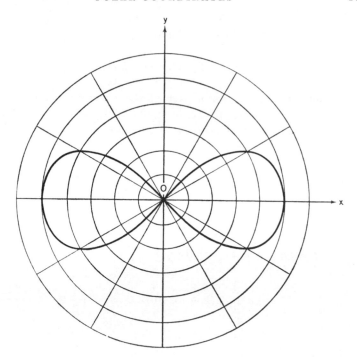

FIG. 145.

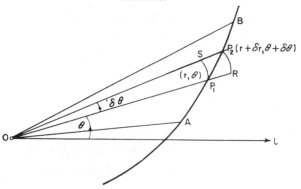

FIG. 146.

Then, in the case shown in Fig. 146, area of circular sector $OP_2R >$ area of sector $OP_2P_1 >$ area of circular sector OP_1S. The area of the sector OAB is equal to the common value of $\lim\limits_{\delta\theta \to 0} \sum\limits_{\theta=\alpha}^{\theta=\beta} \frac{1}{2} r^2 \delta\theta$ and

$$\lim_{\delta\theta\to 0} \sum_{\theta=\alpha}^{\theta=\beta} \tfrac{1}{2}(r+\delta r)^2\,\delta\theta \text{ if these limits exist,}$$

i.e.,　　　　　　　　Area of Sector $= \int_\alpha^\beta \tfrac{1}{2} r^2\,\mathrm{d}\theta.$　　　　　(15.9)

This result is obtained for a part of the curve for which r increases steadily as θ increases. A similar result holds for a part of the curve for which r decreases steadily as θ increases. Then, by addition, the result (15.9) holds in general.

Examples. (i) Calculate the area of one loop of the curve $r = a\sin 3\theta$. The curve was sketched in example (ii), § 15.2, Fig. 142. One loop is completed in the range $0 \leq \theta \leq \dfrac{1}{3}\pi$. Hence the area of one loop is

$$\int_0^{\pi/3} \tfrac{1}{2} r^2\,\mathrm{d}\theta = \int_0^{\pi/3} \tfrac{1}{2} a^2 \sin^2 3\theta\,\mathrm{d}\theta$$

$$= \tfrac{1}{2} a^2 \int_0^{\pi/3} \tfrac{1}{2}(1-\cos 6\theta)\,\mathrm{d}\theta = \tfrac{1}{2} a^2\left[\frac{1}{2}\theta - \frac{\sin 6\theta}{12}\right]_0^{\pi/3}$$

$$= \pi a^2/12.$$

(ii) Calculate the area between the loops of the limaçon $r = a(1 + 2\cos\theta)$.

The inner loop is traced by the radius vector for $\dfrac{2}{3}\pi \leq \theta \leq \dfrac{4}{3}\pi$, (Fig. 147). Hence the area enclosed by the inner loop is

$$\int_{2\pi/3}^{4\pi/3} \tfrac{1}{2} a^2(1 + 2\cos\theta)^2\,\mathrm{d}\theta$$

$$= \tfrac{1}{2} a^2 \int_{2\pi/3}^{4\pi/3} (3 + 2\cos 2\theta + 4\cos\theta)\,\mathrm{d}\theta = \tfrac{1}{2} a^2\left[3\theta + \sin 2\theta + 4\sin\theta\right]_{2\pi/3}^{4\pi/3}$$

$$= \tfrac{1}{2} a^2(2\pi - 3\sqrt{3}).$$

The outer loop is traced in two parts $0 \leq \theta \leq \dfrac{2}{3}\pi$ and $\dfrac{4}{3}\pi \leq \theta \leq 2\pi$ and the areas of the two parts are equal. Hence the area enclosed by the outer loop is

$$2\int_0^{2\pi/3} \tfrac{1}{2} r^2\,\mathrm{d}\theta = a^2\left[3 + \sin 2\theta + 4\sin\theta\right]_0^{2\pi/3}$$

$$= a^2\left(2\pi + \frac{3}{2}\sqrt{3}\right).$$

Thus the area between the loops is $\tfrac{1}{2} a^2(2\pi + 6\sqrt{3})$.

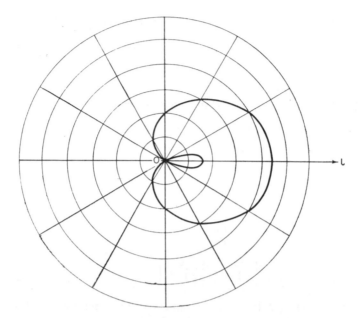

Fig. 147.

Exercises 15.6

1. Calculate the area enclosed by the cardioid $r = a(1 + \cos\theta)$.

2. Calculate the area of one loop of the lemniscate $r^2 = a^2 \cos 2\theta$.

3. Calculate the area of one loop of $r = a \cos 4\theta$.

4. Calculate the total area enclosed by the curve $r = a \sin n\theta$ where n is a positive integer.

5. The area enclosed by the first revolution of the spiral $r = a\theta$ and the initial line from $r = 0$ to $r = 2a\pi$ is divided into four parts by the lines $\theta = 0$, $\theta = \frac{1}{2}\pi$, $\theta = \pi$ and $\theta = 3\pi/2$. Calculate the ratios of these four areas.

6. For the reciprocal spiral $r = a/\theta$, prove that the area enclosed by the lines $\theta = \beta$, $\theta = \alpha$ and the arc of the curve for $\beta \leq \theta \leq \alpha$ where $0 < \beta < \alpha < 2\pi$ varies as the increase of r from $\theta = \beta$ to $\theta = \alpha$.

7. Calculate the *total* area swept out by the radius vector for the spiral $r = a\mathrm{e}^\theta$ from $\theta = 0$ to $\theta = 4\pi$.

8. Transform the equation of the parabola $y^2 = 4ax$ to a polar equation with the focus at the pole and the axis of x as initial line. Hence find the area enclosed between the curve and a focal chord which makes an angle $\frac{1}{3}\pi$ with the positive x-direction.

9. Calculate the area of one of the inner loops bounded by the two curves $r = a\,(1 + \cos\theta)$, $r = a\,(1 - \cos\theta)$.

10. Calculate the area bounded by the curve $r = a\tan\tfrac{1}{2}\theta$ and the lines $\theta = \tfrac{1}{2}\pi$, $\theta = \dfrac{3}{2}\pi$.

15.7 The length of an arc in polar coordinates

If $P(r, \theta)$ is a point on a polar curve and if the cartesian coordinates of P referred to the pole as origin and the initial line as the positive x-axis are (x, y) then

$$x = r\cos\theta, \; y = r\sin\theta.$$

$$\therefore \frac{\mathrm{d}x}{\mathrm{d}\theta} = -r\sin\theta + \frac{\mathrm{d}r}{\mathrm{d}\theta}\cos\theta,$$

$$\frac{\mathrm{d}y}{\mathrm{d}\theta} = r\cos\theta + \frac{\mathrm{d}r}{\mathrm{d}\theta}\sin\theta.$$

$$\therefore \left(\frac{\mathrm{d}s}{\mathrm{d}\theta}\right)^2 = \left(\frac{\mathrm{d}x}{\mathrm{d}\theta}\right)^2 + \left(\frac{\mathrm{d}y}{\mathrm{d}\theta}\right)^2 = r^2 + \left(\frac{\mathrm{d}r}{\mathrm{d}\theta}\right)^2. \tag{15.10}$$

This result (15.10) can be obtained by reference to Fig. 149 and reference to this figure probably affords the best way of remembering the result. In the triangle PQM, $MQ \fallingdotseq \delta r$, $PM \fallingdotseq r\delta\theta$ and $PQ \fallingdotseq \delta s$, whence, using

$$PQ^2 = MQ^2 + MP^2,$$

$$\delta s^2 \fallingdotseq \delta r^2 + r^2\delta\theta^2,$$

i.e.,

$$\left(\frac{\delta s}{\delta\theta}\right)^2 \fallingdotseq r^2 + \left(\frac{\delta r}{\delta\theta}\right)^2,$$

whence (15.10) follows.

Similarly,

$$\frac{\mathrm{d}x}{\mathrm{d}r} = \cos\theta - r\sin\theta\,\frac{\mathrm{d}\theta}{\mathrm{d}r},$$

$$\frac{\mathrm{d}y}{\mathrm{d}r} = \sin\theta + r\cos\theta\,\frac{\mathrm{d}\theta}{\mathrm{d}r}.$$

$$\therefore \left(\frac{\mathrm{d}x}{\mathrm{d}r}\right)^2 + \left(\frac{\mathrm{d}y}{\mathrm{d}r}\right)^2 = 1 + \left(r\,\frac{\mathrm{d}\theta}{\mathrm{d}r}\right)^2. \tag{15.11}$$

From Eqns. (14.5) and (15.10) the length of the arc of $f(r, \theta) = 0$ from $\theta = \theta_1$ to $\theta = \theta_2$, if s is measured so as to increase with θ, is

$$\int_{\theta_1}^{\theta_2} \sqrt{\left\{r^2 + \left(\frac{\mathrm{d}r}{\mathrm{d}\theta}\right)^2\right\}}\,\mathrm{d}\theta. \tag{15.12}$$

Similarly, if s is measured so as to increase with r, the length of the arc from $r = r_1$ to $r = r_2$ is

$$\int_{r_1}^{r_2} \sqrt{\left\{1 + \left(r\frac{\mathrm{d}\theta}{\mathrm{d}r}\right)^2\right\}}\,\mathrm{d}r. \tag{15.13}$$

Example. Calculate the length of the arc of the spiral $r = a\theta$ from $\theta = 0$ to $\theta = 2\pi$.

$$\text{Length of arc} = \int_0^{2\pi} \sqrt{(a^2\theta^2 + a^2)}\,\mathrm{d}\theta$$

$$= a \int_0^{\sinh^{-1}2\pi} \sqrt{(1 + \sinh^2 t)} \cdot \cosh t\,\mathrm{d}t = a \int_0^{\sinh^{-1}2\pi} \cosh^2 t\,\mathrm{d}t$$

$$= a\left[\frac{t}{2} + \frac{\sinh 2t}{4}\right]_0^{\sinh^{-1}2\pi} = \tfrac{1}{2}\,a[\log\{2\pi + \sqrt{(1 + 4\pi^2)}\} + 2\pi\sqrt{(1 + 4\pi^2)}].$$

15.8 Volumes of revolution and areas of surfaces of revolution in polar coordinates

These problems are best considered from first principles using the theorems of Pappus.

If the area bounded by the curve $r = f(\theta)$ and the radius vectors OA, OB ($\theta = \alpha$ and $\theta = \beta$, respectively, from Fig. 146) rotates through 360° about the initial line (Ox), then, by the first theorem of Pappus, the volume δV generated by the elementary sector OP_1P_2 is given by

$$\delta V = \text{area } OP_1P_2 \times 2\pi p$$

where p is the distance of the centroid of the sector from Ox. But the area $OP_1P_2 \doteqdot \tfrac{1}{2}r^2\delta\theta$ and since OP_1P_2 is approximately a triangle with centroid at the intersection of the medians, $p \doteqdot \dfrac{2}{3}r\sin\theta$.

$$\therefore\ \delta V \doteqdot \tfrac{1}{2}r^2\delta\theta \times 2\pi\frac{2}{3}r\sin\theta.$$

Hence, by addition, V, the total volume generated, is given by

$$V = \frac{2\pi}{3}\int_\alpha^\beta r^3 \sin\theta\,\mathrm{d}\theta, \tag{15.14}$$

provided that the area is not cut by Ox.

The area of the surface generated by revolution of the arc AB about the x-axis is given by

$$S = \int_{s_1}^{s_2} 2\pi y \, ds$$

and, if the conditions for a change of variable are satisfied, this becomes

$$S = \int_{\theta=\theta_1}^{\theta=\theta_2} 2\pi r \sin\theta \frac{ds}{d\theta} \, d\theta = \int_{\theta=\theta_1}^{\theta=\theta_2} 2\pi r \sin\theta \sqrt{\left\{ r^2 + \left(\frac{dr}{d\theta}\right)^2 \right\}} \, d\theta. \quad (15.15)$$

Centroids and seconds moments of area and of volume are best found from first principles. The theorems of Pappus are used where it is expedient to do so.

Example. For the area of that part of the cardioid $r = a(1 + \cos\theta)$ lying in the first quadrant, calculate

(1) the volume of revolution about the initial line,
(2) the surface area of the solid obtained by the rotation of the curve,
(3) the distances of the centroid of the area from Ox and Oy.

The curve is symmetrical about Ox and is sketched in Fig. 148.

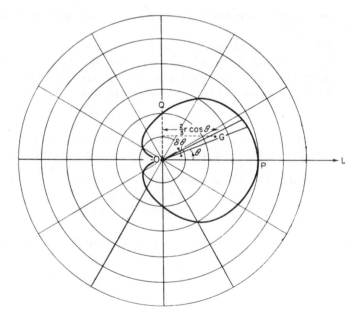

FIG. 148.

(1) The volume of revolution

$$V = \frac{2\pi}{3} \int_0^{\pi/2} r^3 \sin\theta \, d\theta$$

$$= \frac{2\pi a^3}{3} \int_0^{\pi/2} (1 + \cos\theta)^3 \sin\theta \, d\theta$$

$$= \frac{2\pi a^3}{3} \left[-\frac{1}{4}(1+\cos\theta)^4 \right]_0^{\pi/2} = \frac{5\pi a^3}{2} .$$

(2) The surface area of the shell

$$S = \int_{\theta=0}^{\theta=\pi/2} 2\pi y \, ds$$

$$= 2\pi \int_{\theta=0}^{\theta=\pi/2} r\sin\theta \sqrt{\left\{ r^2 + \left(\frac{dr}{d\theta}\right)^2 \right\}} d\theta.$$

Here s is measured from $P\,(2a, 0)$ to $Q(a, \tfrac{1}{2}\pi)$, i.e., in the direction increasing with θ.

$$\therefore S = 2\pi a \int_0^{\pi/2} (1 + \cos\theta)\sin\theta \sqrt{\{a^2 + 2a^2\cos\theta + a^2\cos^2\theta + a^2\sin^2\theta\}} \, d\theta$$

$$= 2\pi a^2 \int_0^{\pi/2} (1 + \cos\theta)\sin\theta \sqrt{(2 + 2\cos\theta)} \, d\theta$$

$$= 16\pi a^2 \int_0^{\pi/2} \cos^4 \tfrac{1}{2}\theta \sin\tfrac{1}{2}\theta \, d\theta$$

$$= 32\pi a^2 \left[\frac{-\cos^5 \tfrac{1}{2}\theta}{5} \right]_0^{\pi/2}$$

$$= \frac{32\pi a^2}{5} \left(1 - \frac{1}{\sqrt{2}} \right).$$

(3) The area A of the cardioid in the first quadrant is given by

$$A = \int_0^{\pi/2} \tfrac{1}{2} r^2 \, d\theta = \tfrac{1}{2} a^2 \int_0^{\pi/2} (1 + \cos\theta)^2 \, d\theta$$

$$= \tfrac{1}{2} a^2 \int_0^{\pi/2} (1 + 2\cos\theta + \cos^2\theta) \, d\theta$$

$$= \tfrac{1}{2} a^2 \left(\frac{\pi}{2} + 2 + \frac{\pi}{4} \right) = \frac{(3\pi + 8)a^2}{8} .$$

Hence, if the cartesian coordinates of the centroid, G, of the given area are $\bar{x}$, $\bar{y}$, then by the first theorem of Pappus,

$$V = A \times 2\pi\bar{y}.$$

$$\therefore \bar{y} = 10a/(3\pi + 8).$$

To find $\bar{x}$ we take first moments about Oy. The first moment δM of the elementary sector OP_1P_2 of Fig. 146 about Oy is given by

$$\delta M \doteqdot \tfrac{1}{2} r^2 \delta\theta \times \frac{2}{3} r\cos\theta.$$

By addition the first moment of the whole area is

$$M = \frac{1}{3} \int_0^{\pi/2} r^3 \cos\theta \, d\theta$$

$$= \frac{a^3}{3} \int_0^{\pi/2} (1 + \cos\theta)^3 \cos\theta \, d\theta$$

$$= \frac{a^3}{3} \int_0^{\pi/2} (\cos\theta + 3\cos^2\theta + 3\cos^3\theta + \cos^4\theta) \, d\theta$$

$$= \frac{a^3}{3} \left[1 + \frac{3\pi}{4} + 3 \cdot \frac{2}{3} + \frac{3}{4} \cdot \frac{1}{2} \cdot \frac{\pi}{2} \right]$$

$$= \frac{(16 + 5\pi) a^3}{16}.$$

But $M = A\bar{x}$.

$$\therefore \bar{x} = \frac{(16 + 5\pi) a}{(16 + 6\pi)} \, .$$

Exercises 15.8

1. Find the length of the arc of the spiral $r = ae^{k\theta}$ from $\theta = 0$ to $\theta = 2\pi$.

2. Find the total length of the cardioid $r = a(1 + \cos\theta)$.

3. Find the length of the arc of the curve $r = a\sin^3 \dfrac{1}{3}\theta$ from $\theta = 0$ to $\theta = 3\pi$.

4. Find the length of the arc of $r = a/\theta$ from $r = \dfrac{3a}{4}$ to $r = \dfrac{12a}{5}$.

5. Find the volume generated by the revolution of the cardioid $r = a(1 + \cos\theta)$ about its axis, i.e., about the initial line.

6. Calculate the distance of the centroid of one loop of the lemniscate $r^2 = a^2 \cos 2\theta$ from the pole.

7. Calculate the volume of rotation of the area bounded by the curve $r^2 = a^2 \cos\theta$ about the initial line.

8. Calculate the area of the surface of revolution of the curve $r = e^\theta$ from $\theta = 0$ to $\theta = \pi$ about the initial line.

15.9 The angle between the tangent and the radius vector

Fig. 149. P is the point (r, θ) and Q the point $(r + \delta r, \theta + \delta\theta)$. The angle between OP and the tangent to the curve at P is φ and PM is perpendicular to OQ. In $\triangle OPM$, $MP = r \sin \delta\theta$ and $OM = r \cos \delta\theta$. Hence from $\triangle PMQ$,

$$\sin OQP = \frac{MP}{PQ} = \frac{r \sin \delta\theta}{PQ} = r\, \frac{\sin \delta\theta}{\delta\theta}\, \frac{\delta\theta}{\delta s}\, \frac{\delta s}{PQ}.$$

We shall again assume that $\lim\limits_{\delta\theta \to 0} \dfrac{\delta s}{PQ} = 1$.

Then $\lim\limits_{\delta\theta \to 0} \dfrac{\sin \delta\theta}{\delta\theta} = 1$ and $\lim\limits_{\delta\theta \to 0} \dfrac{\delta\theta}{\delta s} = \dfrac{d\theta}{ds}$.

$$\therefore\ \lim_{\delta\theta \to 0} \sin OQP = \sin\varphi = r\, \frac{d\theta}{ds}.$$

Also from $\triangle PMQ$

$$\cos OQP = \frac{MQ}{QP} = \frac{OQ - OM}{QP}$$

$$= \frac{r + \delta r - r \cos \delta\theta}{\delta s} \cdot \frac{\delta s}{PQ} = \left\{ \frac{r(1 - \cos \delta\theta)}{\delta s} + \frac{\delta r}{\delta s} \right\} \frac{\delta s}{PQ}.$$

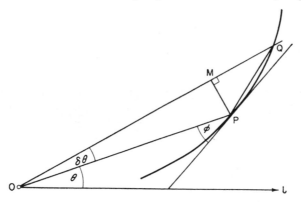

FIG. 149.

$$\therefore \lim_{\delta\theta\to 0} \cos O\,Q\,P = \lim_{\delta\theta\to 0} \left\{ \frac{r(1 - \cos \delta\theta)}{\delta s} \right\} + \frac{\mathrm{d}\,r}{\mathrm{d}\,s}.$$

But
$$\lim_{\delta\theta\to 0} \frac{r(1 - \cos \delta\theta)}{\delta s} = \lim_{\delta\theta\to 0} \frac{r(1 - \cos \delta\theta)}{\delta\theta} \cdot \frac{\delta\theta}{\delta s}$$

$$= \lim_{\delta\theta\to 0} \frac{r \sin \tfrac{1}{2}\,\delta\theta}{\tfrac{1}{2}\,\delta\theta} \cdot \sin \tfrac{1}{2}\,\delta\theta \cdot \frac{\delta\theta}{\delta s}$$

$$= 0.$$

$$\therefore \cos \varphi = \frac{\mathrm{d}\,r}{\mathrm{d}\,s}.$$

$$\therefore \tan \varphi = r\,\frac{\mathrm{d}\,\theta}{\mathrm{d}\,s} \Big/ \frac{\mathrm{d}\,r}{\mathrm{d}\,s} = r\,\frac{\mathrm{d}\,\theta}{\mathrm{d}\,r} = r \Big/ \frac{\mathrm{d}\,r}{\mathrm{d}\,\theta}. \qquad (15.16)$$

Examples. (i) For the cardioid $r = a(1 - \cos\theta)$, discuss the shape of the curve at $\theta = 0$ and at $\theta = \pi$. See Fig. 141 and Example (i) § 15.3.

$$r\,\frac{\mathrm{d}\,\theta}{\mathrm{d}\,r} = \frac{a(1 - \cos\theta)}{a \sin \theta} = \frac{a\,2 \sin^2 \tfrac{1}{2}\,\theta}{a\,2 \sin \tfrac{1}{2}\,\theta \cos \tfrac{1}{2}\,\theta} = \tan \tfrac{1}{2}\,\theta.$$

Hence at $\theta = 0$ the curve is parallel to the initial line and at $\theta = \pi$ the curve is at right angles to the initial line.

(ii) Show that in the *equiangular spiral* $r = a e^{\theta \cot \alpha}$, the angle between the tangent and the radius vector is constant and equal to α.

$$r = a\, e^{\theta \cot \alpha}.$$

$$\therefore \frac{\mathrm{d}\,r}{\mathrm{d}\,\theta} = a \cot \alpha\, e^{\theta \cot \alpha}.$$

$$\therefore \tan \varphi = r\,\frac{\mathrm{d}\,\theta}{\mathrm{d}\,r} = a\, e^{\theta \cot \alpha}/(a \cot \alpha\, e^{\theta \cot \alpha}) = \tan \alpha.$$

$$\therefore \qquad \varphi = \alpha.$$

(iii) Find the angle between the tangent to the limaçon $r = a(1 + 2 \cos\theta)$ at the point P on the curve where $\theta = \pi/6$ and the initial line. Calculate the angle between this tangent and the tangent at the other point Q on the curve which lies on the line OP. (O is the pole).

Fig. 150. At P,

$$\tan \varphi = r\,\frac{\mathrm{d}\,\theta}{\mathrm{d}\,r} = \left[\frac{a(1 + 2 \cos \theta)}{-2a \sin \theta} \right]_{\theta\,=\,\pi/6} = -(1 + \sqrt{3}).$$

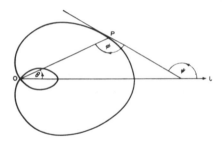

FIG. 150.

$$\tan \psi = \tan(\theta + \varphi) = \frac{(1/\sqrt{3}) - 1 - \sqrt{3}}{1 + (1/\sqrt{3}) + 1} = -\frac{4 + 3\sqrt{3}}{11}.$$

But Q is the point on the curve at which $\theta = 7\pi/6$. Hence, at Q,

$$\tan \varphi = \left[\frac{a(1 + 2\cos\theta)}{-2a\sin\theta}\right]_{\theta = 7\pi/6} = 1 - \sqrt{3},$$

and

$$\tan \psi = \tan(\theta + \varphi) = \frac{1 - \sqrt{3} + (1/\sqrt{3})}{1 - (1/\sqrt{3}) + 1} = \frac{4 - 3\sqrt{3}}{11}.$$

Therefore the angle between the tangent at P and the tangent at Q is

$$\tan^{-1}\left[\frac{\dfrac{4 - 3\sqrt{3}}{11} + \dfrac{4 + 3\sqrt{3}}{11}}{1 - \dfrac{(4 - 3\sqrt{3})(4 + 3\sqrt{3})}{121}}\right] = \tan^{-1}\left(\frac{2}{3}\right).$$

15.10 The tangential polar equation—Curvature

In Fig. 151 ON is the perpendicular from the pole on to the tangent at $P(r, \theta)$ to the curve $f(r, \theta) = 0$. If $ON = p$,

$$p = r \sin\varphi.$$

$$\therefore \frac{1}{p^2} = \frac{1}{r^2}\operatorname{cosec}^2\varphi = \frac{1}{r^2}(1 + \cot^2\varphi).$$

$$\therefore \frac{1}{p^2} = \frac{1}{r^2} + \frac{1}{r^4}\left(\frac{dr}{d\theta}\right)^2. \qquad (15.17)$$

If θ is eliminated between this equation and the polar equation, the tangential-polar, pedal, or $p - r$ equation is obtained.

G

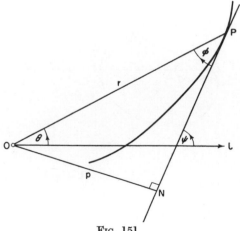

FIG. 151.

Example. Find the $p - r$ equation for the curve $r^2 = a^2 \sec 2\theta$.

$$r^2 = a^2 \sec 2\theta.$$

$$\therefore 2r\frac{\mathrm{d}r}{\mathrm{d}\theta} = 2a^2 \sec 2\theta \tan 2\theta.$$

$$\therefore \left(\frac{\mathrm{d}r}{\mathrm{d}\theta}\right)^2 = \frac{a^4 \sec^2 2\theta \tan^2 2\theta}{r^2} = \frac{r^4\left(\dfrac{r^4}{a^4} - 1\right)}{r^2}.$$

$$\therefore \frac{1}{r^4}\left(\frac{\mathrm{d}r}{\mathrm{d}\theta}\right)^2 = \frac{r^2}{a^4} - \frac{1}{r^2}.$$

$$\therefore \frac{1}{p^2} = \frac{1}{r^2} + \frac{r^2}{a^4} - \frac{1}{r^2} = \frac{r^2}{a^4}.$$

Hence the $p - r$ equation is $pr = a^2$.

Curvature. Calculation of curvature directly from a polar equation is an involved process and a better method is to obtain the $p - r$ equation and then to calculate the curvature from the formula given below. From Fig. 151,

$$\varkappa = \frac{\mathrm{d}\psi}{\mathrm{d}s} = \frac{\mathrm{d}(\theta + \varphi)}{\mathrm{d}s}.$$

But $\dfrac{\mathrm{d}\theta}{\mathrm{d}s} = \dfrac{1}{r}\sin\varphi$ and $\dfrac{\mathrm{d}r}{\mathrm{d}s} = \cos\varphi$ from   § 15.9.

$$\therefore \varkappa = \frac{1}{r} \sin \varphi + \cos \varphi \frac{d \varphi}{d r}$$

$$= \frac{1}{r} \left(\sin \varphi + r \cos \varphi \frac{d \varphi}{d r} \right)$$

$$= \frac{1}{r} \frac{d}{d r} (r \sin \varphi).$$

$$\therefore \varkappa = \frac{1}{r} \frac{d p}{d r} \qquad (15.18)$$

and
$$\varrho = \left| r \frac{d r}{d p} \right|. \qquad (15.19)$$

Example. Calculate the radius of curvature of the lemniscate $r^2 = a^2 \cos 2\theta$ at the point $\theta = 0$.

$$r^2 = a^2 \cos 2\theta.$$

$$\therefore 2 r \frac{d r}{d \theta} = - 2 a^2 \sin 2\theta.$$

$$\therefore \frac{d r}{d \theta} = \frac{- a^2 \sin 2\theta}{r}.$$

$$\therefore \left(\frac{d r}{d \theta} \right)^2 = \frac{a^4 \sin^2 2\theta}{r^2} = \frac{a^4 - r^4}{r^2}.$$

$$\therefore \frac{1}{p^2} = \frac{1}{r^2} + \frac{a^4 - r^4}{r^6} = \frac{a^4}{r^6}.$$

Hence the $p - r$ equation is $r^3 = a^2 p$.

$$\therefore \varkappa = \frac{1}{r} \frac{d p}{d r} = \frac{3 r}{a^2}.$$

At $\theta = 0$, $r = a$; hence the radius of curvature there is $a/3$.

Exercises 15.10

Throughout these exercises φ is used to denote the angle between the tangent and the radius vector and ψ to denote the angle between the initial line and the tangent.

1. Calculate φ for each of the following curves at the point named:

(i) $r = a/(1 + \cos \theta)$; $\left(\frac{2 a}{3}, \frac{1}{3} \pi \right)$,

(ii) $r = a \sin 3\theta$; $(a/\sqrt{2}, \pi/12)$,

(iii) $r = a(\sec\theta - \cos\theta); \left(\dfrac{3a}{2}, \dfrac{1}{3}\pi\right),$

(iv) $r = a\theta; (\theta = \alpha),$

(v) $r^2 - 12r\cos\left(\theta - \dfrac{1}{4}\pi\right) + 5 = 0; (\theta = 7\pi/12).$

2. Prove that, for the spiral $r = a\theta$, $\cos\varphi = a/\sqrt{(a^2 + r^2)}$.

3. Calculate the angle between the two branches of $r^2 = a^2\cos2\theta$ at the pole.

4. Show that tangents to the equiangular spiral $r = \epsilon^{\theta\cot\alpha}$ are parallel to the initial line at points where $\theta = n\pi - \alpha$.

5. For the curve $r = a\tan\frac{1}{2}\theta$ show that the branches corresponding to the ranges $0 \leq \theta \leq \pi$ and $\pi \leq \theta \leq 2\pi$ each touch the initial line at the pole and calculate the angle between the branches at their other point of intersection. Calculate also the θ-coordinates of the points at which the tangent to the curve is parallel to the line $\theta = \frac{1}{2}\pi$. Sketch the curve.

6. Calculate the angle of intersection of the cardioids $r = a(1 + \cos\theta)$ and $r = a(1 - \cos\theta)$.

7. Prove that the curves $r = a(1 + \cos\theta)$ and $r = a/(1 + \cos\theta)$ intersect the initial line and also the line $\theta = \frac{1}{2}\pi$ at equal angles.

8. Calculate ψ for the curve $r = \sin2\theta$ at the point $\theta = \dfrac{1}{4}\pi$ and find the coordinates of the points on the curve at which the tangent is parallel to the initial line.

9. Prove that the tangents to the cardioid $r = a(1 + \cos\theta)$ at the points $\theta = \alpha$, $\theta = \alpha + \dfrac{2}{3}\pi$, $\theta = \alpha + \dfrac{4}{3}\pi$ are parallel.

10. Sketch the polar curve $r^2 = a^2\sec2\theta$ and find an expression for φ in terms of θ.

Discuss the significance of the values of φ and of ψ when $\theta = \dfrac{1}{4}\pi$ and $\theta = \dfrac{3}{4}\pi$.

11. Calculate the angle between the tangents at the pole to $r = 1 + 2\cos\theta$.

12. Find the intrinsic equation of the cardioid $r = a(1 + \cos\theta)$. [The intrinsic equation is the $s - \psi$ equation.]

13. Find the $p - r$ equation for the spiral $r = a\theta$ and hence find the curvature at $\theta = 1$.

14. Find the $p - r$ equation for the reciprocal spiral $r = a/\theta$ and hence find the curvature at $\theta = 1$.

15. Show that, for the curve $r^2\cos2\theta = a^2$, the radius of curvature at any point is proportional to r^3.

16. Find the $p - r$ equation of the curve $a = r\cosh\theta$ and hence find the curvature at $\theta = 0$.

17. For the parabola $r = a/(1 + \cos\theta)$ prove that $\varrho = a\operatorname{cosec}^3\varphi$.

18. Find the values of θ at the points on the curve $r^2 = a^2\sin2\theta$ at which $\varphi = \dfrac{1}{4}\pi$.

Miscellaneous Exercises XV

1. Use the relations $x = r \cos \theta$, $y = r \sin \theta$ to find cartesian equations for the loci

 (i) $r(3 \cos \theta + 4 \sin \theta) = 1$,

 (ii) $r = 3 \cos \theta + 4 \sin \theta$.

Show that one of these loci is a circle and the other a straight line. Find the points where the loci cut the cartesian axes, determine the radius and the coordinates of the centre of the circle and sketch both loci on the same diagram. (N.)

2. Plot, using values of θ at 30° intervals from 0° to 360°, the curve whose equation in polar coordinates is

$$r = 5 + 4 \cos \theta.$$

Show that all chords PQ drawn through the pole O are of length 10 units. Calculate the values of θ for which O trisects the chord PQ. (N.)

3. Sketch the curve which has the polar equation $r = 16/(5 + 3 \cos \theta)$ showing the pole and the initial line in the sketch.

Express the equation in cartesian coordinates x and y, where $x = r \cos \theta$, $y = r \sin \theta$. Show that the curve is an ellipse and find the lengths of its semi-axes and its eccentricity. (N.)

4. Plot the portion of the curve whose equation in polar coordinates is $r = \sin 2\theta$, for values of θ which lie between 0 and $\frac{1}{2} \pi$.

Prove that the point on the curve whose perpendicular distance from the initial line $(\theta = 0)$ is greatest lies at an extremity of the chord through the origin which

makes an angle $\cos^{-1} \sqrt{\dfrac{1}{3}}$ with the initial line. (N.)

5. Find the values of θ for the points at which the curve

$$r^3 = a^3 \cos 3\theta$$

is cut at an angle of 45° by the radius vector drawn from the pole. Draw a rough sketch of the curve. (N.)

6. Show that the area of the sector of the ellipse $x^2/a^2 + y^2/b^2 = 1$ between two radius vectors OP, OQ is given by the integral

$$\frac{1}{2} \int_{\alpha}^{\beta} \frac{a^2 \, b^2 \, \mathrm{d} \theta}{b^2 \cos^2 \theta + a^2 \sin^2 \theta}$$

where α, β denote the angles which OP, OQ respectively make with the axis Ox, and $\beta > \alpha$.

The major axes of two concentric equal ellipses intersect at right angles. Prove that the area common to both ellipses is equal to $4ab \tan^{-1}(b/a)$, where a, b are the lengths of the semi-axes of each ellipse $(a > b)$. (N.)

7. Construct, correct to one place of decimals, a table of values for r, where $r = 1 + 2 \cos \theta$, taking values of θ at intervals of 30° from 0° to 360°. Hence give a sketch of the curve whose polar equation is $r = 1 + 2 \cos \theta$, showing it to be a closed curve with an interior loop.

A straight line OPQ is drawn through the pole O making a positive acute angle α with the initial line so as to meet the curve again at two points P and Q on the same side as O. Calculate the length of PQ. (N.)

8. Show that the radius of curvature at a point on the curve $r = a(1 + \cos\theta)$ is $\dfrac{4}{3}\, a \cos\tfrac{1}{2}\theta$.

If C is the centre of curvature at the point P of the curve and A is the point $r = \dfrac{2}{3}\, a$, $\theta = 0$, prove that CA is parallel to the line joining P to the origin. (N.)

9. Show that at any point of the curve given in polar coordinates by the equation $r^2 = a^2 \cos 2\theta$, the radius of curvature is $3a^2/r$.

If (x, y) are the cartesian coordinates of the centre of curvature at the point where $\theta = \alpha$, show that $y + x\tan^3\alpha = 0$. (N.)

10. Plot the curve whose equation in polar coordinates is $r = 1 + \cos\theta$, and on the same diagram draw the straight line $r = \sec\theta$. Find by calculation the values of r and θ for the points where the curve and the line intersect. (N.)

11. (a) Prove that $r = 4\sec(\theta - 30°)$ is the equation of a straight line in polar coordinates. State the polar coordinates of the foot of the perpendicular on the line from the pole.

(b) A point P moves so that its distance from the pole is equal to its perpendicular distance from the straight line $r = 2a\sec\theta$. Show that the polar equation of the locus of P is $r = a\sec^2\tfrac{1}{2}\theta$.

(c) Eliminate r and θ from the equations

$$x = r\cos\theta, \quad y = r\sin\theta, \quad r(1 + \cos\theta) = 2. \tag{N.}$$

12. Sketch the curve $r = a(2 + \tan^2\theta)$ and prove that the perpendicular p from the origin to the tangent at any point of this curve is given by

$$p^2 \{ar^2 + 4(r - a)^2(r - 2a)\} = ar^4. \tag{N.}$$

13. The polar equation of a plane curve is $r = f(\theta)$, where $0 \leqq \theta \leqq \pi$ and $f(\theta) \geqq 0$. Assuming Pappus' theorem, or otherwise, prove that the volume enclosed when the curve is rotated through four right angles about the line $\theta = 0$ is

$$\frac{2\pi}{3} \int_0^\pi r^3 \sin\theta \, d\theta.$$

Evaluate this integral in the case of the curve

$$r = a(1 + \sin\tfrac{1}{2}\theta). \tag{O.C.}$$

14. Show that the equation of the tangent to the conic $l/r = 1 + e\cos\theta$ at the point whose vectorial angle is α is

$$l/r = e\cos\theta + \cos(\theta - \alpha).$$

Prove that if p be the length of the perpendicular from the pole of coordisnate upon the tangent at the point whose radius vector is r,

$$l^2/p^2 = 2l/r + e^2 - 1.$$

15. Find the area of that part of the cardioid $r = a + a \cos\theta$ which lies outside the circle $r = 3a \cos\theta$. (N.)

16. Sketch the curve $r = a \sin 2\theta$. If $0 < \theta < \frac{1}{2}\pi$ and the tangent to the curve is perpendicular to the line $\theta = 0$, show that

$$\tan\theta \tan 2\theta = 2.$$ (N.)

17. Find the area of one loop of the curve

$$r = a \cos 4\theta.$$ (O.C.)

18. A small bead P is threaded on a thin rod. The rod rotates in a plane with uniform angular velocity 2 radians/sec about one end A, and P is made to move along the rod and away from A so that its velocity relative to the rod at any instant is proportional to the distance AP. Initially P is at unit distance from A and has a velocity relative to the rod equal to 2 units. Taking A as the pole and the initial position of the rod as the initial line,

(i) show that the bead is moving parallel to the initial line when $\theta = \dfrac{3}{4}\pi$,

(ii) find the values of θ in the range $0 \leq \theta \leq 2\pi$ for which the bead is moving in a direction perpendicular to the initial line,

(iii) find the equation of the path of the bead and sketch the path for values of θ from 0 to 2π. (N.)

19. A loop of the curve $r^2 = a^2 \cos 2\theta$ rotates round the line $\theta = \frac{1}{2}\pi$. Prove that the area of the surface generated is $2\sqrt{2\pi a^2}$, and find the volume contained by it. (O.C.)

20. Find the area of the parabola $r = 2a/(1 + \cos\theta)$ between the radii $\theta = \dfrac{\pi}{4}$ and $\theta = \dfrac{3\pi}{4}$.

Find the distance of the centre of gravity of the same area from the axis. (O.C.)

21. Obtain the polar equation of the curve whose parametric equations are

$$x = 2a \cos t - 2a \cos^2 t,$$

$$y = 2a \sin t - 2a \sin t \cos t.$$

Trace the curve and show that it encloses an area of $6\pi a^2$ sq. units. (N.)

22. Show that the area of the cardioid $r = a(1 + \cos\theta)$ between the radii $\theta = 0$ and $\theta = \pi$ is $\dfrac{3}{4}\pi a^2$; and show that the ordinate of the centre of gravity of the area is $16a/9\pi$. (O.C.)

23. Sketch the curve (the cardioid) whose parametric equations are

$$x = r \cos\theta, \; y = r \sin\theta \quad \text{where} \quad r = a(1 - \cos\theta).$$

Show that the tangents to the curve at the three points at which $\theta = \alpha$, $\theta = \alpha + 2\pi/3$, $\theta = \alpha + 4\pi/3$ are parallel to one another, α being any angle.

A rectangle is drawn containing the curve and having two of its sides tangential to the curve and parallel to the axis of x, and the other two sides tangential to the curve and parallel to the axis of y. Find the area of the rectangle. (N.)

24. Sketch the curve $r = a \sin^2\theta$ and show that the initial line is the tangent at the pole. Find the area enclosed by either loop of the curve. Find also the position of the centroid of the upper loop. Hence show that if the upper loop is rotated through an angle of 2π radians about the tangent at the point $r = a$, $\theta = \pi/2$, the volume of the solid so formed is

$$2\pi a^3 \left(\frac{3\pi}{16} - \frac{32}{105} \right). \tag{L.}$$

25. A lamina is bounded by a curve $r = f(\theta)$ which is symmetrical about the initial line of the polar coordinates. Prove that the distance of its centre of mass from the origin is

$$\bar{x} = \frac{2 \int r^3 \cos \theta \, d\theta}{3 \int r^2 \, d\theta}.$$

Such a lamina has the shape of one loop of the curve $r = a \cos 3\theta$, where $-\pi/6 \leq \theta \leq \pi/6$.

Find the area of the loop and the distance of its centre of mass from the origin. (L.)

CHAPTER XVI

COMPLEX NUMBERS

16.1 The number system

In the process of the development of the subject which we call Algebra the number system has been successively extended *by definition* from the *natural numbers*, or *counting numbers* 1, 2, 3 ... to (i) the *positive fractions*, (ii) the *directed numbers* or *integers* 0, +1, −1, +2, −2 etc., (iii) the *rational numbers* p/q where p is an integer and q is any integer other than 0, and (iv) the *irrational numbers* which include among others surds and such irrational numbers as e and π. The integers constitute an infinite ordered sequence in which comparison between one number and another of the form $a > b$ is possible, and within which it is possible to designate the *next number* to any given number of the sequence. The whole system of numbers, including the irrational numbers, admits of comparisons of the form $a > b$, but with the system thus completed it is no longer possible to designate the next number to any given number. The aggregate of all the numbers, rational and irrational, classified here is called the *Arithmetical Continuum*. The members of the whole class of numbers are called *real* numbers.

In Fig.152, O is a fixed point on a straight line. Points to the right of O such as A_r represent positive numbers and points to the left of O such as A_{-s} represent negative numbers. With this analogy, the straight line is considered as being composed of all the numbers of the arithmetical continuum.

$$A_{-s} \qquad O \qquad A_r$$

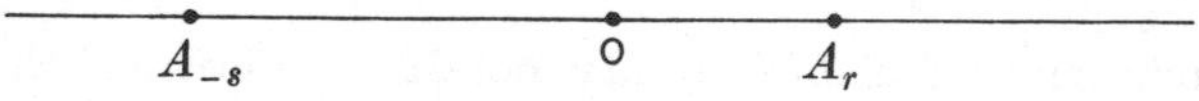

FIG. 152.

16.2 Definition of complex number

The solution of the equation $ax + b = 0$, where a and b are real numbers and $a \neq 0$, is possible in all cases with the system of real numbers, but the solution of the equation

$$ax^2 + bx + c = 0,$$

where a, b and c are real numbers and $a \neq 0$, is limited with the system of real numbers to the cases in which $b^2 \geq 4ac$. In the historical development of the subject this position is analogous to the one in which, for example, a definition of the irrational numbers which we call surds was required, in order to enable solutions to be stated for this same equation in cases in which $b^2 - 4ac$ is not a perfect square. We now add to the number system a new class of numbers which enable us to state a solution to this equation in the case $b^2 < 4ac$.

Suppose the symbol i represents a number which possesses the property of combining with itself and with the real numbers according to the laws of Algebra, and suppose that $i^2 = -1$. *A complex number* is represented in the form $a + ib$ where a and b are real numbers.

By definition a complex number satisfies the following laws.

(1) *Addition.* $(a + ib) + (c + id) = (a + c) + i(b + d)$.   (16.1)

(2) *Subtraction.* $(a + ib) - (c + id) = (a - c) + i(b - d)$.   (16.2)

(3) *Multiplication.* $(a + ib)(c + id) = (ac - bd) + i(ad + bc)$. (16.3)

(4) *Division.* $\dfrac{a + ib}{c + id} = x + iy$ where $(c + id)(x + iy) = (a + ib)$. (16.4)

We also postulate that $a + ib = 0$ if and only if $a = 0$, $b = 0$. It follows from (16.2) that,

(5) if $x + iy = x_1 + iy_1$, then $x = x_1$ and $y = y_1$.   (16.5)

Complex numbers, thus defined, can be manipulated alongside real numbers according to the ordinary processes of Algebra. It is important to observe that complex numbers do not possess *magnitude* in the sense that real numbers do; it is not possible to compare two complex numbers by saying that one is greater, or less, than the other.

Notation

(i) In some fields of scientific study, notably in Electrical Engineering, it is customary to use j to denote $\sqrt{(-1)}$ and to write a complex number in the form $a + bj$ where we here write $a + ib$.

(ii) By accepted convention a is called the *real* part and b is called the *imaginary* part of the complex number $a + ib$. This unfortunately misleading nomenclature is nevertheless useful and widely used since it is frequently necessary to refer to the separate parts of a complex number.

(iii) The single letter z is used to denote the complex variable $x + iy$.

(iv) $R(z)$ and $I(z)$ are used to denote, respectively, the real and imaginary parts of the complex number z.

Examples. (i) Solve the equation $x^2 + 2x + 4 = 0$.

$$x = \frac{-2 \pm \sqrt{(4 - 16)}}{2}$$

$$= -1 \pm \sqrt{(-3)}$$

$$= -1 + i\sqrt{3} \text{ or } -1 - i\sqrt{3}.$$

(ii) $(3 + 4i)(2 - 5i) = 6 - 7i + 20 = 26 - 7i$.

(iii) $(2 - 3i)^5 = 2^5 + 5(-3i) \cdot 2^4 + 10(-3i)^2 \cdot 2^3 + 10(-3i)^3 \cdot 2^2 +$
$+ 5(-3i)^4 \cdot 2 + (-3i)^5$

$$= 32 - 240i + 720i^2 - 1080i^3 + 810i^4 - 243i^5$$

$$= 32 - 240i - 720 + 1080i + 810 - 243i$$

$$= 122 + 597i.$$

(iv) $\dfrac{1 + 2i}{1 - 3i} = \dfrac{(1 + 2i)(1 + 3i)}{(1 - 3i)(1 + 3i)} = \dfrac{1 + 5i - 6}{1 + 9} = \dfrac{1}{10}(-5 + 5i) = \tfrac{1}{2}(-1 + i)$.

Here the division is simplified by the multiplication of numerator and denominator by $(1 + 3i)$. The complex number $(a - ib)$ is called the *conjugate* of the complex number $(a + ib)$ and the relation

$$(a + ib)(a - ib) = a^2 + b^2$$

enables us to make a denominator real in much the same way that we have hitherto rationalized a denominator.

(v) Factorize $x^2 + 2x + 10$.

$x^2 + 2x + 10 \equiv (x + 1)^2 + 9 \equiv (x + 1)^2 + 3^2$
$\equiv (x + 1 + 3i)(x + 1 - 3i)$.

(vi) Form the equation whose roots are $\alpha + i\beta$, $\alpha - i\beta$, γ.

The equation is $\{x - (\alpha + i\beta)\}\{x - (\alpha - i\beta)\}(x - \gamma) = 0$, i.e.,

$x^3 - (\alpha + i\beta + \alpha - i\beta + \gamma)x^2 + \{(\alpha + i\beta)(\alpha - i\beta) + (\alpha + i\beta)\gamma +$
$+ (\alpha - i\beta)\gamma\}x - (\alpha + i\beta)(\alpha - i\beta)\gamma = 0$,

or $x^3 - (2\alpha + \gamma)x^2 + (\alpha^2 + \beta^2 + 2\alpha\gamma)x - (\alpha^2\gamma + \beta^2\gamma) = 0$.

(vii) Express $\dfrac{x+1}{x^2 + 2x + 17}$ as the sum of two partial fractions with linear denominators.

$$\frac{x+1}{x^2 + 2x + 17} \equiv \frac{x+1}{(x+1)^2 + 16} \equiv \frac{x+1}{(x+1+4i)(x+1-4i)}.$$

Hence using the cover-up rule we find

$$\frac{x+1}{x^2 + 2x + 17} \equiv \frac{1}{2(x+1+4i)} + \frac{1}{2(x+1-4i)}.$$

(viii) If $(a + ib)^2 = 3 + 4i$ where a and b are real, calculate a and b.

$$(a + ib)^2 = 3 + 4i.$$

$$\therefore a^2 - b^2 + 2aib = 3 + 4i.$$

Thus, using (16.5), $a^2 - b^2 = 3$, $ab = 2$.
This process is called "*equating real and imaginary parts*".

$$\therefore a^2 - \frac{4}{a^2} = 3,$$

i.e., $$a^4 - 3a^2 - 4 = 0,$$

or $$(a^2 - 4)(a^2 + 1) = 0.$$

Hence, since a is real, $a = 2$, $b = 1$ or $a = -2$, $b = -1$.

Exercises 16.2

1. Solve each of the following equations in x:
(i) $x^2 + x + 1 = 0$,   (ii) $x^2 - x + 3 = 0$,   (iii) $2x^2 - 2x + 1 = 0$,
(iv) $x^2 - 2x\cos\theta + 1 = 0$,   (v) $x^2\cos^2\theta + x\sin2\theta + 1 = 0$,   (vi) $x^4 + 1 = 0$.

2. Express each of the following in the form $a + ib$:
(i) $(3 + 5i) + (7 - 2i)$,   (ii) $(4 + 2i) - (1 - 5i)$,   (iii) $(5 + 3i)^2$,
(iv) $(2 + 3i)(3 - 7i)$,   (v) $(2 + 4i)(5 - 2i)$,   (vi) $(3 + i)/(5 - i)$,
(vii) $(2 + 4i)(3 - 2i)/(1 - 2i)$,   (viii) $(1 - i)^4$,   (ix) $(1 + 2i)^5$,   (x) $1/(3 - 4i)^2$.

3. If $z = x + iy$, express each of (i) z^2, (ii) $1/z$, (iii) $(z + 1)/(z - 1)$ in the form $X + iY$.

4. Calculate the square roots of each of the following:
(i) $7 - 24i$, (ii) $15 - 8i$, (iii) $5 + 12i$, (iv) $(4 + 3i)/(12 + 5i)$, (v) $2i$, (vi) $-2i$.

5. In each of the following cases, form the equation with the given roots.
(i) $2i$, $-2i$;   (ii) $(2 + 3i)$, $(2 - 3i)$;   (iii) $(1 + i)$, $(1 - i)$, 1, -1;   (iv) $(2 + i)$, $(2 - i)$, 0, 0 (a repeated root);   (v) $(1 + i)$, $(1 - i)$, $(2 + i)$, $(2 - i)$.

6. Simplify
$$\frac{a+bi}{c+di} + \frac{a-bi}{c-di}.$$

7. Express each of the following as the product of linear factors:
(i) $x^2 + 2x + 5$, (ii) $x^2 - 4x + 5$, (iii) $4x^2 + 4x + 2$, (iv) $x^3 - 3x^2 + 4x - 2$,
(v) $x^2 + 2ax + (a^2 + b^2)$.

8. Express $\dfrac{x}{x^2 - 2x + 10}$ in partial fractions with denominators linear in x.

9. Assume $\dfrac{1}{x^2+1} \equiv \dfrac{Ai+B}{x+i} + \dfrac{Ci+D}{x-i}$, where A, B, C, D are real constants,
and hence obtain partial fractions of $\dfrac{1}{x^2+1}$.

10. Prove that if $(a + ib)^3$, where a and b are real and $a \neq 0, b \neq 0$, is a number of the form $(x + 0\,i)$, then $b^2 = 3a^2$ and find the condition that $(a + ib)^3$ is a number of the form $(0 + iy)$.

11. Solve $z^2 - 2iz + 1 = 0$.

12. Solve $iz^2 + 2z + i = 0$.

13. Solve $z^2 - 3z + 10i = 0$.

14. Solve $2\left(z + \dfrac{1}{z}\right) = 3 + i$.

15. Solve $z^2 - 2iz \sec\theta - 1 = 0$.

16.3 The cube roots of unity

If $x^3 - 1 = 0$, then
$$(x - 1)(x^2 + x + 1) = 0$$
so that $x = 1$ or $x = \frac{1}{2}(-1 + \sqrt{3}i)$ or $x = \frac{1}{2}(-1 - \sqrt{3}i)$.
These three values are the three cube roots of $1 + 0i$ in the field of complex numbers. We have

$$\left\{\tfrac{1}{2}(-1 + \sqrt{3}i)\right\}^2 = \frac{1}{4}(-2 - 2\sqrt{3}i) = \tfrac{1}{2}(-1 - \sqrt{3}i),$$

$$\left\{\tfrac{1}{2}(-1 - \sqrt{3}i)\right\}^2 = \frac{1}{4}(-2 + 2\sqrt{3}i) = \tfrac{1}{2}(-1 + \sqrt{3}i).$$

Each of the complex cube roots of unity is thus the square of the other. The three roots are usually denoted by $1, \omega, \omega^2$.

(i) Because $1, \omega, \omega^2$ are the roots of $x^3 - 1 = 0$, a cubic equation in x in which the coefficient of x^2 is zero,
$$\therefore 1 + \omega + \omega^2 = 0.$$
(This result also follows directly from the values of ω, ω^2 given above.)

(ii) $\omega^{3n+1} = \omega$, $\omega^{3n+2} = \omega^2$ for $n = 0, 1, 2, \ldots$.

(iii) The cube roots of any real number N (say) are $N^{1/3}$, $N^{1/3}\omega$ and $N^{1/3}\omega^2$ where $N^{1/3}$ denotes the real cube root of N.

Examples. (i) Show that

$$a^3 + b^3 + c^3 - 3abc \equiv (a + b + c)(a + \omega b + \omega^2 c)(a + \omega^2 b + \omega c).$$
$$(a + \omega b + \omega^2 c)(a + \omega^2 b + \omega c) \equiv a^2 + \omega^2 ab + \omega ac + \omega ab + \omega^3 b^2$$
$$+ \omega^2 bc + \omega^2 ac + \omega^4 bc + \omega^3 c^2$$
$$\equiv a^2 + b^2 + c^2 + ab(\omega^2 + \omega) + bc(\omega^2 + \omega^4) + ca(\omega + \omega^2)$$
$$\equiv a^2 + b^2 + c^2 - ab - bc - ca.$$

Multiplication by $(a + b + c)$ now gives the required result.

(ii) Find the sum of

$$1 - \omega + \omega^2 - \omega^3 + \ldots + (-1)^{n-1}\omega^{n-1}$$

and express the answer in its simplest form

  (i) when $n = 3m$,

  (ii) when $n = 3m + 1$,

  (iii) when $n = 3m + 2$, $(m = 0, 1, 2 \ldots)$

considering separately the cases in which m is odd and the cases in which m is even.

If $S_n = 1 - \omega + \omega^2 - \omega^3 + \ldots + (-1)^{n-1}\omega^{n-1}$,

which is a geometric progression of n terms and common ratio $-\omega$, then

$$S_n = \frac{1 - (-\omega)^n}{1 + \omega}.$$

When m is odd.

  (i) If $n = 3m$, $(-\omega)^n = -(\omega^3)^m = -1$ and $S_n = \dfrac{2}{1 + \omega} = -2\omega$.

  (ii) If $n = 3m + 1$, $(-\omega)^n = \omega^{3m}$. $\omega = \omega$ and $S_n = \dfrac{1 - \omega}{1 + \omega}$.

  (iii) If $n = 3m + 2$, $(-\omega)^n = -\omega^{3m}$. $\omega^2 = -\omega^2$ and $S_n = \dfrac{1 + \omega^2}{1 + \omega} = \dfrac{1}{\omega}$.

When m is even.

  (i) If $n = 3m$, $(-\omega)^n = (\omega^3)^m = 1$ and $S_n = 0$.

  (ii) If $n = 3m + 1$, $(-\omega)^n = -\omega^{3m}$. $\omega = -\omega$ and $S_n = \dfrac{1 + \omega}{1 + \omega} = 1$.

  (iii) If $n = 3m + 2$, $(-\omega)^n = \omega^{3m}$. $\omega^2 = \omega^2$

  and $S_n = \dfrac{1 - \omega^2}{1 + \omega} = 1 - \omega$.

16.4 Conjugate pairs of complex roots

We have stated in § 16.2 Example (iv) that the numbers $(a + ib)$ and $(a - ib)$ are called conjugate complex numbers and that $(a^2 + b^2) = (a + ib)(a - ib)$. We now establish some important results involving conjugate complex numbers.

(*Notation.* If $z = x + iy$, its conjugate $x - iy$ is written $\bar{z}$.)

From Equation (16.3) which defines multiplication for complex numbers it follows that

$$(a_1 + ib_1)(a_2 + ib_2) = (a_1a_2 - b_1b_2) + i(a_1b_2 + a_2b_1), \tag{16.6}$$

$$(a_1 - ib_1)(a_2 - ib_2) = (a_1a_2 - b_1b_2) - i(a_1b_2 + a_2b_1). \tag{16.7}$$

Equation (16.7) could be obtained directly from Equation (16.6) by a simple replacement of i by $-i$ and by a similar reference to the definition of division for complex numbers in equation (16.4) we could show that if $\dfrac{a_1 + ib_1}{a_2 + ib_2} = x + iy$, then $\dfrac{a_1 - ib_1}{a_2 - ib_2} = x - iy$.

These results can be extended, by the method of mathematical induction, to the product (or combination of product and quotient) of n complex numbers.

In a similar way if the definitions of addition and subtraction in Equations (16.1) and (16.2) are generalized it follows that the result of any operation involving addition and subtraction with complex numbers can be obtained from the result of the corresponding operation for the conjugate complex numbers by changing i into $-i$.

It follows from this that from any identity connecting complex numbers, another identity can be obtained by changing i into $-i$.

In particular if $f(x + iy)$ is a rational function of $(x + iy)$ *with real coefficients* and

$$f(x + iy) = X + iY, \tag{16.8}$$

where X, Y are functions of x, y, then

$$f(x - iy) = X - iY. \tag{16.9}$$

We now show that *if a polynomial equation with real coefficients has complex roots, then these roots occur in conjugate pairs.*

For, if $\alpha + i\beta$ is a root of $f(x) = 0$ where $f(x)$ is a polynomial in x with real coefficients and α and β are real,

$$f(\alpha + i\beta) = A + iB = 0.$$

$$\therefore A = 0, B = 0.$$

But from (16.8) and (16.9) $f(\alpha - i\beta) = A - iB = 0$.

$$\therefore \alpha - i\beta \text{ is a root of } f(x) = 0.$$

Example. Given that $1 - 2i$ is a root of the equation

$$f(x) \equiv x^4 + 11x^2 - 10x + 50 = 0,$$

find the other roots.

If $(1 - 2i)$ is one root, $(1 + 2i)$ is also a root.

$$\therefore x^2 - (1 + 2i + 1 - 2i)x + (1 - 2i)(1 + 2i) = x^2 - 2x + 5$$

is a quadratic factor of $f(x)$.

$$\therefore x^4 + 11x^2 - 10x + 50 \equiv (x^2 - 2x + 5)(x^2 + 2x + 10).$$

The solution of $x^2 + 2x + 10 = 0$ is $x = -1 + 3i$ or $x = -1 - 3i$. Hence the other roots of $f(x) = 0$ are $(1 + 2i)$, $(-1 + 3i)$, $(-1 - 3i)$.

Exercises 16.4

Where appropriate in these exercises give answers in terms of ω.

1. Solve the equation $x^3 + 1 = 0$.

2. Express $x^3 + 6x^2 + 12x + 7$ in the form $(x + a)^3 + b = 0$ and hence solve the equation $x^3 + 6x^2 + 12x + 7 = 0$.

3. Use the methods of the calculus to show that, if p and q are real, the equation

$$x^3 - px + q = 0$$

has   (i) three real distinct roots if $4p^3 > 27q^2$,

    (ii) a pair of real coincident roots if $4p^3 = 27q^2$,

    (iii) two complex roots and one real root if $4p^3 < 27q^2$.

4. Sketch the graph of $y = x^3 - 6x - 4$. Transform the equation $x^3 - 6x - 4 = 0$ by means of the substitutions $x = w + z$ and $wz = 2$ into the form $w^3 + z^3 = a$ where a is a constant. Hence calculate the possible values of w^3.

5. Prove that $1 + \dfrac{1}{\omega} + \dfrac{1}{\omega^2} = 0$.

6. Prove that $1 + \omega^4 + \omega^8 = 0$.

7. Prove that $(1 - \omega + \omega^2)(1 + \omega - \omega^2)(1 - \omega - \omega^2) = 8$.

8. Prove that if $x = a + b$, $y = a\omega + b\omega^2$, $z = a\omega^2 + b\omega$ then $xyz = a^3 + b^3$.

9. Prove that $(1 - \omega)^5 = -9(2 + \omega)$.

10. Form the equation whose roots are $-1, 1 + \omega, 1 + \omega^2$.

11. Find the sum of

$$1 + \omega + \omega^2 + \omega^3 + \ldots + \omega^{3n}.$$

12. Solve the equation $8x^3 - 44x^2 + 86x - 65 = 0$ given that one root is $\frac{1}{2}(3 - 2i)$.

13. Form the equation of the fourth degree in x, two of the roots of which are $3 + i$ and $1 + 3i$.

14. If two of the roots of the equation $x^3 + ax - 4c^3 = 0$ are complex and the real root is $x = 2c$, find the complex roots in terms of c only.

15. The sum of the four complex roots of the equation

$$x^5 - 5x^3 + 14x + 20 = 0$$

is. 2. State the value of the real root and, given that one of the complex roots is $2 + i$, find the other roots.

16. Solve the equation $x^4 - x^3 - x + 1 = 0$ given that two of its roots are complex and the other two are equal.

17. If $\alpha + i\beta$ is a root of the equation $x^3 + px + q = 0$, prove that (i) $2\alpha(\alpha^2 + \beta^2) = q$, (ii) $3\alpha^2 - \beta^2 = -p$, (iii) α is a root of the equation

$$8x^3 + 2px - q = 0.$$

16.5 The geometry of complex numbers

In § 16.1 we discussed the arithmetical continuum of real numbers and their *representation* by the points in a line. A complex number $x + iy$ can be considered as a *number pair* which is uniquely defined by the numbers x and y, and as such it is represented uniquely by the point (x, y) in the plane of the cartesian coordinate axes Ox and Oy.

The number can be considered as represented by

(i) the point (x, y),

or (ii) *displacements* x and y in directions parallel to the axes from the origin,

or (iii) the resultant *displacement* (which is a vector).

The point is not represented by any length in the plane. Indeed, *a complex number cannot be represented by a length.*

It must be stressed that this is an arbitrarily chosen method of *representing* complex numbers geometrically, and it is not a property of complex numbers which is deduced from the definition. The representation is justified because every point in the plane represents a complex number and every complex number is represented uniquely by a point in the plane. Methods of carrying out the four processes of addition, subtraction, multiplication and division within this geometrical framework can now be evolved. The diagram of the coordinate plane, when it is used to represent complex numbers in this way, is sometimes called the Argand Diagram [after Jean Robert Argand (1768 – 1822) of Geneva, who first expressed these ideas clearly]. By convention Ox is called the *real axis* and Oy is called the *imaginary axis*.

Addition. Fig. 153. If $z_1 = x_1 + iy_1$ is represented by P, and $z_2 = x_2 + iy_2$ by Q, then $z_1 + z_2$ is represented by R where $OPRQ$ is a parallelogram.

$$\text{For } \Delta\, OPM \equiv \Delta\, QRN.$$

$$\therefore QN = OM = x_1 \text{ and } RN = PM = y_1.$$

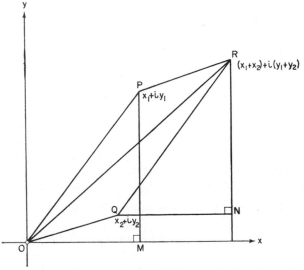

FIG. 153.

Thus R is $\{(x_1 + x_2),\ (y_1 + y_2)\}$ and hence R represents the complex number $z_1 + z_2$.

Justification for this construction when the complex numbers z_1 and z_2 are represented by the vectors $\overrightarrow{OP}$ and $\overrightarrow{OQ}$ (Volume I, Chapter IV, § 4.7) is as follows:

$$\overrightarrow{PR} = \overrightarrow{OQ}.$$

But
$$\overrightarrow{OP} + \overrightarrow{OQ} = \overrightarrow{OP} + \overrightarrow{PR} = \overrightarrow{OR}.$$

$$\therefore \overrightarrow{OR} = z_1 + z_2.$$

Subtraction. Fig.154. If z_1 is represented by P and z_2 by Q, $z_1 - z_2$ is represented by R, where $OQPR$ is a parallelogram.

$$\text{For } \Delta\, QOM \equiv \Delta\, PRN.$$

$$\therefore RN = OM = x_2 \text{ and } PN = QM = y_2.$$

Thus R is $\{(x_1 - x_2),\, (y_1 - y_2)\}$ and hence R represents the complex number $z_1 - z_2$. *With vector representation*

$$\overrightarrow{OP} - \overrightarrow{OQ} = \overrightarrow{OP} + \overrightarrow{QO} = \overrightarrow{OP} + \overrightarrow{PR} = \overrightarrow{OR}.$$

$$\therefore z_1 - z_2 = \overrightarrow{OR}.$$

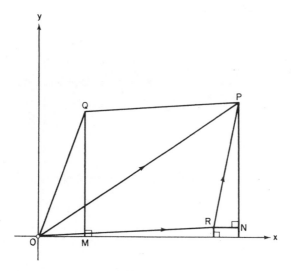

FIG. 154.

Note that with vector representation $\overrightarrow{OR} = \overrightarrow{QP}$. It is often convenient to use $\overrightarrow{QP}$ to represent $z_1 - z_2$.

Multiplication and Division. These processes are more conveniently carried out when a system of polar coordinates is used for representing complex numbers (§ 16.7). One result is noted here

$$i(a + ib) = -b + ia. \tag{16.10}$$

In Fig. 155, P is (a, b) and Q is $(-b, a)$.

$$\Delta\, OPN \equiv \Delta\, QOM.$$

$$\therefore\ \angle\, PON = \angle\, OQM.$$

$$\therefore\ \angle\, QOP = \tfrac{1}{2}\pi.$$

The effect of multiplication by i has been to rotate the vector representing the complex number through an angle $\tfrac{1}{2}\pi$. The effect of multiplication by i^2 is thus to rotate the vector successively through two right angles, that is through an angle π. This is analogous to the effect of multiplication of a real number by -1 when the field of real numbers is represented by the points of a line.

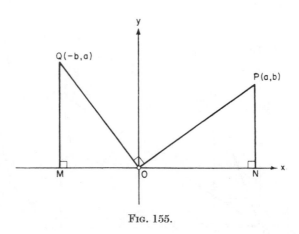

FIG. 155.

Exercises 16.5

For these questions use perpendicular axes drawn on graph paper.

1. Mark points on the Argand diagram representing the following complex numbers,

$A(4 + 3i)$; $B(3 - 4i)$; $C(-3 + 4i)$; $D(-3 - 4i)$; $E(1 + 0i)$; $F(0 + i)$.

2. Draw diagrams to illustrate each of the following operations with complex numbers, stating the answer from your diagram and verifying it by calculation.

   (i) $(3 + 4i) + (2 - 3i)$, (ii) $(2 - i) - (3 - 2i)$, (iii) $(3 + 2i) - (1 - 0i)$,
   (iv) $i(2 - 3i) - i(3 - 2i)$, (v) $i^3(4 - i)$.

16.6 The polar coordinate form of a complex number—Modulus and Argument

If the complex number $z(= x + iy)$ is represented by the point $P(x, y)$ in the coordinate plane, then, in the usual notation for polar coordinates with the origin as pole and the x-axis as initial line, $x = r\cos\theta$, $y = r\sin\theta$ where $r = \sqrt{(x^2 + y^2)}$ and $\tan\theta$ is defined by $\sin\theta = y/r$, $\cos\theta = x/r$. The polar form of the complex number is now $r(\cos\theta + i\sin\theta)$. This form is sometimes shortened in calculations either to $r\operatorname{cis}\theta$ or to $r \angle \theta$.

r is a scalar and it is represented by the *length* of OP.

r is called the *modulus* of the complex number z and it is written thus $r = |z|$.

If θ is the unique angle in the range $-\pi < \theta \leq \pi$ which is defined by $\sin\theta = y/r$, $\cos\theta = x/r$, then θ is called the *principal value of the argument* of the complex number. The angles $\theta + 2n\pi$ are also defined by $\sin\theta = y/r$, $\cos\theta = x/r$ and thus the *argument of a complex number is a many valued function*. The argument of a complex number is written thus, $\arg z$.

It follows from the definition of equal complex numbers that two complex numbers are equal if, and only if, their moduli are equal and the principal values of their arguments are equal.

Notation. The word *amplitude* (amp) was originally used instead of the word *argument* (arg). Its use is now largely discontinued because of the possibilities of confusion with the idea of amplitude in wave motion. *Amplitude* is still used however by some authors.

Examples. (i) The complex number $(3 - 4i) = \sqrt{(3^2 + 4^2)}(\cos\theta + i\sin\theta)$ where $\sin\theta = -\dfrac{4}{5}$ and $\cos\theta = \dfrac{3}{5}$,

i.e. $\tan\theta = -\dfrac{4}{3}$ (θ in the fourth quadrant).

$\therefore 3 - 4i = 5\{\cos(360n° - 53°\,8') + i\sin(360n° - 53°\,8')\}$.

$\therefore |3 - 4i| = 5$, $\arg(3 - 4i) = 360n° - 53°\,8'$.

(ii) Find the modulus and principal value of the argument of $(3 + 4i)/(3 - 4i)$.

$$\text{Let } z = \frac{3 + 4i}{3 - 4i}.$$

Then $z = \dfrac{(3 + 4i)^2}{(3 + 4i)\,(3 - 4i)} = \dfrac{-7 + 24i}{25} = \dfrac{1}{25}\cdot 25\,(\cos\theta + i\sin\theta)$

$$= \cos\theta + i\sin\theta,$$

where $\tan\theta = 24/(-7)$ and θ is in the second quadrant.

$\therefore\ |z| = 1$ and the principal value of arg $z \doteqdot 1.85^c$.

(iii) Find the locus of the point representing z in the coordinate plane if $|z + 2 - i| = 1$.

For points on the locus $|(x + 2) + i(y - 1)| = 1$,

$$\text{i.e.,}\ (x + 2)^2 + (y - 1)^2 = 1.$$

Hence the locus is a circle with centre $(-2, 1)$ and radius 1.

Alternatively. The equation $|z + 2 - i| = 1$ can be written $|z - (-2 + i)| = 1$ expressing the fact that the point z moves so that its distance from the point $-2 + i$ is unity. Hence the locus of the point z is a circle with centre $-2 + i$ and of radius 1.

(iv) If $z = \cos\theta + i\sin\theta$, $w = (1 - z)/1 + z)$ show that the point representing w on the Argand diagram lies on the imaginary axis and find its coordinates.

Discuss the limiting positions of this point as $\theta \to \pi - 0$ and as $\theta \to \pi + 0$.

$w = \dfrac{(1 - \cos\theta) - i\sin\theta}{(1 + \cos\theta) + i\sin\theta} = \dfrac{\big\{(1 - \cos\theta) - i\sin\theta\big\}\big\{(1 + \cos\theta) - i\sin\theta\big\}}{\big\{(1 + \cos\theta) + i\sin\theta\big\}\big\{(1 + \cos\theta) - i\sin\theta\big\}}$

$\quad = \dfrac{1 - \cos^2\theta - \sin^2\theta - i\,(\sin\theta + \sin\theta\cos\theta + \sin\theta - \sin\theta\cos\theta)}{1 + 2\cos\theta + \cos^2\theta + \sin^2\theta}$

$\quad = \dfrac{-2i\sin\theta}{2 + 2\cos\theta} = \dfrac{-i\sin\theta}{1 + \cos\theta}\,.$

Thus w is represented by the point $\left(0, \dfrac{-\sin\theta}{1 + \cos\theta}\right)$ on the imaginary axis. Also

$$w = \dfrac{-i\sin\theta}{1 + \cos\theta} = \dfrac{-2i\sin\tfrac{1}{2}\theta\cos\tfrac{1}{2}\theta}{2\cos^2\tfrac{1}{2}\theta} = -i\tan\tfrac{1}{2}\theta\,.$$

$$\text{As }\theta \to \pi - 0,\ w \to 0 - \infty\,i\,.$$

$$\text{As }\theta \to \pi + 0,\ w \to 0 + \infty\,i\,.$$

16.7 Products and quotients

If $z_1 = r_1(\cos\theta_1 + i\sin\theta_1)$ and $z_2 = r_2(\cos\theta_2 + i\sin\theta_2)$, then

$$z_1 z_2 = r_1 r_2 (\cos\theta_1 + i\sin\theta_1)(\cos\theta_2 + i\sin\theta_2)$$

$$= r_1 r_2 (\cos\theta_1 \cos\theta_2 - \sin\theta_1 \sin\theta_2) + i(\sin\theta_1 \cos\theta_2 + \cos\theta_1 \sin\theta_2).$$

$$\therefore\ z_1 z_2 = r_1 r_2 \{\cos(\theta_1 + \theta_2) + i\sin(\theta_1 + \theta_2)\}. \tag{16.11}$$

Also

$$\frac{z_1}{z_2} = \frac{r_1(\cos\theta_1 + i\sin\theta_1)}{r_2(\cos\theta_2 + i\sin\theta_2)}$$

$$= \frac{r_1(\cos\theta_1 + i\sin\theta_1)(\cos\theta_2 - i\sin\theta_2)}{r_2(\cos^2\theta_2 + \sin^2\theta_2)}$$

$$= \frac{r_1}{r_2}\{\cos(\theta_1 - \theta_2) + i\sin(\theta_1 - \theta_2)\}. \tag{16.12}$$

It should be noted that these results (16.11) and (16.12) do not necessarily give the principal values of the argument of the product or quotient.

These results are summarised as follows:

The modulus of the product of two complex numbers is equal to the product of the moduli of the numbers.

One value of the argument of the product of two complex numbers is equal to the sum of the arguments of the numbers.

The modulus of the quotient of two complex numbers is equal to the quotient of the moduli of the numbers.

One value of the argument of the quotient of two complex numbers is equal to the difference of the arguments of the numbers.

Example. If the complex numbers z_1, z_2 and z_3 are represented in the Argand diagram by the points P, Q and R respectively, and the angles of the triangle PQR at Q and R are each $\frac{1}{2}(\pi - \alpha)$, prove that

$$(z_3 - z_2)^2 = 4(z_3 - z_1)(z_1 - z_2)\sin^2\tfrac{1}{2}\alpha.$$

Let $\theta_1, \theta_2, \theta_3$ be the angles made with Ox by $\overrightarrow{QR}$, $\overrightarrow{PR}$ and $\overrightarrow{PQ}$ (Fig. 156.) Then, from triangles RAB and QAC,

$$\theta_2 + \theta_3 = \theta_1 + (\tfrac{1}{2}\pi - \tfrac{1}{2}\alpha) + \theta_1 + (\tfrac{1}{2}\alpha + \tfrac{1}{2}\pi) = \pi + 2\theta_1.$$

But

$$\overrightarrow{OR} - \overrightarrow{OQ} = \overrightarrow{QR}.$$

$$\therefore z_3 - z_2 = \overrightarrow{QR}.$$

$$\therefore (z_3 - z_2)^2 = \{QR(\cos\theta_1 + i\sin\theta_1)\}^2$$

$$= QR^2(\cos 2\theta_1 + i\sin 2\theta_1).$$

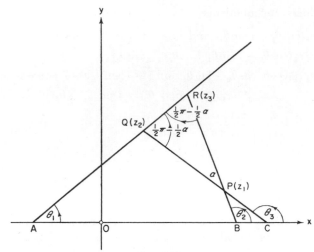

Fig. 156.

Also $$\overrightarrow{OQ} - \overrightarrow{OP} = \overrightarrow{PQ}.$$

$$\therefore (z_2 - z_1) = PQ\,(\cos\theta_3 + i\sin\theta_3)\,.$$

Similarly $$(z_3 - z_1) = PR\,(\cos\theta_2 + i\sin\theta_2)\,.$$

$$\therefore (z_3 - z_1)(z_2 - z_1) = PQ \cdot PR\,(\cos\theta_3 + i\sin\theta_3)(\cos\theta_2 + i\sin\theta_2)$$

$$= PQ \cdot PR\,\{\cos(\theta_3 + \theta_2) + i\sin(\theta_3 + \theta_2)\}$$

$$= -PQ \cdot PR\,(\cos2\theta_1 + i\sin2\theta_1)\,.$$

$$\text{But}\quad PQ = PR = \frac{RQ}{2\sin\frac{1}{2}\alpha}\,.$$

$$\therefore PQ \cdot PR = \frac{RQ^2}{4\sin^2\frac{1}{2}\alpha}\,.$$

$$\therefore (z_3 - z_1)\,(z_1 - z_2) = \frac{RQ^2\,(\cos2\theta_1 + i\sin2\theta_1)}{4\sin^2\frac{1}{2}\alpha}\,.$$

$$\therefore (z_3 - z_2)^2 = 4\,(z_3 - z_1)\,(z_1 - z_2)\sin^2\tfrac{1}{2}\alpha\,.$$

Multiplication and Division in the Argand Diagram

In Fig. 157 P is the point z_1, Q is the point z_2 and A is the point $(1, 0)$. The triangle OQR is constructed similar to the triangle OAP, i.e. so that $\angle A = \angle Q$ and $\angle P = \angle R$. Let R represent the complex number z_3.

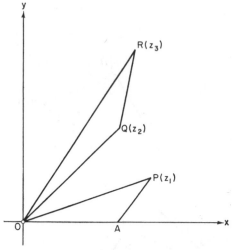

F IG. 157.

Then $\angle ROA = \angle POA + \angle QOA.$

$$\therefore \arg z_3 = \arg z_1 + \arg z_2.$$

Also $$OR/OQ = OP/OA.$$

$$\therefore OR \cdot OA = OP \cdot OQ.$$

$$\therefore |z_3| = |z_1 z_2|.$$

$$\therefore z_3 = z_1 z_2$$

because we have proved that $|z_3| = |z_1 z_2|$ and $\arg z_3 = \arg z_1 z_2.$

In Fig. 158 P is the point z_1, Q is the point z_2 and A is the point $(1, 0)$. The triangle OAR is constructed similar to the triangle OQP, i.e. so that $\angle A = \angle Q$, $\angle R = \angle P$. Let R represent z_3.

Then $$\arg z_3 = \angle AOR = \angle QOP$$

$$= \arg z_1 - \arg z_2 = \arg\left(\frac{z_1}{z_2}\right).$$

Also $$OR/OA = OP/OQ.$$

$$\therefore |z_3| = \frac{|z_1|}{|z_2|} = \left|\frac{z_1}{z_2}\right|.$$

$$\therefore z_3 = z_1/z_2.$$

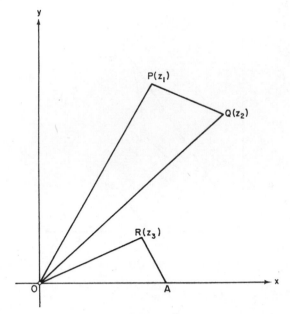

FIG. 158.

Example. Show on an Argand diagram the locus of the point P, representing the complex number z, if

(a) $\dfrac{z - i}{z - 1}$ is purely imaginary,

(b) $\dfrac{z - i}{z - 1}$ is real.   **(L.)**

$$\frac{z - i}{z - 1} = \frac{x + i(y - 1)}{(x - 1) + iy} \cdot \frac{(x - 1) - iy}{(x - 1) - iy}$$

$$= \frac{x^2 + y^2 - x - y + i(1 - x - y)}{x^2 + y^2 - 2x + 1} .$$

The denominator of this fraction is real.

(a) Then $\dfrac{z - i}{z - 1}$ is purely imaginary if

$$x^2 + y^2 - x - y = 0.$$

In this case the locus of P is the circle $x^2 + y^2 - x - y = 0$.

(b) $\dfrac{z-i}{z-1}$ is purely real if

$$1 - x - y = 0.$$

Then the locus of P is the straight line $x + y - 1 = 0$.

These results could have been reached by geometrical considerations as shown in Fig. 159. If P is the point z, A the point $(1 + 0i)$, B the

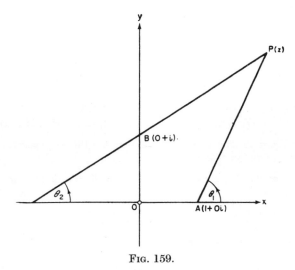

FIG. 159.

point $(0 + i)$ and if $\overrightarrow{AP}$ and $\overrightarrow{BP}$ make angles of θ_1, θ_2 respectively with Ox, then

$$z - i = \overrightarrow{BP}, \quad z - 1 = \overrightarrow{AP}.$$

$$\therefore \arg\left(\frac{z-i}{z-1}\right) = \arg(z-i) - \arg(z-1) = \theta_2 - \theta_1.$$

(a) If $\dfrac{z-i}{z-1}$ is purely imaginary, then $\arg\left(\dfrac{z-i}{z-1}\right) = \pm\tfrac{1}{2}\pi$.

$$\text{But } \angle APB = \pm(\theta_2 - \theta_1).$$

$$\therefore \angle APB = \tfrac{1}{2}\pi.$$

Hence the locus of P is a circle on AB as diameter when $\dfrac{z-i}{z-1}$ is purely imaginary.

(b) If $\dfrac{z-i}{z-1}$ is real, then

$$\arg \frac{z-i}{z-1} = 0 \text{ or } \pi.$$

$$\therefore\ \theta_2 - \theta_1 = 0 \text{ or } \pi.$$

Hence the locus of P is the line AB when $\dfrac{z-i}{z-1}$ is real.

Exercises 16.7

1. Express each of the following in the form

$$r(\cos\theta + i\sin\theta)$$

where θ is the principal value of the argument.

(i) $5 + 12i$, (ii) $3 + 4i$, (iii) 1, (iv) i, (v) ω (one of the cube roots of unity), (vi) ω^2, (vii) $(2 + i)^2$, (viii) $(3 + 4i)(5 - 12i)$, (ix) $(2 + i)/(1 - 2i)$, (x) $1/(2 + 3i)$.

2. Draw a figure to show on the Argand diagram the conjugate complex numbers $r(\cos\theta + i\sin\theta)$ and $r(\cos\theta - i\sin\theta)$.

3. Find the locus of the point representing z in the coordinate plane in each of the following cases:

(i) $|z + i| = 1$,　(ii) $\left|\dfrac{1}{z}\right| = a$,　(iii) $|z + 1| = |z + i|$,

(iv) $\arg(z + 1) = \dfrac{1}{4}\pi$,　(v) $\arg z = \alpha$.

4. Use the Argand diagram to prove

$$\text{(i)}\quad |z_1 + z_2| \leqq |z_1| + |z_2|,$$

$$\text{(ii)}\quad |z_1 - z_2| \geqq |z_1| - |z_2|.$$

Prove these theorems analytically by putting z_1 and z_2 into polar form.

5. If $|z| = 2$, what is the locus of P when it represents the complex number $z + 2$?

6. If P represents the point z_1 and Q the point z_2 on the Argand diagram, find the locus of the mid-point of PQ when

(i) $|z_1|$ and $|z_2|$ are constant and $\arg z_1 - \arg z_2$ is constant, (ii) $|z_1| = |z_2|$ and $\arg(z_1 - z_2)$ is constant, (iii) z_1 is constant and $\arg z_2 - \arg z_1 = \frac{1}{2}\pi$.

7. Simplify each of the following:

(i) $(\cos\theta + i\sin\theta)(\cos\theta - i\sin\theta)$,

(ii) $2\left(\cos\dfrac{1}{4}\pi + i\sin\dfrac{1}{4}\pi\right) \times 3\left(\cos\dfrac{1}{3}\pi - i\sin\dfrac{1}{3}\pi\right)$,

(iii) $\left\{3\left(\cos\dfrac{1}{3}\pi + i\sin\dfrac{1}{3}\pi\right)\right\}^3$,

(iv) $4\left(\cos\dfrac{3}{4}\pi + i\sin\dfrac{3}{4}\pi\right)\Big/2\left(\cos\dfrac{1}{3}\pi - i\sin\dfrac{1}{3}\pi\right)$,

(v) $(\cos\tfrac{1}{2}\theta + i\sin\tfrac{1}{2}\theta)^3\Big/\left(\cos\dfrac{1}{3}\theta - i\sin\dfrac{1}{3}\theta\right)^2$.

8. If $z = \cos\theta + i\sin\theta$, express each of $\dfrac{1}{z}$; $z + \dfrac{1}{z}$; $z^2 + \dfrac{1}{z^2}$; and $z^3 + \dfrac{1}{z^3}$ in polar form.

9. In the Argand diagram, the points P and Q represent the complex numbers $(1 + i)$ and $(4 + 5i)$ respectively. Find the complex numbers which correspond to the other two vertices of an equilateral triangle with P as centre and one vertex at Q.

10. Identify the loci on the Argand diagram given by

(i) $|z - 1| + |z + 1| = 4$,

(ii) $|z - 1| + |z + 1| = 2$,

(iii) $\arg(z - 1) - \arg(z + 1) = \tfrac{1}{2}\pi$,

(iv) $\arg(z - 1) + \arg(z + 1) = \pi$,

(v) $\dfrac{|z - 1|}{|z + 1|} = 2$. (O.C.)

11. Prove that if P_1, P_2, P_3, P_4 represent the points z_1, z_2, z_3, z_4 on the Argand diagram and if

$$\frac{(z_1 - z_2)(z_3 - z_4)}{(z_1 - z_4)(z_3 - z_4)}$$

is real, $P_1 P_2 P_3 P_4$ is cyclic.

12. (a) If x and y are real, solve the equation

$$\frac{iy}{1 + ix} - \frac{3y + 4i}{3x + y} = 0.$$

(b) If $z = x + iy$ where x and y are real, show that

(i) when $\dfrac{z + i}{z + 2}$ is real, the locus of the point (x, y) is a straight line,

(ii) when $\dfrac{z + i}{z + 2}$ is purely imaginary, the locus of the point (x, y) is a circle of radius $\tfrac{1}{2}\sqrt{5}$. State the coordinates of the centre of the circle. (L.)

13. If a, b, c are real, and $ac - b^2 > 0$, show that the roots of the equation $az^2 + 2bz + c = 0$ form a pair of conjugate complex numbers and write down the complex number representing the vector joining their positions in the Argand diagram.

If this line forms the diagonal of a square, find the numbers representing the other diagonal and the sides. (L.)

14. Two complex numbers z and z' are connected by the relation

$$z' = \frac{2+z}{2-z}.$$

Show that as the point which represents z describes the axis of y, from the negative end to the positive end, the point which represents z' describes completely the circle $x^2 + y^2 = 1$ in the counterclockwise direction.   (O.C.)

15. (i) Write down the values of $|z|$ and arg z, where $z = x + iy$. Illustrate by means of an Argand diagram.

The numbers c and p are given, c being real and p being complex, with $p = a + ib$; $\bar{z}$ and $\bar{p}$ denote the conjugates of z and p respectively. Prove that, if

$$z\bar{z} - \bar{p}z - p\bar{z} + c = 0,$$

then the point on the Argand diagram which represents z lies on a certain circle whose centre and radius should be determined.

(ii) Prove that, if

$$x + iy = \frac{1}{\lambda + i\mu},$$

then the points on the Argand diagram defined by making λ constant lie on a circle, and the points defined by making μ constant also lie on a circle.

Prove also that, whatever the values of the constants, the centres of the two systems of circles obtained lie on two fixed perpendicular lines.   (O.C.)

16.8 De Moivre's Theorem

This theorem states that if n is a rational number, then

$$(\cos\theta + i\sin\theta)^n = \cos n\theta + i\sin n\theta. \tag{16.13}$$

(i) *When n is a positive integer.* We prove the theorem in this case by mathematical induction. Suppose the theorem to be true for exponent n,

i.e., $(\cos\theta + i\sin\theta)^n = \cos n\theta + i\sin n\theta.$

Then

$$(\cos\theta + i\sin\theta)^{n+1} = (\cos n\theta + i\sin n\theta)(\cos\theta + i\sin\theta)$$

$$= (\cos n\theta \cos\theta - \sin n\theta \sin\theta) + i(\sin n\theta \cos\theta + \cos n\theta \sin\theta)$$

$$= \cos(n+1)\theta + i\sin(n+1)\theta.$$

Hence, if the theorem is true for the exponent n, it is true for the exponent $n + 1$. But the theorem is (identically) true for the exponent 1. Therefore it is true for exponents $1 + 1 = 2$, $2 + 1 = 3$ and so on for all positive integral values of n.

(ii) *When n is a negative integer.* If $n = -m$, where m is a positive integer, then

$$(\cos\theta + i\sin\theta)^n = \frac{1}{(\cos\theta + i\sin\theta)^m}$$

$$= \frac{1}{\cos m\theta + i\sin m\theta}$$

$$= \frac{\cos m\theta - i\sin m\theta}{\cos^2 m\theta + \sin^2 m\theta}$$

$$= \cos(-m\theta) + i\sin(-m\theta)$$

$$= \cos n\theta + i\sin n\theta$$

since $-m = n$. This proves the theorem when n is a negative integer.

(iii) *When n is a rational fraction.* Let $n = p/q$ where p is an integer and q is a positive integer. In this case $(\cos\theta + i\sin\theta)^n$ is a many valued function of θ.

We shall prove first that *one* of the values of $(\cos\theta + i\sin\theta)^n$ is $(\cos n\theta + i\sin n\theta)$. From (i), we have

$$\left(\cos\frac{p\theta}{q} + i\sin\frac{p\theta}{q}\right)^q = \cos p\theta + i\sin p\theta,$$

and from (i) or (ii)

$$\cos p\theta + i\sin p\theta = (\cos\theta + i\sin\theta)^p.$$

Hence $\left(\cos\dfrac{p\theta}{q} + i\sin\dfrac{p\theta}{q}\right)$ is a value of $(\cos\theta + i\sin\theta)^{p/q}$.

Thus we have proved De Moivre's theorem for all *rational* values of n.

Notation $z = a^{p/q}$ means that z satisfies the equation $z^q = a^p$ and therefore has many values; $\sqrt[q]{a^p}$ means the principal value of $a^{p/q}$.

The values of $(\cos\theta + i\sin\theta)^{p/q}$ where p and q are integers and q is positive.

If a value of $(\cos\theta + i\sin\theta)^{p/q}$ is $k(\cos\varphi + i\sin\varphi)$, then

$$k^q(\cos\varphi + i\sin\varphi)^q = (\cos\theta + i\sin\theta)^p.$$

$$\therefore\ k^q(\cos q\varphi + i\sin q\varphi) = \cos p\theta + i\sin p\theta.$$

But if two complex numbers are equal, their moduli are also equal.

$$\therefore\ k^q = 1.$$

$\therefore\ k = 1$ for all positive integral values of q.

$$\therefore\ \cos q\varphi + i\sin q\varphi = \cos p\theta + i\sin p\theta.$$

Then, equating real and imaginary parts,

$$\cos q\varphi = \cos p\theta,$$

$$\sin q\varphi = \sin p\theta.$$

$$\therefore \ q\varphi = p\theta + 2m\pi,$$

i.e.,
$$\varphi = \frac{p}{q}\theta + \frac{2m}{q}\pi \quad \text{where } m \text{ is an integer.}$$

If now we give to m the successive values $0, 1, 2, \ldots, (q-1)$, we have

$$(\cos\theta + i\sin\theta)^{p/q} = \cos\left(\frac{p\theta}{q}\right) + i\sin\left(\frac{p\theta}{q}\right),$$

$$\text{or} \quad \cos\left(\frac{p\theta + 2\pi}{q}\right) + i\sin\left(\frac{p\theta + 2\pi}{q}\right),$$

$$\text{or} \quad \cos\left(\frac{p\theta + 4\pi}{q}\right) + i\sin\left(\frac{p\theta + 4\pi}{q}\right),$$

$$\cdot \quad \cdot \quad \cdot \quad \cdot \quad \cdot \quad \cdot \quad \cdot \quad \cdot$$

$$\cdot \quad \cdot \quad \cdot \quad \cdot \quad \cdot \quad \cdot \quad \cdot \quad \cdot$$

$$\cdot \quad \cdot \quad \cdot \quad \cdot \quad \cdot \quad \cdot \quad \cdot \quad \cdot$$

$$\text{or} \quad \cos\left(\frac{p\theta + 2(q-1)\pi}{q}\right) + i\sin\left(\frac{p\theta + 2(q-1)\pi}{q}\right).$$

These values of $(\cos\theta + i\sin\theta)^{p/q}$ are all distinct because the difference between the arguments of any two of them is less than 2π. There are no further values of $(\cos\theta + i\sin\theta)^{p/q}$, for if $m = aq + b$ where a and b are positive integers and $b < q$, then $(\cos\theta + i\sin\theta)^{p/q}$ takes the value

$$\cos\left\{\frac{p\theta + 2(aq+b)\pi}{q}\right\} + i\sin\left\{\frac{p\theta + 2(aq+b)\pi}{q}\right\},$$

i.e.,
$$\cos\left\{\frac{p\theta + 2b\pi}{q} + 2a\pi\right\} + i\sin\left\{\frac{p\theta + 2b\pi}{q} + 2a\pi\right\},$$

and this value is equal to the value of

$$\cos\left(\frac{p\theta + 2b\pi}{q}\right) + i\sin\left(\frac{p\theta + 2b\pi}{q}\right)$$

which is one of those listed above. We have shown, therefore, that $(\cos\theta + i\sin\theta)^{p/q}$ when p and q are integers has q values and these values are given above.

It is worth repeating that De Moivre's theorem, as stated and proved above, refers only to a real, rational exponent. The theorem can be shown to be true for an irrational exponent, but in this book we shall not extend the proof beyond the case of a real rational exponent.

Examples. (i) Solve the equation $z^5 + 1 = 0$.

$$z^5 = -1.$$

$$\therefore z = (-1)^{\frac{1}{5}} = (\cos \pi + i \sin \pi)^{\frac{1}{5}}.$$

$$\therefore z = \cos \frac{\pi}{5} + i \sin \frac{\pi}{5},$$

$$\text{or} \ \cos \frac{3\pi}{5} + i \sin \frac{3\pi}{5},$$

$$\text{or} \ \cos \frac{5\pi}{5} + i \sin \frac{5\pi}{5} = -1 + i\,0,$$

$$\text{or} \ \cos \frac{7\pi}{5} + i \sin \frac{7\pi}{5},$$

$$\text{or} \ \cos \frac{9\pi}{5} + i \sin \frac{9\pi}{5}.$$

(ii) Solve the equation

$$z^6 + 2z^3 + 2 = 0.$$

This equation can be written

$$(z^3 + 1)^2 + 1 = 0$$

so that $z^3 = -1 - i$ or $-1 + i$.

$$\therefore z^3 = \sqrt{2} \left\{ \cos\left(-\frac{3}{4}\pi\right) + i \sin\left(-\frac{3}{4}\pi\right) \right\}$$

$$\text{or} \ z^3 = \sqrt{2}\left(\cos \frac{3}{4}\pi + i \sin \frac{3}{4}\pi \right).$$

$$\therefore z \doteqdot 1.12 \left\{ \cos\left(-\frac{3}{4}\pi\right) + i \sin\left(-\frac{3}{4}\pi\right) \right\}^{1/3}$$

$$\text{or} \ z \doteqdot 1.12 \left(\cos \frac{3}{4}\pi + i \sin \frac{3}{4}\pi \right)^{1/3}.$$

H

$$\therefore z \doteqdot 1.12\left\{\cos\left(-\frac{1}{4}\pi\right) + i\sin\left(-\frac{1}{4}\pi\right)\right\} \text{ or } z \doteqdot 1.12\left(\cos\frac{1}{4}\pi + i\sin\frac{1}{4}\pi\right)$$

$$\text{or } 1.12\left(\cos\frac{5\pi}{12} + i\sin\frac{5\pi}{12}\right) \qquad \text{or } 1.12\left(\cos\frac{11\pi}{12} + i\sin\frac{11\pi}{12}\right)$$

$$\text{or } 1.12\left(\cos\frac{13\pi}{12} + i\sin\frac{13\pi}{12}\right) \qquad \text{or } 1.12\left(\cos\frac{19\pi}{12} + i\sin\frac{19\pi}{12}\right).$$

(iii) Expand $(\cos\theta + i\sin\theta)^4$ by the binomial theorem. Hence deduce expressions for (a) $\cos 4\theta$ in terms of $\cos\theta$ and (b) $\sin 4\theta$ in terms of $\sin\theta$ and $\cos\theta$. Deduce an expression for $\tan 4\theta$ in terms of $\tan\theta$.

By the binomial theorem

$$(\cos\theta + i\sin\theta)^4 = \cos^4\theta + 4i\cos^3\theta\sin\theta - 6\cos^2\theta\sin^2\theta$$
$$- 4i\cos\theta\sin^3\theta + \sin^4\theta.$$

But by De Moivre's theorem

$$(\cos\theta + i\sin\theta)^4 = \cos 4\theta + i\sin 4\theta.$$

Equating real and imaginary parts, we have

$$\cos 4\theta = \cos^4\theta - 6\cos^2\theta\sin^2\theta + \sin^4\theta,$$
$$= \cos^4\theta - 6\cos^2\theta(1 - \cos^2\theta) + (1 - \cos^2\theta)^2$$
$$= 8\cos^4\theta - 8\cos^2\theta + 1,$$

and $\qquad \sin 4\theta = 4\cos^3\theta\sin\theta - 4\cos\theta\sin^3\theta.$

$$\therefore \tan 4\theta = \frac{4\cos^3\theta\sin\theta - 4\cos\theta\sin^3\theta}{\cos^4\theta - 6\cos^2\theta\sin^2\theta + \sin^4\theta}$$

$$= \frac{4\tan\theta - 4\tan^3\theta}{1 - 6\tan^2\theta + \tan^4\theta}.$$

(iv) Prove that $\qquad \cos^4\theta = \frac{1}{8}(\cos 4\theta + 4\cos 2\theta + 3).$

$$\sin^4\theta = \frac{1}{8}(\cos 4\theta - 4\cos 2\theta + 3).$$

If $z = \cos\theta + i\sin\theta$, then $\dfrac{1}{z} = \cos\theta - i\sin\theta,$

$$\left(z^n + \frac{1}{z^n}\right) = 2\cos n\theta,$$

$$\left(z^n - \frac{1}{z^n}\right) = 2i\sin n\theta.$$

$$\therefore \ 16\cos^4\theta = (2\cos\theta)^4 = \left(z + \frac{1}{z}\right)^4$$

$$= z^4 + 4z^2 + 6 + 4/z^2 + 1/z^4$$

$$= \left(z^4 + \frac{1}{z^4}\right) + 4\left(z^2 + \frac{1}{z^2}\right) + 6$$

$$= 2\cos 4\theta + 8\cos 2\theta + 6.$$

$$\therefore \ \cos^4\theta = \frac{1}{8}(\cos 4\theta + 4\cos 2\theta + 3).$$

Also
$$16\sin^4\theta = (2i\sin\theta)^4 = \left(z - \frac{1}{z}\right)^4$$

$$= z^4 - 4z^2 + 6 - 4/z^2 + 1/z^4$$

$$= \left(z^4 + \frac{1}{z^4}\right) - 4\left(z^2 + \frac{1}{z^2}\right) + 6$$

$$= 2\cos 4\theta - 8\cos 2\theta + 6.$$

$$\therefore \ \sin^4\theta = \frac{1}{8}(\cos 4\theta - 4\cos 2\theta + 3).$$

(v) Prove that $\cos\dfrac{\pi}{7} + \cos\dfrac{3\pi}{7} + \cos\dfrac{5\pi}{7} = \dfrac{1}{2}$.

The roots of $x^7 + 1 = 0$ are $-1, \ \cos\dfrac{\pi}{7} + i\sin\dfrac{\pi}{7}$,

$$\cos\frac{3\pi}{7} + i\sin\frac{3\pi}{7}, \quad \cos\frac{5\pi}{7} + i\sin\frac{5\pi}{7}, \quad \cos\frac{9\pi}{7} + i\sin\frac{9\pi}{7},$$

$$\cos\frac{11\pi}{7} + i\sin\frac{11\pi}{7}, \quad \cos\frac{13\pi}{7} + i\sin\frac{13\pi}{7}.$$

Since the coefficient of x^6 in the equation is zero, the sum of the roots of the equation is zero. The sum of the real parts of the roots is therefore zero.

$$\therefore \ \left(\cos\frac{\pi}{7} + \cos\frac{13\pi}{7}\right) + \left(\cos\frac{3\pi}{7} + \cos\frac{11\pi}{7}\right)$$

$$+ \left(\cos\frac{5\pi}{7} + \cos\frac{9\pi}{7}\right) - 1 = 0.$$

But $\cos\dfrac{\pi}{7} = \cos\dfrac{13\,\pi}{7}$ etc.

$$\therefore\ 2\cos\frac{\pi}{7} + 2\cos\frac{3\,\pi}{7} + 2\cos\frac{5\,\pi}{7} = 1 .$$

$$\therefore\ \cos\frac{\pi}{7} + \cos\frac{3\,\pi}{7} + \cos\frac{5\,\pi}{7} = \frac{1}{2} .$$

Exercises 16.8

1. Simplify $(\cos\theta - i\sin\theta)^4/(\cos\theta + i\sin\theta)^5$.

2. Simplify $\left(\cos\dfrac{1}{4}\pi + i\sin\dfrac{1}{4}\pi\right)^3\left(\cos\dfrac{3}{4}\pi - i\sin\dfrac{3}{4}\pi\right)^2$.

3. Simplify $\dfrac{\left(\cos\dfrac{1}{3}\pi + i\sin\dfrac{1}{3}\pi\right)^2\left(\cos\dfrac{2}{3}\pi + i\sin\dfrac{2}{3}\pi\right)^3}{\left(\cos\dfrac{1}{4}\pi + i\sin\dfrac{1}{4}\pi\right)^2\left(\cos\dfrac{3}{4}\pi + i\sin\dfrac{3}{4}\pi\right)^4}$.

4. Find the square roots of each of the following in the form $r(\cos\theta + i\sin\theta)$

   (i) i, (ii) $3 + 4i$, (iii) $5 - 12i$, (iv) $-7 + 24i$, (v) $(1 + i)/(1 - i)$.

5. Find the cube roots of each of the following in the form $r(\cos\theta + i\sin\theta)$

   (i) 1, (ii) i, (iii) $1 + i$, (iv) $3 + 4i$.

6. Draw a diagram to show the positions of the points representing the five fifth-roots of unity in the coordinate plane.

7. Solve the equation $x^4 - 2x^2 + 5 = 0$.

8. Solve the equation $x^6 + 1 = 0$.

9. Express $\cos 5\theta$ in terms of powers of $\cos\theta$ only and obtain an expression for $\tan 5\theta$ in terms of $\tan\theta$.

10. Obtain expressions for $\cos n\theta$ and for $\sin n\theta$ in terms of $\sin\theta$ and $\cos\theta$.

11. Express $\sin^6\theta$ and $\cos^6\theta$ each in terms of cosines of multiples of θ.

12. Use the method suggested by question 11 to obtain the following integrals

   (i) $\int \sin^6\theta\,d\theta$, (ii) $\int \cos^8\theta\,d\theta$, (iii) $\int \sin^4\theta\cos^2\theta\,d\theta$.

13. Express $\dfrac{(\cos\alpha + i\sin\alpha)(\cos\beta + i\sin\beta)}{(\cos\gamma + i\sin\gamma)(\cos\delta + i\sin\delta)}$ in the form $(\cos\theta + i\sin\theta)$.

14. Solve the equation $(x + 1)^6 = (x - 1)^6$.

15. Find all the fourth roots of $28 + 96i$

16. If $\alpha = \cos\dfrac{2}{5}\pi + i\sin\dfrac{2}{5}\pi$, prove that

$$1 + \alpha + \alpha^2 + \alpha^3 + \alpha^4 = 0.$$

17. Prove that the roots of the equation $x^n = (x-1)^n$, where n is a positive integer, are $\frac{1}{2}\{1 + i\cot(r\pi/n)\}$ where $r = 1, 2, 3, \ldots, (n-1)$.

18. Show that $(-1 + i\sqrt{3})^n + (-1 - i\sqrt{3})^n$ is a real number and find its value in terms of trigonometrical functions.

19. Show that $\cos n\theta$ and $\sin n\theta/\sin\theta$ are expressible as polynomials in $\cos\theta$ and obtain these polynomials in the case $n = 6$. (L.)

20. Show that $z^2 - 2z\cos 2\pi/7 + 1$ is a factor of $z^7 - 1$, and write down the two other real quadratic factors of $z^7 - 1$. (L.)

16.9 The exponential form of a complex number

We have shown in § 16.8 that, for a rational n,

$$(\cos\theta + i\sin\theta)^n = \cos n\theta + i\sin n\theta,$$

i.e., if
$$f(\theta) = \cos\theta + i\sin\theta,$$

then
$$\{f(\theta)\}^n = f(n\theta).$$

This is a property of a function which obeys the index laws, as for example if $f(x) = a^x$, then

$$\{f(x)\}^n = a^{nx} = f(nx).$$

It is reasonable, therefore, to expect $\cos\theta + i\sin\theta$ to be associated with such a function as a^θ.

In Volume I, § 9.1 we obtained the results

$$\cos\theta = 1 - \frac{\theta^2}{2!} + \frac{\theta^4}{4!} - \frac{\theta^6}{6!} + \cdots,$$

$$\sin\theta = \theta - \frac{\theta^3}{3!} + \frac{\theta^5}{5!} - \frac{\theta^7}{7!} + \cdots,$$

whence we have

$$\cos\theta + i\sin\theta = 1 + i\theta - \frac{\theta^2}{2!} - \frac{i\theta^3}{3!} + \frac{\theta^4}{4!} + \frac{i\theta^5}{5!} - \frac{\theta^6}{6!} - \frac{i\theta^7}{7!} + \cdots$$

$$= 1 + i\theta + \frac{(i\theta)^2}{2!} + \frac{(i\theta)^3}{3!} + \frac{(i\theta)^4}{4!} + \frac{(i\theta)^5}{5!} + \frac{(i\theta)^6}{6!} + \frac{(i\theta)^7}{7!} + \cdots$$

assuming that rearrangement of the sum of the two infinite series is permissible.

This is of the same form as the expansion

$$e^x = 1 + x + \frac{x^2}{2!} + \frac{x^3}{3!} + \cdots$$

where x is a *real variable*, but since the expansion was obtained for a real variable only it is not possible to *deduce* that $\cos\theta + i\sin\theta = e^{i\theta}$; indeed $e^{i\theta}$ has not yet been defined.

If we write $\exp z$ for $1 + z + \dfrac{z^2}{2!} + \dfrac{z^3}{3!} + \ldots$ where z is a complex variable, then

$$\exp(i\theta) = \cos\theta + i\sin\theta$$

and
$$\{\exp(i\theta)\}^n = \exp(ni\theta).$$

Also from Eqn. (16.11)

$$\exp(i\theta) \times \exp(i\varphi) = \exp\{i(\theta + \varphi)\}$$

and from Eqn. (16.12)

$$\exp(i\theta) \div \exp(i\varphi) = \exp\{i(\theta - \varphi)\}.$$

We therefore *define* $e^{i\theta}$ as $\exp(i\theta)$ so that

$$e^{i\theta} = 1 + i\theta + \frac{(i\theta)^2}{2!} + \frac{(i\theta)^3}{3!} + \cdots$$

and $e^{i\theta}$, thus defined, obeys the laws which e^x, where x is real, obeys. Then

$$e^{i\theta} = \cos\theta + i\sin\theta. \tag{16.14}$$

It follows that
$$\begin{aligned}
e^{i(\theta + 2n\pi)} &= e^{i\theta}e^{i2n\pi} \\
&= e^{i\theta}(\cos 2n\pi + i\sin 2n\pi) \\
&= e^{i\theta}(n = 0, \pm 1, \pm 2 \cdots).
\end{aligned}$$

Also, since

$$(\cos\theta + i\sin\theta)^{-1} = (\cos\theta - i\sin\theta) = \cos(-\theta) + i\sin(-\theta),$$
$$e^{-i\theta} = 1/e^{i\theta}.$$

We can summarize these postulates, and the deductions we have made from them here, in the conclusion that complex numbers, expressed by definition as powers of e, obey the same laws of computation as do real numbers similarly expressed.

Examples. (i) $3 + 4i = 5(\cos\theta + i\sin\theta)$ where $\sin\theta = 4/5$, $\cos\theta = 3/5$ and therefore $\theta \doteqdot 0.927^c$.

$$\therefore 3 + 4i \doteqdot 5e^{(0.927)i}.$$

(θ *must* now be expressed in radians, since this is the assumption involved in the use of the series for sine and cosine.)

(ii)
$$\frac{1 + i}{1 - i} = \frac{\sqrt{2}\,e^{-\frac{1}{4}\pi i}}{\sqrt{2}\,e^{\frac{1}{4}\pi i}} = e^{\frac{1}{2}\pi i} = i.$$

(iii) Evaluate the integrals $I_1 = \int\limits_0^{\pi/2} e^{2\theta} \cos\theta\, d\theta$ and $I_2 = \int\limits_0^{\pi/2} e^{2\theta} \sin\theta\, d\theta$.

We make the assumption here that the rules of integration are applicable to complex quantities.

$$I_1 + i I_2 = \int\limits_0^{\pi/2} e^{2\theta}\, e^{i\theta}\, d\theta = \int\limits_0^{\pi/2} e^{(2+i)\theta}\, d\theta$$

$$= \left[\frac{e^{(2+i)\theta}}{2+i}\right]_0^{\pi/2} = \frac{1}{5}\left[(2-i)\, e^{(2+i)\theta}\right]_0^{\pi/2}$$

$$= \frac{1}{5}\left[e^{2\theta}\,(2-i)\,(\cos\theta + i\sin\theta)\right]_0^{\pi/2}$$

$$= \frac{1}{5}\left[e^{2\theta}\{(2\cos\theta + \sin\theta) - i\,(\cos\theta - 2\sin\theta)\}\right]_0^{\pi/2}.$$

Equating real and imaginary parts, we find

$$I_1 = \frac{1}{5}\left[e^{2\theta}\,(2\cos\theta + \sin\theta)\right]_0^{\pi/2} = \frac{1}{5}\,e^{\pi} - \frac{2}{5},$$

$$I_2 = -\frac{1}{5}\left[e^{2\theta}\,(\cos\theta - 2\sin\theta)\right]_0^{\pi/2} = \frac{2}{5}\,e^{\pi} + \frac{1}{5}.$$

16.10 Exponential values of sine and cosine

Since, for real values of θ,

$$e^{i\theta} = \cos\theta + i\sin\theta,$$

$$e^{-i\theta} = \cos\theta - i\sin\theta,$$

$$\therefore\ \cos\theta = \tfrac{1}{2}(e^{i\theta} + e^{-i\theta})$$

$$\text{(16.15)}$$

$$\sin\theta = \frac{1}{2i}(e^{i\theta} - e^{-i\theta})$$

for real values of θ. These values of $\sin\theta$ and $\cos\theta$ are analagous to the definitions $\sinh x = \tfrac{1}{2}(e^x - e^{-x})$ and $\cosh x = \tfrac{1}{2}(e^x + e^{-x})$ for real values of x.

The reason for the close relation between the properties of the two sets of functions is now clear.

We *define* $\sin z$ and $\cos z$ for complex values of z by the equations

$$\sin z = \frac{e^{iz} - e^{-iz}}{2i}, \quad \cos z = \frac{e^{iz} + e^{-iz}}{2}$$

and we similarly extend the definition of $\sinh x$ and $\cosh x$ to include

$$\sinh z = \tfrac{1}{2}(e^z - e^{-z}), \quad \cosh z = \tfrac{1}{2}(e^z + e^{-z}).$$

These definitions include the special cases (in which z is purely real or purely imaginary) considered earlier.

Now we have

$$\cosh iz = \cos z, \quad \sinh iz = i \sin z \tag{16.16}$$

and it can be proved that

$$\cos iz = \cosh z, \quad \sin iz = i \sinh z. \tag{16.17}$$

Formulae which we have established for both the circular and hyperbolic functions of real variables can now be extended to the corresponding functions of complex variables and can be shown to be true by simple processes of algebra. Formulae for hyperbolic functions can be obtained from the corresponding formulae for circular functions by replacing cos by cosh and sin by i sinh. This is equivalent to the rule given in Chapter XII whereby the formulae for hyperbolic functions were obtained from the corresponding formulae for circular functions by changing the sign in front of every product of two sines.

Examples. (i) Prove that

$$\cos(z + w) \equiv \cos z \cos w - \sin z \sin w,$$

where z and w are complex variables, and deduce the corresponding formula for $\cosh(z + w)$.

$$\cos(z + w) \equiv \tfrac{1}{2}\left\{e^{i(z+w)} + e^{-i(z+w)}\right\} \equiv \tfrac{1}{2}(e^{iz}e^{iw} + e^{-iz}e^{-iw})$$

$$\equiv \tfrac{1}{2}\{(\cos z + i\sin z)(\cos w + i\sin w) + (\cos z - i\sin z)(\cos w - i\sin w)\}$$

$$\therefore \cos(z + w) \equiv \cos z \cos w - \sin z \sin w.$$

By a similar algebraic process we could obtain

$$\cos i(z + w) \equiv \cos iz \cos iw - \sin iz \sin iw.$$

$$\therefore \cosh(z + w) \equiv \cosh z \cosh w - i^2 \sinh z \sinh w$$

$$\equiv \cosh z \cosh w + \sinh z \sinh w.$$

(ii) Evaluate $\text{Cos}^{-1}1.5$.

If $\text{Cos}^{-1}1.5 = x + iy$, where x and y are real, then

$$\text{Cos}(x + iy) = \text{Cos}\,x\,\text{Cos}\,iy - \text{Sin}\,x\,\text{Sin}\,iy = 1.5.$$

$$\therefore\ \text{Cos}\,x\,\text{Cosh}\,y - i\,\text{Sin}\,x\,\text{Sinh}\,y = 1.5.$$

Equating real and imaginary parts

$$\text{Cos}\,x\,\text{Cosh}\,y = 1.5, \tag{1}$$

$$\text{Sin}\,x\,\text{Sinh}\,y = 0. \tag{2}$$

Hence from Eqn. (2)

$$\text{Sin}\,x = 0,\ \ \text{i.e.,}\ \ x = n\pi,$$

$$\text{or Sinh}\,y = 0,\ \ \text{i.e.,}\ \ y = 0.$$

But $y = 0$ involves $\text{Cos}\,x = 1.5$ from Eqn. (1), which is contrary to the hypothesis that x is real. Also $x = n\pi$ when n is an odd integer involves $\text{Cosh}\,y = -1.5$, which is contrary to the hypothesis that y is real.

$$\therefore\ x = 2m\pi\ (m = 0,\ \pm1,\ \pm2,\ \ldots)$$

and

$$y = \text{Cosh}^{-1}1.5 = \pm\,\log\left(\frac{3 + \sqrt{5}}{2}\right).$$

$$\therefore\ \text{Cos}^{-1}1.5 = 2m\pi \pm i\,\log\left(\frac{3 + \sqrt{5}}{2}\right) \qquad (m = 0,\ \pm1,\ \pm2,\ \ldots).$$

(iii)

$$\frac{e^{2i\theta} - 1}{e^{2i\theta} + 1} = \frac{e^{i\theta} - e^{-i\theta}}{e^{i\theta} + e^{-i\theta}} = \frac{2i\sin\theta}{2\cos\theta} = i\tan\theta.$$

(iv) Find the real and imaginary parts of $\tan(x + iy)$.

If

$$u + iv = \tan(x + iy), \tag{1}$$

then

$$u - iv = \tan(x - iy). \tag{2}$$

From (1) and (2)

$$2u = \tan(x + iy) + \tan(x - iy)$$

$$= \frac{\sin(x + iy)\cos(x - iy) + \sin(x - iy)\cos(x + iy)}{\cos(x + iy)\cos(x - iy)}$$

$$= \frac{\sin 2x}{\frac{1}{2}(\cos 2x + \cos 2iy)}$$

$$= \frac{2\sin 2x}{\cos 2x + \cosh 2y}.$$

$$\therefore u = \frac{\sin 2x}{\cos 2x + \cosh 2y}.$$

Also from (1) and (2) $2iv = \tan(x + iy) - \tan(x - iy)$

which leads to

$$v = \frac{\sinh 2y}{\cos 2x + \cosh 2y}.$$

This example is an illustration of an important technique which is used to find the real and imaginary parts of $f(z)$ where $z = x + iy$.

For, if $w = u + iv = f(z),$

then $\overline{w} = u - iv = \overline{f(z)},$

and so $2u = f(z) + \overline{f(z)}, \quad 2iv = f(z) - \overline{f(z)}.$

It should be observed that $\overline{f(z)} = f(\overline{z})$ only if the numerical coefficients in $f(z)$ are all real.

(v) Find the sum of the series

$$\sin\theta + \sin 2\theta + \sin 3\theta + \cdots + \sin n\theta.$$

If $C = \cos\theta + \cos 2\theta + \cdots + \cos n\theta,$

$$S = \sin\theta + \sin 2\theta + \cdots + \sin n\theta,$$

then $C + iS = e^{i\theta} + e^{i2\theta} + \cdots + e^{in\theta}$

$$= \frac{e^{i\theta} - e^{i(n+1)\theta}}{1 - e^{i\theta}}$$

and, using the method of example (iv),

$$C - iS = \frac{e^{-i\theta} - e^{-i(n+1)\theta}}{1 - e^{-i\theta}}.$$

$$\therefore 2iS = \frac{e^{i\theta} - e^{i(n+1)\theta}}{1 - e^{i\theta}} - \frac{e^{-i\theta} - e^{-i(n+1)\theta}}{1 - e^{-i\theta}}$$

$$= \frac{(e^{i\theta} - e^{-i\theta}) + (e^{in\theta} - e^{-in\theta}) - (e^{i(n+1)\theta} - e^{-i(n+1)\theta})}{1 - (e^{i\theta} + e^{-i\theta}) + e^{i\theta}e^{-i\theta}}$$

$$= \frac{2i\{\sin\theta + \sin n\theta - \sin(n+1)\theta\}}{2 - 2\cos\theta}$$

$$\therefore S = \frac{\sin\theta + \sin n\theta - \sin(n+1)\theta}{2(1 - \cos\theta)}.$$

Exercises 16.10

1. Express each of the following in the form $re^{i\theta}$ (giving the principal value of θ):

(i) $1 + i$, (ii) $1 - i$, (iii) i, (iv) $\dfrac{1 - i}{1 + i}$, (v) $1 - i\sqrt{3}$.

2. Express each of the following in the form $a + ib$:

(i) $e^{-i\pi/4}$, (ii) $e^{i\pi}$, (iii) $e^{-i\pi}$, (iv) $e^{1 + i\pi/3}$, (v) $e^{x + iy}$.

3. Express $1 + e^{i\theta}$ in the form $re^{i\varphi}$.

4. Express $e^{i\theta} + e^{-i\theta}$ in the form $re^{i\varphi}$.

5. Express $\sin(x - iy) + \cos(x + iy)$ in the form $a + ib$.

6. Express (i) $\sinh(1 - i)$, (ii) $\cosh(1 - i)$ in the form $a + ib$.

7. Use the exponential form of $\sin z$ and $\cos z$ to prove each of the following formulae and, in each case, write down the corresponding formula for hyperbolic functions:

(i) $\sin 2z = 2 \sin z \cos z$, (ii) $\sin 3z = 3 \sin z - 4 \sin^3 z$,

(iii) $\cos z + \cos w = 2 \cos\frac{1}{2}(z + w) \cos\frac{1}{2}(z - w)$,

(iv) $\tan 3z = (3 \tan z - \tan^3 z)/(1 - 3 \tan^2 z)$,

(v) $\tan^2 z = \dfrac{1 - \cos 2z}{1 + \cos 2z}$.

8. Find the real and imaginary parts of $\tanh(x + iy)$.

9. Express $\cos^6\theta$ as an exponential function and hence obtain $\int \cos^6\theta \, d\theta$ in terms of multiple angles.

10. Evaluate $\int_0^{\pi/4} \sin^8\theta \, d\theta$.

11. Show that all solutions of the equation

$$\sin z = 2i \cos z$$

are given by $z = (\frac{1}{2} \pm n)\pi + \frac{1}{2}i \log 3$ where n is zero or any positive integer. (L.)

12. Evaluate each of the following in the form $a + ib$ where a and b are real numbers:

(i) $\mathrm{Sin}^{-1}2$, (ii) $\mathrm{Cos}^{-1}3$, (iii) $\mathrm{Sec}^{-1}(-\frac{1}{2})$.

13. Find the polar equation of the curve in the Argand diagram described by the point z when it varies so that ze^z is real.

Sketch that part of the curve which is such that $\arg(ze^z) = 0$ and $-\pi < \arg z < \pi$ indicating the asymptotes. (L.)

14. If $\tan\left(\dfrac{1}{4}\pi + iv\right) = re^{i\theta}$ where v, r and θ are all real, prove that $r = 1$, $\tan\theta = \sinh 2v$ and $\tanh\theta = \tan\frac{1}{2}\theta$. (L.)

15. Find the real and imaginary parts of $\cos z$, where $z = x + iy$. Hence show that one solution of the equation

$$\cos z = \frac{1}{2\sqrt{2}} \{e(1-i) + e^{-1}(1+i)\}$$

is

$$z = \frac{1}{4}\pi + i.$$

16. Prove that $(e^{i\alpha} + e^{2i\alpha} + e^{4i\alpha})$ is one root of $x^2 + x + 2 = 0$, where $\alpha = 2\pi/7$. Hence show that

$$\cos\alpha + \cos 2\alpha + \cos 4\alpha = -\tfrac{1}{2},$$

$$\sin\alpha + \sin 2\alpha + \sin 4\alpha = \tfrac{1}{2}\sqrt{7}. \qquad \text{(O.C.)}$$

17. Find the general solution of the equation $\sinh z = 2\cosh z$, where z is complex.   (L.)

18. If x, y, u, v are real and $\cosh(x+iy) = \tan(u+iv)$ prove that

$$\cosh 2x + \cos 2y = 2\left\{\frac{\cosh 2v - \cos 2u}{\cosh 2v + \cos 2u}\right\}. \qquad \text{(L.)}$$

19. If $u + iv = \coth(x+iy)$ show that

$$v = \frac{-\sin 2y}{(\cosh 2x - \cos 2y)}.$$

Hence show that

$$iv = \frac{1}{1 - e^{2(x-iy)}} - \frac{1}{1 - e^{2(x+iy)}},$$

and deduce that, if $x < 0$, v can be expressed as the infinite series

$$-2\sum_{r=1}^{\infty} e^{2rx}\sin 2ry. \qquad \text{(L.)}$$

20. Prove that

$$\sin\alpha - \sin(\alpha+\beta) + \sin(\alpha+2\beta) - \cdots - \sin\{\alpha + (2n-1)\beta\}$$
$$= -\sin n\beta \cos\{\alpha + (n-\tfrac{1}{2})\beta\}\sec\tfrac{1}{2}\beta.$$

$A_1, A_2, \ldots, A_{2n}$ are the vertices of a regular polygon of $2n$ sides. Prove that the sum of the lengths of $A_1A_3, A_1A_5, \ldots, A_1A_{2n-1}$ differs from the sum of the lengths $A_1A_2, A_1A_4, \ldots, A_1A_{2n}$ by $a\sec^2(\pi/4n)$ where $2a$ is the length of a side of the polygon.   (L.)

21. Show that $\displaystyle\sum_{r=0}^{\infty} \frac{\cos r\theta}{r!} = e^{\cos\theta}\cos(\sin\theta)$.   (L.)

22. Prove that

$$(1+x)^{2n} - (1-x)^{2n} = 4nx\prod_{r=1}^{n-1}\left(x^2 + \tan^2\frac{r\pi}{2n}\right).$$

Hence, or otherwise, prove that

$$\prod_{r=1}^{n-1}\cos\frac{r\pi}{2n} = 2^{1-n}\sqrt{n}. \qquad \text{(L.)}$$

23. Sum the infinite series

$$x \sin \theta + \frac{x^3 \sin 3\theta}{3!} + \frac{x^5 \sin 5\theta}{5!} + \cdots,$$

where x and θ are real. (L.)

Miscellaneous Exercises XVI

1. (a) If ω is a complex cube root of unity, prove that
$$(a + \omega b + \omega^2 c)^3 - (a + \omega^2 b + \omega c)^3 = 3(\omega^2 - \omega)(b - c)(c - a)(a - b).$$

(b) State the moduli and arguments of all the roots of the equation
$$z^4 = \sqrt{3} - i. \tag{N.}$$

2. (a) Show that the argument of the ratio $(z - 2)/(z + 2)$ is constant if
$$z = ai + \sqrt{(4 + a^2)}\, e^{i\theta},$$

where a is a real constant and θ is a real variable.

(b) State a geometrical construction for finding the point corresponding to $1/(z - i)$, given the point z on the Argand diagram. (N.)

3. (a) Show that in the Argand diagram the points representing the roots of the equation
$$z^8 + 12z^4 - 64 = 0$$

are the vertices of a square and the mid-points of the sides of that square.

(b) If $S_n = 8(1 + z + z^2 + \cdots + z^{n-1})$, where $z = -\frac{1}{2}i$, mark in the Argand diagram the points representing S_1, S_2, S_3, S_4, S_5 and also the point representing the sum of the infinite series $8(1 + z + z^2 + \cdots)$. (N.)

4. If P_1, P_2, P_3 are the points in the Argand diagram that correspond respectively to the complex numbers
$$z_1, \quad z_2, \quad (1 + \tfrac{1}{2}i) z_2 - \tfrac{1}{2}i z_1,$$

show that $P_1 P_2 P_3$ is a right angled triangle. Also find in terms of z_1, z_2 the complex number that corresponds to P_4 if $P_1 P_2 P_3 P_4$ is a rectangle. (N.)

5. (a) If $\alpha = \cos(2\pi/n) + i \sin(2\pi/n)$, where n is an integer greater than 2, show that in the Argand diagram the points corresponding to
$$0, \ 1, \ 1 + \alpha, \ 1 + \alpha + \alpha^2, \ \ldots, 1 + \alpha + \alpha^2 \cdots + \alpha^{n-2}$$

are the vertices of a regular polygon.

Obtain the complex number corresponding to the centre of this polygon.

(b) If $\omega = \cos 2\pi/3 + i \sin 2\pi/3$ and $(z_1 + \omega z_2 + \omega^2 z_3)^3 = (z_1 + \omega^2 z_2 + \omega z_3)^3$, show that
$$(z_2 - z_3)(z_3 - z_1)(z_1 - z_2) = 0. \tag{N.}$$

6. (a) If $w = 1 + z$, express the modulus and argument of w in terms of θ when $z = \cos\theta + i \sin\theta$. Hence, or otherwise, show that if $-\pi < \theta < \pi$,
$$\theta = 2(\sin\theta - \tfrac{1}{2}\sin 2\theta + \tfrac{1}{3}\sin 3\theta - \ldots).$$

(b) If $w = e^z$, sketch the curve in the Argand diagram for w corresponding to

$$z = t(\cos\alpha + i\sin\alpha)$$

where t is a variable parameter, α is a constant. (N.)

7. (a) Find the modulus of the complex number

$$\frac{z - 3i}{1 + 3iz}$$

when $|z| = 1$.

(b) Find the region of the Argand diagram in which z must lie when

$$\left|\frac{z - 3 + 2i}{1 - 3z - 2iz}\right| > 1.$$ (N.)

8. (a) Solve the equation

$$z^{10} + \sqrt{2}z^5 + 1 = 0,$$

expressing the argument of each root in degrees.

(b) If $\dfrac{|z_1|}{3} = \dfrac{|z_2|}{4} = \dfrac{|z_1 - z_2|}{5}$, show that $16z_1 + 9z_2^2 = 0$. (N.)

9. If ω is a complex cube root of unity, prove that

$$(a + \omega b + \omega^2 c)^3 + (a + \omega^2 b + \omega c)^3 = (2a - b - c)(2b - c - a)(2c - a - b).$$ (N.)

10. Solve completely the equation

$$x^6 + 2x^3 + 4 = 0,$$

giving each root in the form $r(\cos\theta + i\sin\theta)$, where $r > 0$ and $0 \leq \theta \leq 2\pi$. (N.)

11. (i) Show that one root of the equation

$$2z^4 = 1 + i\sqrt{3}$$

is $\dfrac{(\sqrt{3} + 1) + i(\sqrt{3} - 1)}{2\sqrt{2}}$, and determine in similar numerical form the other roots.

(ii) If $z_3 - z_1 = \frac{1}{2}(z_2 - z_1)(1 + i\sqrt{3})$, show that

$$z_1 - z_2 = \tfrac{1}{2}(z_3 - z_2)(1 + i\sqrt{3}).$$

Prove that, if the points in the Argand diagram corresponding to the complex numbers α, β, γ form an equilateral triangle, then

$$\alpha^2 + \beta^2 + \gamma^2 - \beta\gamma - \gamma\alpha - \alpha\beta = 0.$$ (N.)

12. (i) Show that $x^2 - (1 + \sqrt{3})x + 2$ is a factor of $x^{12} + 64$, and obtain in similar numerical form the other quadratic factors.

(ii) The complex numbers that correspond to the points A, B, C, D are z_1, z_2, z_3, z_4. Interpret geometrically each of the following relations:

(a) $z_1 - z_2 + z_3 - z_4 = 0$,

(b) $z_1 + 2iz_2 - z_3 - 2iz_4 = 0$.

What is the nature of the quadrilateral $ABCD$ if both relations are satisfied? (N.)

13. Show that

$$\cos 7\theta = 64 \cos^7\theta - 112 \cos^5\theta + 56 \cos^3\theta - 7 \cos\theta.$$

Prove that

(i) $\cos \dfrac{\pi}{7} \cos \dfrac{3\pi}{7} \cos \dfrac{5\pi}{7} = -\dfrac{1}{8}$,

(ii) $\cos^4 \dfrac{\pi}{7} + \cos^4 \dfrac{3\pi}{7} + \cos^4 \dfrac{5\pi}{7} = \dfrac{13}{16}$. (N.)

14. (a) If $|z| < 1$, show that $|z - 3 - 4i| > 4$.

(b) If the point $z(\equiv x + iy)$ describes the circle $(x - 1)^2 + y^2 = 1$, find the equation of the path described by the point $1/z$.

(c) Show that the roots of the equation

$$z^6 - 4z^3 + 8 = 0$$

lie on a circle of radius $\sqrt{2}$. Make a diagram showing the circle and the six roots. (N.)

15. (a) Find the modulus and argument of each root of the equation

$$z^4 - iz^2 \sqrt{3} - 1 = 0.$$

(b) Solve the simultaneous equations

$$|z - 1| = 2\sqrt{2}, \quad |z - 1 - i| = |z|. \tag{N.}$$

16. Express $\cos(\alpha + i\beta)$ in the form $A + iB$, where A, B are real if α, β are real.

Prove that the general solution of the equation $\cos z = 2$ is

$$z = 2n\pi \pm i \log_e (2 + \sqrt{3}),$$

where n is an integer. (N.)

17. (a) Solve the equation $x^2 + 6x + 13 = 0$, giving the solution in the form $a \pm ib$, where a and b are real.

(b) If $\omega^3 = 1$ and $\omega \neq 1$, indicate on a diagram the possible positions of the point representing ω.

Prove that   (i) $1 + \omega + \omega^2 = 0$,
 (ii) $(1 + \omega)^3 = -1$,
 (iii) $(1 - \omega + \omega^2)(1 + \omega - \omega^2) = 4$. (N.)

18. Verify   that $(a + ib)(c + id) = (ac - bd) + i(bc + ad)$. Write down a similar product of factors for $(ac - bd) - i(bc + ad)$, and hence express $(ac - bd)^2 + (bc + ad)^2$ as a product of real factors. (N.)

19. (a) Show how a complex number z can be represented by a point on an Argand diagram. If $\bar{z}$ is the conjugate of z, show on the same diagram the points which represent $\bar{z}$, $z + \bar{z}$, $z - \bar{z}$ and $\sqrt{(z\bar{z})}$.

(b) If a, b, c, d are real numbers, and if the equation

$$x^2 + (a + ib)x + (c + id) = 0$$

has a real root, show that $abd = d^2 + b^2c$. (N.)

20. (a) Find the modulus and argument of the complex number z given by

$$\text{(i)} \quad z = \frac{3\left(\cos\dfrac{1}{3}\pi + i\sin\dfrac{1}{3}\pi\right)}{2\left(\cos\dfrac{2}{3}\pi + i\sin\dfrac{2}{3}\pi\right)}, \quad \text{(ii)} \quad \frac{3z+2}{4z-1} = i.$$

(b) Verify that $z = \cos\dfrac{1}{3}\pi + i\sin\dfrac{1}{3}\pi$ is a root of the equation

$$z^5 + z = 1.$$

Write down another root of the equation and obtain the cubic equation satisfied by the remaining three roots.   (N.)

21. Find the sum of the first n terms of the series $1 + z + z^2 + z^3 + \cdots$. Show that, if $z = \frac{1}{2}(\cos\theta + i\sin\theta)$, the sum to infinity is

$$\frac{2(2 - \cos\theta) + 2i\sin\theta}{5 - 4\cos\theta}.$$ (N.)

22. Prove that $\sin(x + iy) = re^{i\theta}$ where

$$r = \sqrt{\{\tfrac{1}{2}(\cosh 2y - \cos 2x)\}} \quad \text{and} \quad \theta = \tan^{-1}(\cot x \tanh y).$$

23. (i) Complex numbers z_1 and z_2 are given by the formulae

$$z_1 = R_1 + i\omega L, \quad z_2 = R_2 - \frac{i}{\omega C},$$

and z is given by the formula

$$\frac{1}{z} = \frac{1}{z_1} + \frac{1}{z_2}.$$

Find the value of ω for which z is a real number.

(ii) Use De Moivre's theorem to prove that, if

$$2\cos\theta = x + \frac{1}{x},$$

then

$$2\cos n\theta = x^n + \frac{1}{x^n}.$$

Hence, or otherwise, solve the equation

$$5x^4 - 11x^3 + 16x^2 - 11x + 5 = 0.$$ (O.C.)

24. Solve the equation

$$(\cos\theta + i\sin\theta)(\cos 2\theta + i\sin 2\theta)\cdots(\cos n\theta + i\sin n\theta) = 1.$$ (O.C.)

25. Prove that, if $x + iy = c\cot(u + iv)$, then

$$\frac{x}{\sin 2u} = \frac{-y}{\sinh 2v} = \frac{c}{\cosh 2v - \cos 2u},$$

and show that, if x, y are coordinates of a point in a plane, then for a given value of v the point lies on the circle

$$x^2 + y^2 + 2cy\coth 2v + c^2 = 0.$$

Also verify that, if a_1, a_2 denote the radii of the circles for the values v_1, v_2 of v, and d denotes the distance between their centres, then

$$\frac{a_1^2 + a_2^2 - d^2}{2 a_1 a_2} = \cosh 2 \,(v_1 - v_2). \tag{O.C.}$$

26. Prove that the roots of the equation

$$x^{n-1} + x^{n-2} + x^{n-3} + \cdots + 1 = 0$$

are the values of

$$\cos \frac{2 r \pi}{n} + i \sin \frac{2 r \pi}{n},$$

where $r = 1, 2, 3, \ldots, n - 1$.

If $\alpha = \cos \dfrac{2 \pi}{13} + i \sin \dfrac{2 \pi}{13}$, prove that

$$\alpha + \alpha^3 + \alpha^4 + \alpha^9 + \alpha^{10} + \alpha^{12} \quad \text{and} \quad \alpha^2 + \alpha^5 + \alpha^6 + \alpha^7 + \alpha^8 + \alpha^{11}$$

are the roots of the quadratic

$$X^2 + X - 3 = 0. \tag{O.C.}$$

27. If

$$\alpha = \cos \frac{2 \pi}{7} + i \sin \frac{2 \pi}{7}.$$

prove that the quadratic whose roots are

$$\alpha + \alpha^2 + \alpha^4, \quad \alpha^3 + \alpha^5 + \alpha^6$$

is

$$x^2 + x + 2 = 0.$$

Deduce or prove otherwise that

$$\sin \frac{\pi}{7} \sin \frac{2 \pi}{7} \sin \frac{3 \pi}{7} = \frac{\sqrt{7}}{8}. \tag{O.C.}$$

28. Explain briefly how complex numbers may be represented in a diagram.

If $Z = \dfrac{z - 1}{z + 1}$, and if z describes the circle of unit radius whose centre is the origin, prove that the locus of Z is also a circle, and find its centre and its radius. (O.C.)

29. Prove that with a suitable choice of α, the roots of the equation

$$z^n = 1 \quad \text{are} \quad 1, \alpha, \alpha^2, \ldots, \alpha^{n-1}.$$

Show that the fifteenth roots of unity, other than cube and fifth roots, give the zeros of the quotient of $z^{10} + z^5 + 1$ on division by $z^2 + z + 1$. Deduce that the roots of the equation

$$x^4 - x^3 - 4x^2 + 4x + 1 = 0 \quad \text{are} \quad x = 2 \cos \frac{2 r \pi}{15}, \ (r = 1, 2, 4, 7). \tag{O.C.}$$

30. Show that, if $i^2 = -1$,

$$\begin{vmatrix} 2x & 2(i-1) & 1+i \\ 1+i & x+i & 1 \\ 1-i & 1 & x-i \end{vmatrix} = 2x\,(x^2 + 1). \tag{L.}$$

31. If ω is a complex cube root of unity, evaluate the determinant

$$\begin{vmatrix} 1 & \omega & \omega^2 \\ \omega & \omega^2 & 1 \\ \omega^2 & 1 & \omega \end{vmatrix}.$$

32. (i) If $z = x + iy$, where $i^2 = -1$, find x and y when

$$\frac{z}{2+i} + \frac{25}{11i+2} = \frac{2z}{1+i}.$$

(ii) Express $\cos\left\{\frac{1}{4}\pi(1-i)\right\}$ in the form $a + ib$, where a and b are real, and find its modulus. (L.)

33. (i) If $\dfrac{z-i}{z-1}$ is purely imaginary, show that the locus of $z(= x + iy)$ in the Argand diagram is a circle whose centre represents the complex number $(1 + i)/2$ and whose radius is $1/\sqrt{2}$.

(ii) Show that the roots of the equation

$$z^4 + z^2 + 1 = 0$$

may be expressed in the form α, α^2, α^4, α^5 where $\alpha = e^{i\pi/3}$. (L.)

34. (i) If $z = x + iy$, find x and y when

$$\frac{2z}{1+i} - \frac{2z}{i} = \frac{5}{2+i}.$$

(ii) On an Argand diagram z is a representative point and $w = \dfrac{z-2}{z-i}$. Show that, when the point represented by w moves along the real axis, z traces the line through 2 and i.

Find the locus of z when w moves along the imaginary axis. (L.)

35. If z_1, z_2, z_3 are complex numbers, interpret geometrically the complex numbers $z_1 - z_2$, $(z_1 - z_2)/(z_3 - z_2)$.

Triangles BCX, CAY, ABZ are described on the sides of a triangle ABC. If the points A, B, C, X, Y, Z in the Argand diagram represent the complex numbers a, b, c, x, y, z respectively, and

$$\frac{x-c}{b-c} = \frac{y-a}{c-a} = \frac{z-b}{a-b},$$

show that the triangles BCX, CAY, ABZ are similar.

Prove also that the centroids of ABC, XYZ are coincident. (L.)

36. (i) Use the relation

$$e^{i\theta} = \cos\theta + i\sin\theta$$

to prove that

$$\cos 5\theta = 16\cos^5\theta - 20\cos^3\theta + 5\cos\theta.$$

(ii) Two complex variables z and w are connected by the relation

$$zw + z - w + 1 = 0.$$

If $z = 2 + e^{i\theta}$, prove that $w = -2 + i\tan\frac{1}{2}\theta$, and describe the loci of the points z and w as θ varies from $-\pi$ to π. (L.)

37. If $\sin (x + iy) = a + ib$, where x, y, a, b are real, show that $\sin x \cosh y = a$, $\cos x \sinh y = b$.

Hence, or otherwise, find the value of x between $-\frac{1}{2}\pi$ and $\frac{1}{2}\pi$ if $a = 1 = b$. (L.)

38. If $z = \dfrac{\sqrt{3} + i}{1 - i}$, where $i = \sqrt{(-1)}$, express z in the form $re^{i\theta}$ and evaluate $z^2 + \dfrac{1}{z^2}$ in the form $a + ib$. (L.)

39. If $\cos (x + iy) = 1 + i$, where x and y are real, prove that

$$\cos x = \sqrt{\left(\frac{3 - \sqrt{5}}{2}\right)}.$$ (L.)

40. Express the complex number $1 + i$ in the form $r(\cos\theta + i \sin\theta)$. Hence, or otherwise, prove that, n being any positive integer,

$$(1 + i)^n + (1 - i)^n = 2 \left(2^{n/2} \cos \frac{n\pi}{4}\right).$$

If
$$(1 + x)^n = p_0 + p_1 x + p_2 x^2 + \cdots + p_n x^n,$$

prove that
$$p_0 - p_2 + p_4 - \cdots = 2^{n/2} \cos \frac{n\pi}{4},$$

and
$$p_1 - p_3 + p_5 - \cdots = 2^{n/2} \sin \frac{n\pi}{4}.$$ (L.)

41. Express the left-hand side of the equation

$$\cos 6\varphi + 6 \cos 4\varphi - 9 \cos 3\varphi + 15 \cos 2\varphi - 27 \cos\varphi + 14 = 0$$

as a polynomial in $\cos\varphi$, and hence, or otherwise, find all angles φ between $0°$ and $360°$ inclusive satisfying the equation. (L.)

42. (i) In the Argand diagram, a square $OABC$ (lettered anticlockwise) has a vertex at the origin O. If the centre of the square represents the complex number z, find the complex numbers represented by A, B and C.

(ii) Use De Moivre's theorem to find the roots of the equation $(1 - x)^5 = x^5$ in the form $a + ib$. (Numerical evaluation is not required.) (L.)

43. (i) If $u + iv = \log \sin (x + iy)$, show that

$$2e^{2u} = \cosh 2y - \cos 2x.$$

(ii) Show that the roots of the equation

$$(1 + z)^{2n+1} = (1 - z)^{2n+1}$$

are $z = \pm i \tan \{r\pi/(2n + 1)\}$, $(r = 0, 1, 2, \ldots, n)$. (L.)

44. Show that

$$x^{2n} - 2a^n x^n \cos n\theta + a^{2n} = \prod_{r=0}^{n-1} \left\{x^2 - 2ax \cos \left(\theta + \frac{2r\pi}{n}\right) + a^2\right\}.$$

Deduce that $\displaystyle\prod_{r=0}^{n-1} \sin \left\{\frac{(4r + 1)\pi}{4n}\right\} = 2^{-(2n-1)/2}.$ (L.)

45. Show that, if n is a positive integer,

$$(-1)^n 2^{2n} \sin^{2n+1}\theta = \sin(2n+1)\theta - \left(\frac{2n+1}{1}\right) \sin(2n-1)\theta + \cdots$$

$$+ (-1)^r \left(\frac{2n+1}{r}\right) \sin(2n+1-2r)\theta + \cdots + (-1)^n \left(\frac{2n+1}{n}\right) \sin\theta.$$

Hence find all the solutions of the equation

$$16 \sin^5\theta = \sin 5\theta. \tag{L.}$$

46. If 1, ω, ω^2 are the three cube roots of unity, prove that

$$\begin{vmatrix} \theta+1 & \omega & \omega^2 \\ \omega & \theta+\omega^2 & 1 \\ \omega^2 & 1 & \theta+\omega \end{vmatrix} \equiv \theta^3. \tag{L.}$$

CHAPTER XVII

DIFFERENTIAL EQUATIONS

17.1 Formation of differential equations

Definition. An ordinary differential equation is an equation which states a relationship between an independent variable x, a dependent variable y and one or more of the derivatives $\dfrac{dy}{dx}$, $\dfrac{d^2y}{dx^2}$, $\dfrac{d^3y}{dx^3}$, $\cdots$. If the equation contains the first derivative $\dfrac{dy}{dx}$ only, it is called a differential equation of the first order. The order of the equation is fixed by the highest derivative that occurs in it. Thus $\dfrac{d^2y}{dx^2} + P\dfrac{dy}{dx} + Qy = R$, where P, Q and R are functions of x, is a differential equation of the second order. The "degree" of an equation is fixed by the highest power of the differential coefficient of highest order when the equation has been made rational and free of fractions in the *differential coefficients.* Thus

$$\left(\frac{d^2y}{dx^2}\right)^2 = a^2\left\{1 + \left(\frac{dy}{dx}\right)^2\right\}^3$$

is a differential equation of the second degree. *Linear differential equations* involve y and its derivatives only in the first degree.

Derivation of differential equations. An equation such as $xy = c^2$ represents a *family* of curves, in this case rectangular hyperbolas, particular members of which are obtained by giving appropriate values to the arbitrary constant c. Geometrically $c\sqrt{2}$ is the distance from the origin of the point at which a member of the family is cut by the line $x = y$.

For any curve of the family we have

$$x y = c^2 \quad \text{and} \quad \frac{dy}{dx} = -\frac{c^2}{x^2}.$$

Therefore, eliminating c,

$$\frac{dy}{dx} = -\frac{y}{x}.$$

This is the differential equation which represents the whole family of curves. It can be interpreted geometrically as representing a property common to all the curves, as in this case, for example, it represents the property. "For the rectangular hyperbola $xy = c^2$, the tangent to the curve at any point and the line joining that point to the origin are equally inclined to the axis of x."

Examples. (i) A family of curves possesses the property that, for each of its members, the sum of the subtangent and the subnormal is constant and equal to a. Write down the differential equation for the family. Fig. 160.

$$TN + NG = a.$$

$$\therefore \; \frac{y}{\dfrac{dy}{dx}} + y\,\frac{dy}{dx} = a.$$

$$\therefore \; y\left(\frac{dy}{dx}\right)^2 - a\frac{dy}{dx} + y = 0$$

is the required differential equation.

(ii) Find the differential equation for the curves

$$y = A \cosh nx + B \sinh nx.$$

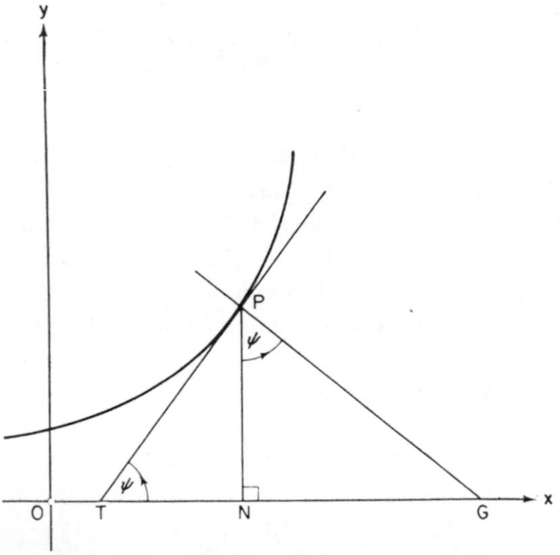

FIG. 160.

In general when there are two arbitrary constants (A and B here) the first and second derivatives are required in order to eliminate the constants.

$$\frac{dy}{dx} = nA \sinh nx + nB \cosh nx.$$

$$\frac{d^2y}{dx^2} = n^2 A \cosh nx + n^2 B \sinh nx.$$

$$\therefore \frac{d^2y}{dx^2} = n^2 y.$$

This is the required differential equation.

(iii) A differential equation can also arise from the mathematical interpretation of a known physical property. In illustration we consider the following problem.

A particle moving in a straight line is attracted towards a fixed point O of the line by a force which is proportional to its distance from O, and the particle is also subject to a constant retarding force. Write down a differential equation which represents the motion of the particle.

In the usual notation, if the mass of the particle is m (using the physical law $P = mf$ for the relationship between force, mass and acceleration for a particle with constant mass),

$$m\frac{d^2s}{dt^2} = -ps - q$$

where p and q are the constants implied in the description of the motion. This is a differential equation of the second order.

The motion could also have been described by the differential equation

$$m v \frac{dv}{ds} = -ps - q,$$

a differential equation of the first order between v and s.

Exercises 17.1

In each of the Questions $1-8$, form the differential equation of the family of curves by eliminating the arbitrary constant(s) in the equation of the family.

1. $x^2 + y^2 = r^2$. (Interpret the differential equation geometrically.)

2. $y^2 = 4kx$. (Interpret geometrically.)

3. $y = Ae^{kx} + Be^{-kx}$.   4. $y = a/x + b$.

5. $y = a \cos nx + b \sin nx$. (Eliminate a and b.)

6. $y = a \log x + b$.   7. $y = a \tan^{-1} x + b$.

8. $y = (A + Bx) e^{-kx}$.

In each of Questions 9–16 form a differential equation of the curves from the given geometrical property.

9. The subnormal is of constant length $\dot{k}$.

10. The curves are parabolas with their axes parallel to the y-axis.

11. The radius of curvature is equal to a.

12. The subnormal varies as the cube of the ordinate.

13. The subnormal varies as the abscissa.

14. The curves are circles which touch both axes.

15. The polar curves are such that the angle between the radius vector and the tangent at any point is equal to half the angle between the radius vector and the initial line.

16. The polar curves are such that the tangent at any point and the radius vector to that point are equally inclined to the initial line.

In each of Questions 17–20, form a differential equation which expresses the physical law stated there.

17. The acceleration of a point moving in a straight line is directed towards a fixed point and is inversely proportional to the square of its distance from the point.

18. The rate of cooling of a body is proportional to its excess temperature over the surroundings.

19. The acceleration of a particle towards a fixed point partly varies as its distance from the point and partly varies as its velocity.

20. The rate of change of the momentum (mass × velocity) of a particle of constant mass moving along a straight line is proportional to the impressed force.

17.2 The solution of a differential equation

The solution of a differential equation in x, y and the differential coefficients of y with respect to x is the functional relationship $f(x, y) = 0$ (say), not involving differential coefficients, which satisfies the differential equation. The solution of a differential equation of the first order will, in general, involve one arbitrary constant, and the number of constants involved in the solution of any differential equation will, in general, correspond to the order of the equation. Such a solution, involving the appropriate number of constants, is called the *general solution* or the *complete primitive*. Any solution which satisfies the differential equation but is not the general solution is called a *particular solution*.

A particular solution which satisfies a given set of *boundary conditions* is obtained by using the boundary conditions to evaluate the constants of the general solution.

An example of the use of boundary conditions is afforded by the problems concerning the bending of beams. If a thin loaded beam is either supported at both ends or clamped at one or both ends and is not subject to forces other than the reactions at the supporting points and the weights of the beam and the loads, it can be shown by the methods of theoretical mechanics that the *downward* (small) deflection y and the horizontal distance x along the beam are related by the differential equation

$$E I \frac{\mathrm{d}^4 y}{\mathrm{d} x^4} = w ,$$

where E and I are physical constants associated with the beam and w is the load per unit length at the point (x, y). From this equation, by using the given boundary conditions, the shape of the beam and the deflection at any point can be obtained by successive integration.

Example. A uniformly loaded beam AB of length l is clamped horizontally at both ends, which are on the same level.

If the equation of the beam is referred to axes Ox, Oy taken horizontally and vertically downwards through A, then

$$E I \frac{\mathrm{d}^4 y}{\mathrm{d} x^4} = w$$

subject to the boundary conditions

$$(i) \quad y = 0 \quad \text{when} \quad x = 0 ,$$

$$(ii) \quad y = 0 \quad \text{when} \quad x = l ,$$

$$(iii) \quad \frac{\mathrm{d} y}{\mathrm{d} x} = 0 \quad \text{when} \quad x = 0 ,$$

$$(iv) \quad \frac{\mathrm{d} y}{\mathrm{d} x} = 0 \quad \text{when} \quad x = l .$$

By repeated integration of $E I \dfrac{\mathrm{d}^4 y}{\mathrm{d}x^4} = w$ with respect to x, we have

$$E I \frac{\mathrm{d}^3 y}{\mathrm{d}x^3} = w\,x + A\,,$$

$$E I \frac{\mathrm{d}^2 y}{\mathrm{d}x^2} = \frac{w\,x^2}{2} + A\,x + B\,,$$

$$E I \frac{\mathrm{d}y}{\mathrm{d}x} = \frac{w\,x^3}{6} + \frac{A\,x^2}{2} + B\,x + C\,,$$

$$E I\,y = \frac{w\,x^4}{24} + \frac{A\,x^3}{6} + \frac{B\,x^2}{2} + C\,x + D\,.$$

From boundary conditions (i) and (iii) $D = 0$, $C = 0$. From boundary conditions (ii) and (iv),

$$\frac{w\,l^4}{24} + \frac{A\,l^3}{6} + \frac{B\,l^2}{2} = 0\,,$$

$$\frac{w\,l^3}{6} + \frac{A\,l^2}{2} + B\,l = 0\,,$$

whence $A = -wl/2$, $B = wl^2/12$ and the shape of the beam is given by

$$E I\,y = \frac{w\,x^4}{24} - \frac{w\,l\,x^3}{12} + \frac{w\,l^2\,x^2}{24}\,,$$

i.e.,
$$y = \frac{w\,x^2\,(l-x)^2}{24\,E I}\,.$$

We note that
$$\frac{\mathrm{d}y}{\mathrm{d}x} = \frac{w\,x\,(l-x)\,(l-2\,x)}{12\,E I}$$

and that therefore $\dfrac{\mathrm{d}y}{\mathrm{d}x} = 0$ when $x = \frac{1}{2}l$. This confirms the expected result that the beam is horizontal at its mid-point.

17.3 First order differential equations with variables separable

The processes of integration which we have discussed in the earlier chapters of this book are examples of the solution of a differential equation of the form

$$\frac{\mathrm{d}y}{\mathrm{d}x} = f(x)\,.$$

The indefinite integral, involving a constant, is the complete solution.

If the differential equation can be written in the form

$$\frac{\mathrm{d}y}{\mathrm{d}x} = \frac{f(x)}{g(y)},$$

it is called a *differential equation of the first order with variables separable.*
In this case

$$g(y)\frac{\mathrm{d}y}{\mathrm{d}x} = f(x),$$

so that, integrating with respect to x,

$$\int g(y)\frac{\mathrm{d}y}{\mathrm{d}x}\,\mathrm{d}x = \int f(x)\,\mathrm{d}x + A,$$

where A is an arbitrary constant, or

$$\int g(y)\,\mathrm{d}y = \int f(x)\,\mathrm{d}x + A.$$

Examples. (i) Solve the equation

$$\frac{\mathrm{d}y}{\mathrm{d}x} = \operatorname{cosec} x \tan y.$$

$$\int \frac{\mathrm{d}y}{\tan y} = \int \operatorname{cosec} x\,\mathrm{d}x.$$

$$\therefore \log \sin y = \log \tan \tfrac{1}{2}x + \log C$$

where C is constant. The form in which the arbitrary constant is written
is also arbitrary. Here it is convenient to write it as $\log C$.

$$\therefore \quad \sin y = C \tan \tfrac{1}{2}x.$$

$$\therefore \qquad y = \operatorname{Sin}^{-1}(C \tan \tfrac{1}{2}x).$$

(ii) $$y(1+x^2)\frac{\mathrm{d}y}{\mathrm{d}x} + 1 = y^2.$$

This equation can be written

$$(1+x^2)\frac{\mathrm{d}y}{\mathrm{d}x} = \frac{y^2-1}{y}.$$

$$\therefore \int \frac{y\,\mathrm{d}y}{y^2-1} = \int \frac{\mathrm{d}x}{1+x^2}.$$

$$\therefore \tfrac{1}{2}\log(y^2-1) + \log C = \tan^{-1}x.$$

$$\therefore x = \tan[\log\{C\sqrt{(y^2-1)}\}].$$

[The solution can alternatively be expressed in the form

$$y^2 = 1 + A e^{2\tan^{-1}x}.]$$

17.4 Homogeneous equations

A function $f(x, y)$ is homogeneous and of degree n if

$$f(\lambda x, \lambda y) = \lambda^n f(x, y).$$

The equation $P \dfrac{\mathrm{d}y}{\mathrm{d}x} = Q$ is said to be homogeneous if P and Q are homogeneous functions of x and y of the same degree.

The substitution of $y = xv$, where v is a function of x only, transforms a homogeneous equation to an equation with variables separable.

Example. Solve
$$(x^2 + y^2) \frac{\mathrm{d}y}{\mathrm{d}x} = x y.$$

Put
$$y = xv.$$

Then
$$\frac{\mathrm{d}y}{\mathrm{d}x} = x \frac{\mathrm{d}v}{\mathrm{d}x} + v$$

and the equation transforms to

$$(x^2 + x^2 v^2)\left(x \frac{\mathrm{d}v}{\mathrm{d}x} + v\right) = x^2 v,$$

i.e.,
$$x \frac{\mathrm{d}v}{\mathrm{d}x} + v = \frac{v}{1 + v^2}.$$

$$\therefore \ x \frac{\mathrm{d}v}{\mathrm{d}x} = \frac{v}{1 + v^2} - v = \frac{-v^3}{1 + v^2}.$$

$$\therefore \ \int \frac{\mathrm{d}x}{x} = - \int \frac{(1 + v^2)\,\mathrm{d}v}{v^3}.$$

$$\therefore \ \log x = \frac{1}{2 v^2} - \log v + \log C.$$

Replacing v by y/x now gives

$$\log x = \frac{x^2}{2 y^2} - \log\left(\frac{y}{x}\right) + \log C.$$

$$\therefore \ \log\left(\frac{x}{C} \cdot \frac{y}{x}\right) = \frac{x^2}{2 y^2}$$

and so the general solution is

$$y = C \, e^{x^2/2 y^2}.$$

Exercises 17.4

1. The gradient of a curve at a point (x, y) on it is $\log x$ and the curve passes through the point (e, e). Find the equation of the curve.

2. A uniformly loaded beam $A B$ of length l rests upon supports at its extremities. Take the horizontal and downward vertical lines through A as axes of x and y respectively and find the equation of the curve assumed by the beam. (It can be assumed that since the ends of the beam are free the curvature is zero at these points).

Obtain the complete solutions of each of the equations in Questions 3–11.

3. $(1 + x^2)\dfrac{dy}{dx} + (1 + y^2) = 0.$ 4. $\dfrac{dy}{dx} + x \sin y = 0.$

5. $\dfrac{dy}{dx} = \dfrac{y}{x} \log y.$ 6. $\sec x \dfrac{dy}{dx} = y.$

7. $\dfrac{dy}{dx} + \dfrac{1}{x} = \dfrac{e^y}{x}.$ 8. $\dfrac{dy}{dx} + \dfrac{x^2}{y} = x^2.$

9. $x y (1 - x^2)\dfrac{dy}{dx} + y^2 = 1.$ 10. $\dfrac{dy}{dx} + e^x = y^2\, e^x.$

11. $\cos y \dfrac{dy}{dx} + \cos x = \cos x \sin y.$

Obtain the particular solution which satisfies the given boundary conditions of each of the differential equations in Questions 12–19.

12. $x y (1 + x^2)\dfrac{dy}{dx} = 1 + y^2; \; x = 1, \; y = 1.$

13. $(x - 1)\dfrac{dy}{dx} = y - 4; \; x = 0, \; y = 0.$

14. $y^2\, e^x + \dfrac{dy}{dx} = y^2; \; x = 0, \; y = 1.$

15. $r\dfrac{d\theta}{dr} = \tan \tfrac{1}{2}\theta; \; r = a, \; \theta = \pi.$

16. $y \sin^2 x \dfrac{dy}{dx} + \cot x = 0; \; x = \tfrac{1}{2}\pi, \; y = 0.$

17. $r\dfrac{d\theta}{dr} = \tan 3\theta; \; r = a, \; \theta = \pi/6.$

18. $\cosh x \dfrac{dy}{dx} = \cos y; \; x = 0, \; y = 0.$

19. $(\sin x + \cos x)\dfrac{dy}{dx} = (\cos x - \sin x) \tan y; \; x = \tfrac{1}{2}\pi, \; y = \tfrac{1}{2}\pi.$

20. Obtain the particular solution of the equation

$$\frac{dy}{dx} = (x + y)^2,$$

given that $y = 1$ when $x = 0$ by using the substitution $x + y = v$.

21. Obtain the general solution of the equation

$$\frac{dy}{dx} = e^{x+y}.$$

22. Use the substitution $y = xv$ to obtain the general solution of the equation

$$(x - y)\frac{dy}{dx} = x + y.$$

23. Obtain the particular solution of the equation

$$2xy - (x^2 - y^2)\frac{dy}{dx} = 0,$$

given that $y = 1$ when $x = 2$.

24. Find the equation of the curve through $(2, 1)$ for which the subnormal is equal to twice the abscissa for all points on the curve.

25. The ordinate at a point P on a curve $y = f(x)$ meets the x-axis at N and the perpendicular from N to the tangent at P meets that tangent at R. If $NR = a$ and if $(0, a)$ is a point on the curve, find the equation of the curve.

26. For the polar curve $r = f(\theta)$, the perpendicular from the pole O on to the tangent at P (r, θ) meets that tangent at R. If OR/PR is constant and equal to k for all points P on the curve, which goes through the point $(1, 0)$, find the equation of the curve.

27. The acceleration of a particle moving in a straight line and starting from rest at unit distance from the origin is given, in the usual notation, by the equation

$$f = 1 + \log s.$$

Calculate v correct to two significant figures when $s = 2$.

17.5 The law of natural growth

The differential equation

$$\frac{dy}{dt} = ky \qquad (17.1)$$

expresses mathematically the statement that the rate of change w.r. to t of a quantity y is proportional to that quantity. A quantity which changes in this way is said to be subject to the law of natural growth or, if k is negative, of natural decay.

The solution of Eqn. (17.1) is

$$\int \frac{dy}{ky} = \int dt,$$

i.e.,

$$t = \frac{1}{k}\log_e y + \frac{1}{k}\log_e C.$$

$$\therefore \ Cy = e^{kt}.$$

$$\therefore \ y = \frac{1}{C}e^{kt}.$$

If $y = y_0$ when $t = 0$, then $\dfrac{1}{C} = y_0$ and hence

$$y = y_0 \, e^{kt}. \tag{17.2}$$

Among many examples involving the laws of natural growth and decay in the physical world, are the following:

(i) *Newton's Law of Cooling*, which states that the rate of change of the temperature of a body is proportional to the excess temperature over the surroundings. If the excess temperature is θ, then

$$\frac{d\theta}{dt} = -k\,\theta.$$

$$\therefore \ \theta = \theta_0 \, e^{-kt},$$

where θ_0 is the excess temperature at time $t = 0$.

(ii) A uniform bar expands when heated so that the rate of change of its length, l, with respect to the temperature, θ, is proportional to its length. Then

$$\frac{dl}{d\theta} = k\,l,$$

where k is called the coefficient of linear expansion of the bar. The solution of this equation is

$$l = l_0 \, e^{k\theta},$$

where l_0 is the length of the bar at temperature $0°$. If $e^{k\theta}$ is expanded in powers of $k\theta$ and squares and higher powers of $k\theta$ are neglected, this result becomes

$$l \doteqdot l_0(1 + k\theta).$$

(iii) *Radioactive decay.* In a mass of radioactive material in which atoms are disintegrating spontaneously, the average rate of disintegration is proportional to the number of atoms present so that

$$\frac{dN}{dt} = -k\,N,$$

where N is the number of atoms present at time t.

$$\therefore N = N_0 \, e^{-kt},$$

where N_0 is the number of atoms present at time $t = 0$.

The rate of decay of a radioactive substance is usually expressed in terms of the *half-life*, i.e., the time taken for one-half of a given mass of substance to decay. The half-life, T, of such a substance is given by

$$\tfrac{1}{2} N_0 = N_0 \, e^{-kT},$$

i.e.,
$$e^{-kT} = \tfrac{1}{2}$$

so that
$$T = \frac{1}{k} \log_e 2 \, .$$

(iv) If an electric current of magnitude i is flowing in a circuit of negligible capacitance, self-inductance L and resistance R, and if a constant E.M.F. E is applied to the circuit, it can be shown that

$$L \frac{d\,i}{d\,t} + R\,i = E.$$

This is an equation with variables separable which can be solved as follows:

$$L \frac{d\,i}{d\,t} = E - R\,i.$$

$$\therefore \int \frac{d\,i}{E - R\,i} = \int \frac{d\,t}{L} \, .$$

$$\therefore -\frac{1}{R} \log \left(E - R\,i \right) = \frac{t}{L} + C.$$

If $i = 0$ when $t = 0$ (i.e., if the time is measured from the instant when the circuit is completed),

$$C = -\frac{1}{R} \log E.$$

$$\therefore \log E - \log (E - R\,i) = \frac{R\,t}{L} \, .$$

$$\therefore \log \left(\frac{E}{E - R\,i} \right) = \frac{R\,t}{L} \, .$$

$$\therefore \frac{E}{E - R\,i} = e^{Rt/L}.$$

$$\therefore R\,i \, e^{Rt/L} = E(e^{Rt/L} - 1).$$

$$\therefore i = \frac{E}{R} \left(1 - e^{-Rt/L} \right).$$

$$\therefore i \to \frac{E}{R} \text{ (its Ohm's Law value) as } t \to \infty.$$

Examples. (i) A radioactive substance disintegrates at a rate proportional to its mass. One-half of a given mass of the substance disintegrates in 136 days. Calculate the time required for five-eighths of the substance to disintegrate. If the original mass of the substance was 100 gm, calculate the mass which has disintegrated after 34 days.

$$\frac{\mathrm{d}M}{\mathrm{d}t} = -kM,$$

where M is the mass of the substance at time t.

The solution of this equation is

$$M = M_0\,e^{-kt}$$

where M_0 is the mass of the substance at time $t = 0$. But $M = \frac{1}{2}M_0$ when $t = 136$.

$$\therefore \tfrac{1}{2}M_0 = M_0 e^{-136k}.$$

$$\therefore 136 = \frac{1}{k}\log 2.$$

$$\therefore k = (\log 2)/136.$$

$$\therefore M = M_0 e^{-(t\log 2)/136}.$$

When $\qquad\qquad M = 3M_0/8,$

$$\frac{3}{8} = e^{-(t\log 2)/136},$$

i.e., $\qquad \dfrac{t\log 2}{136} = \log\dfrac{8}{3}.$

Then $\quad t = \left(136 \times \log\dfrac{8}{3}\right)\Big/\log 2$

$$\doteqdot (136 \times 0.9809)/0.6931.$$

$$\therefore t \doteqdot 192.$$

N	L
136	2.1335
0.9809	$\bar{1}$.9916
	2.1251
0.6931	$\bar{1}$.8408
	2.2843

Hence five-eighths of the mass will disintegrate in approximately 192 days.

If $M_0 = 100 \quad$ and $\quad t = 34$,

$$M = 100\,e^{-(34\log 2)/136} = 100\,e^{-\frac{1}{4}\log 2}.$$

$$\therefore \log_{10} M = \log_{10} 100 - \frac{1}{4}\log 2 \times \log_{10} e$$

$$\doteqdot 2 - \frac{1}{4} \times 0.6931 \times 0.4343$$

$$\doteqdot 2 - 0.0753$$

$$= 1.9247.$$

$$\therefore M \doteqdot 84.1.$$

N	L
0.6931	$\bar{1}$.8408
0.4343	$\bar{1}$.6378
	$\bar{1}$.4786
4	0.6021
	$\bar{2}$.8765

Hence after 34 days, approximately 15.9 gm have disintegrated.

I

(ii) A certain chemical reaction follows Wilhelmy's Law which states that the rate of transformation of the reacting substance is proportional to its concentration. If, initially, the concentration of the reagent was 9·5 gm per litre and if, after five minutes, the concentration was 3·5 gm per litre, find what the concentration was after 2 minutes.

If x gm per litre of the reagent is transformed at time t sec, the differential equation which describes the reaction is

$$\frac{dx}{dt} = k(9 \cdot 5 - x).$$

$$\therefore \int \frac{dx}{9 \cdot 5 - x} = \int k \, dt.$$

$$\therefore -\log(9 \cdot 5 - x) = kt + C.$$

Since $x = 0$ when $t = 0$, $C = -\log 9 \cdot 5$;

$$\therefore \log \left(\frac{9 \cdot 5}{9 \cdot 5 - x} \right) = kt.$$

$$\therefore e^{kt} = \frac{9 \cdot 5}{9 \cdot 5 - x}.$$

$$\therefore x = 9 \cdot 5 (1 - e^{-kt}).$$

But $x = 6$ when $t = 5$.

$$\therefore \frac{6}{9 \cdot 5} = 1 - e^{-5k},$$

$$\therefore e^{-5k} = 1 - \frac{6}{9 \cdot 5} \doteqdot 0 \cdot 3684.$$

$$\therefore -5k \doteqdot \log_e 0 \cdot 3684 = -0 \cdot 9986.$$

$$\therefore k \doteqdot 0 \cdot 19.$$

Hence, when $t = 2$,

$$x \doteqdot 9 \cdot 5 (1 - e^{-0 \cdot 7994})$$

and the concentration at time $t = 2$ is approximately $9 \cdot 5 e^{-0 \cdot 7994}$ gm per litre $= A$ gm per litre (say).

$$\therefore \log_{10} A \doteqdot \log_{10} 9 \cdot 5 - 0 \cdot 7994 \times \log_{10} e$$

$$\doteqdot \log_{10} 9 \cdot 5 - 0 \cdot 7994 \times 0 \cdot 4343$$

$$\doteqdot 0 \cdot 9777 - 0 \cdot 3472$$

$$= 0 \cdot 6305.$$

$\therefore$ concentration $\doteqdot 4 \cdot 27$ gm per litre after **2 minutes**.

Note. In problems such as this example the use of the arbitrary constant can be dispensed with by means of definite integration. The amount x gm of the substance transformed at time 2 sec is given by

$$t = \frac{1}{k} \int_0^2 \frac{dx}{9 \cdot 5 - x} ,$$

subject to the condition $x = 6$ when $t = 5$. The condition is used in order to evaluate k.

Exercises 17.5

1. The temperature of a liquid in a room of constant temperature 30 °C was observed to be 80° and 5 minutes afterwards the temperature of the liquid was observed to be 60°.

Calculate (i) correct to the nearest minute, the time at which the temperature of the liquid was 40°,

(ii) the temperature of the liquid 10 minutes after the first observation.

2. In the usual notation the equation of motion of a particle starting from rest and moving in a straight line is

$$f = 2v + 1 .$$

Calculate (i) the velocity of the particle at time $t = 2$,

(ii) the displacement from the origin of the particle at time $t = 1$.

3. A particle starts from rest and falls under gravity in a resisting medium so that, in the usual notation, its acceleration at time t is given by the formula

$$f = g - kv .$$

Show that the velocity of the particle does not exceed g/k and calculate, in terms of k and a, the time taken from the commencement of the motion for the particle to attain a velocity of $g/(ak)$ where $a > 1$.

4. If the half-life of a radioactive substance is 8 days, calculate the time that will be required for 9 gm out of a mass of 10 gm of the substance to disintegrate.

5. Calculate the current after 0·01 sec in an electric circuit of resistance 8 ohm, self-induction 0·04 henry with a constant E.M.F. of 50 volts if initially the current in the circuit was zero.

6. The velocity v of a particle along a straight line Ox is increasing with respect to its distance x from O at a rate which is proportional to x/v. When the particle is at O its velocity is 10 ft per sec; after it has travelled 10 ft from O its velocity is 20 ft per sec. What further distance will the particle have travelled by the time it attains the velocity of 30 ft per sec?   (N.)

7. In a certain type of chemical reaction the amount x of a substance which has been transformed in time t is such that the rate of increase of x is proportional to the amount of substance that has not yet been transformed. If the original amount of the substance is 100 gm and 60 gm has been transformed after 2 min, find how much will be transformed after 6 min.   (N.)

8. A vessel A contains a mixture of 70% of water and 30% of another liquid L. A second vessel B contains water only and the total amount of liquid in each vessel is the same. At time $t = 0$ the vessels are connected, allowing a flow of liquid from each vessel to the other, the total amount of liquid in each vessel remaining unaltered. The net flow of liquid L is from that vessel in which the concentration of L is greater, and the rate of flow is proportional to the difference between the concentrations in the two vessels. Prove that x, the percentage of liquid L present in A at time t, is given by a differential equation of the form

$$\frac{\mathrm{d}x}{\mathrm{d}t} = -k(x - 15)$$

where k is a positive constant.

Integrate this equation, expressing t as a function of x. If $x = 25$ when $t = 5$, find to the nearest tenth of a unit, the value of t when $x = 18$. (N.)

9. The temperature θ of a cooling liquid is known to decrease at a rate proportional to $(\theta - \alpha)$, where α is the constant temperature of the surrounding medium. Show that $\theta - \alpha$ must be proportional to e^{-kt}, where t is the time and k is a positive constant.

If the constant temperature of the surrounding medium is $15°$ and the temperature of the liquid falls from $60°$ to $45°$ in 4 minutes, find

(i) the temperature after a further 4 minutes,

(ii) the time in which the temperature falls from $45°$ to $30°$. (N.)

10. The natural rate of growth of a colony of micro-organisms in a liquid is k_1 organisms per organism per minute. If the organisms are removed at the rate of k_2 organisms per minute, write down a differential equation to describe the variation in size of the colony.

Initially the colony contains N organisms. Find the number of organisms after t minutes and prove that the colony will be removed in a finite time provided that $N < k_2/k_1$. (N.)

17.6 Linear equations of the first order

Differential equations of the first degree in y and the differential coefficients of y with respect to x are called *linear equations*.

The general linear equation of the first order is

$$\frac{\mathrm{d}y}{\mathrm{d}x} + Py = Q, \tag{17.3}$$

where P and Q are functions of x only.

The method of solution of this equation is suggested by the solution of the equation in which $Q = 0$, i.e., of

$$\frac{\mathrm{d}y}{\mathrm{d}x} + Py = 0$$

which has solution

$$\int \frac{\mathrm{d}y}{y} + \int P\,\mathrm{d}x = 0,$$

i.e., $\log y + \int P\,\mathrm{d}x = \log C,$

i.e., $y\,e^{\int P\,\mathrm{d}x} = C.$

Also we have

$$\frac{\mathrm{d}}{\mathrm{d}x}\left(y\,e^{\int P\,\mathrm{d}x}\right) = \frac{\mathrm{d}y}{\mathrm{d}x}\,e^{\int P\,\mathrm{d}x} + Py\,e^{\int P\,\mathrm{d}x}$$

This gives the information that if the left-hand-side of eqn. (17.3) is multiplied by $e^{\int P\,\mathrm{d}x}$ it becomes the derivative of $y\,e^{\int P\,\mathrm{d}x}$. To solve such equations, therefore, we first multiply throughout the equation by the *integrating factor* $e^{\int P\,\mathrm{d}x}$. Each side of the equation can then be integrated directly with respect to x, the result of the integration on the left-hand-side being $y\,e^{\int P\,\mathrm{d}x}$

Two points should be noted carefully by the student.

(i) The integrating factor frequently reduces to the form $e^{\log f(x)}$ and it should be remembered that $e^{\log f(x)} = f(x)$.

(ii) If the original equation is of the form

$$a(x)\,\frac{\mathrm{d}y}{\mathrm{d}x} + b(x)\,y = c(x),$$

it must be divided throughout by $a(x)$ before the integrating factor is obtained.

Examples. (i) Solve the equation $\dfrac{\mathrm{d}y}{\mathrm{d}x} + y\cot x = \operatorname{cosec} x.$

The integrating factor is $e^{\int \cot x\,\mathrm{d}x} = e^{\log \sin x} = \sin x$. Multiplying throughout by $\sin x$, we have

$$\sin x\,\frac{\mathrm{d}y}{\mathrm{d}x} + y\cos x = 1,$$

i.e., $\dfrac{\mathrm{d}}{\mathrm{d}x}(y\sin x) = 1.$

Integration gives $y\sin x = x + C.$

$$\therefore y = (x + C)\operatorname{cosec} x$$

is the general solution.

(ii) Find the function $y = f(x)$ which satisfies the differential equation

$$x \frac{\mathrm{d}y}{\mathrm{d}x} = y + kx^2 \cos x$$

and the condition $y = 2\pi$ when $x = \pi$.

For what values of k do x and y always have the same sign?

Find a first order differential equation which is satisfied by this function but which does not involve k.   (N.)

The differential equation can be written

$$\frac{\mathrm{d}y}{\mathrm{d}x} - \frac{y}{x} = k \, x \cos x \qquad (1)$$

and the integrating factor is $\mathrm{e}^{-\int \frac{1}{x} \mathrm{d}x} = \mathrm{e}^{-\log x} = \dfrac{1}{x}$.

Hence multiplying Eqn. (1) by $\dfrac{1}{x}$,

$$\frac{1}{x} \frac{\mathrm{d}y}{\mathrm{d}x} - \frac{y}{x^2} = k \cos x,$$

i.e.,
$$\frac{\mathrm{d}}{\mathrm{d}x} \left(\frac{y}{x} \right) = k \cos x.$$

Integration gives

$$\frac{y}{x} = k \sin x + C.$$

Since $y = 2\pi$ when $x = \pi$, $C = 2$ and the particular solution is

$$\frac{y}{x} = k \sin x + 2,$$

i.e.,
$$y = kx \sin x + 2x.$$

x and y always have the same sign so long as $k \sin x + 2 > 0$ for all values of x, i.e., so long as $k > -2$.

$y = kx \sin x + 2x$ will satisfy the differential equation which is the result of eliminating k between the original differential equation and the particular solution.

$$\therefore y = kx \sin x + 2x \text{ satisfies}$$

$$x \frac{\mathrm{d}y}{\mathrm{d}x} = y + \left(\frac{y - 2x}{x \sin x} \right) x^2 \cos x,$$

i.e.,
$$x \sin x \frac{\mathrm{d}y}{\mathrm{d}x} - (\sin x + x \cos x) \, y + 2x^2 \cos x = 0.$$

17.7 Bernoulli's equation $\dfrac{dy}{dx} + Py = Qy^n$

This can be transformed into a linear equation by the substitution

$$z = 1/y^{n-1}.$$

Example. Solve the equation

$$x\frac{dy}{dx} - y = y^3\,e^x.$$

This equation can be written

$$\frac{x}{y^3}\frac{dy}{dx} - \frac{1}{y^2} = e^x.$$

Put $\quad z = \dfrac{1}{y^2}\quad$ so that $\quad \dfrac{dz}{dx} = -\dfrac{2}{y^3}\dfrac{dy}{dx}.$

Then the equation transforms to

$$-\frac{x}{2}\frac{dz}{dx} - z = e^x,$$

i.e. $\quad \dfrac{dz}{dx} + \dfrac{2z}{x} = \dfrac{-2e^x}{x}.$ \hfill (1)

The integrating factor is $e^{\int\frac{2}{x}dx} = x^2$ and Eqn. (1) becomes

$$x^2\frac{dz}{dx} + 2xz = -2xe^x,$$

i.e. $\quad \dfrac{d}{dx}(x^2 z) = -2xe^x.$

Integrating, $\qquad x^2 z = -2xe^x + 2e^x + C.$

$$\therefore \frac{x^2}{y^2} = -2xe^x + 2e^x + C.$$

$$\therefore y = \pm\,x/\sqrt{(2e^x - 2xe^x + C)}.$$

Exercises 17.7

Solve the differential Equations 1−10.

1. $\dfrac{dy}{dx} + \dfrac{y}{x} = e^x.$

2. $\dfrac{dy}{dx} + y\cot x = e^x.$

3. $(1+x^2)\dfrac{dy}{dx} + xy = x.$

4. $\dfrac{dy}{dx} + \dfrac{y}{x-1} = e^x.$

5. $\dfrac{dy}{dx} + ay = b.$

6. $\dfrac{dy}{dx} + y = \cos x.$

7. $\dfrac{dr}{d\theta} + r \tan \theta = \cos \theta.$

8. $\sin x \dfrac{dy}{dx} - y = \sin^2 x.$

9. $\cos x \dfrac{dy}{dx} - y = \cos x.$

10. $\dfrac{dy}{dx} = \dfrac{y + x - 1}{x}.$

11. Make the substitution $y = \dfrac{1}{z}$ and hence obtain the complete solution of the equation $\dfrac{dy}{dx} - y \cot x = y^2 \operatorname{cosec} x.$

12. Solve the equation $\dfrac{dy}{dx} + \dfrac{y}{x} = x^2 y^3$ by means of the substitution $z = 1/y^2.$

13. With the usual notation for the motion of a particle moving in a straight line, the equation of motion of such a particle is $\dfrac{dv}{dt} = -2v + t.$ If $v = 1$ when $t = 0$, calculate v when $t = 1$ and if $s = 0$ when $t = 0$ calculate s when $t = 1$.

14. The current i in an electric circuit of resistance R and self-induction L satisfies the equation

$$L \frac{di}{dt} + Ri = 40 \sin 100\,t.$$

If $i = 0$ when $t = 0$ and if $L = 0\cdot20$ and $R = 20$, find i when $t = 0\cdot01$.

15. Solve the differential equation

$$x \frac{dy}{dx} + 3y = \frac{a^3}{x^2} \cos\left(\frac{x}{a}\right)$$

with the condition $y = 0$ when $x = \dfrac{1}{2}\pi a.$

Show that $|y| \le \dfrac{2a^4}{x^3}$ for $x > 0$. (N.)

16. (a) Solve the differential equation

$$\frac{dy}{dx} + 2y = e^{-2x}.$$

(b) By means of the substitution $y = xz$, where z is a function of x, reduce the differential equation

$$x \frac{dy}{dx} - y = \frac{1}{4} x^2 - y^2$$

to a differential equation involving z and x only. Hence solve it, given that $y = 0$ when $x = \log_e 2.$ (N.)

17. Solve the equation $x \dfrac{dy}{dx} + 2y = x^3$, given that $y = 0$ when $x = 1$. (L.)

18. Solve the differential equation $x \dfrac{dy}{dx} - y = 3x^4$ given that $y = -1$ when $x = 1$. (L.)

19. Solve the equation $\dfrac{\mathrm{d}y}{\mathrm{d}x} - y \cot x = \tan^2 x$ given that $y = 0$ when $x = \dfrac{1}{4}\pi$.

(N.)

20. Find the solution of the equation

$$\left(1 - \frac{\mathrm{d}y}{\mathrm{d}x}\right) \sin x = y \cos x$$

for which $y = \sqrt{2} - 1$ when $x = \dfrac{1}{4}\pi$.   (O.C.)

17.8 Equations of higher orders

The general solution of a differential equation of the second order contains two arbitrary constants.

An equation of the form $\dfrac{\mathrm{d}^n y}{\mathrm{d}x^n} = f(x)$ can be solved by n direct integrations (c.f. § 17.2); each integration introduces one arbitrary constant. The final solution is of the form $y = F(x)$ where $F(x)$ contains n arbitrary constants.

Equations of the form $\dfrac{\mathrm{d}^2 y}{\mathrm{d}x^2} = f(y)$ are transformed by the substitution $\dfrac{\mathrm{d}y}{\mathrm{d}x} = p$ to a first-order equation in p and y with variables separable.

Examples. (i) The equation of a simple harmonic motion is

$$\frac{\mathrm{d}^2 y}{\mathrm{d}t^2} = -\omega^2 y \text{ where } \omega \text{ is constant.}$$

Put $\dfrac{\mathrm{d}y}{\mathrm{d}t} = v$. Then

$$\frac{\mathrm{d}^2 y}{\mathrm{d}t^2} = \frac{\mathrm{d}v}{\mathrm{d}t} = \frac{\mathrm{d}v}{\mathrm{d}y}\frac{\mathrm{d}y}{\mathrm{d}t} = v \frac{\mathrm{d}v}{\mathrm{d}y}.$$

$$\therefore v \frac{\mathrm{d}v}{\mathrm{d}y} = -\omega^2 y.$$

$$\therefore \int v \, \mathrm{d}v = - \int \omega^2 y \, \mathrm{d}y.$$

$$\therefore \frac{v^2}{2} = \frac{a^2 \omega^2}{2} - \frac{\omega^2 y^2}{2}$$

where a is constant and $\dfrac{a^2 \omega^2}{2}$ is the constant of integration.

$$\therefore v = \frac{\mathrm{d}y}{\mathrm{d}t} = \pm \omega \sqrt{(a^2 - y^2)}.$$

$$\therefore \int \frac{\mathrm{d}y}{\sqrt{(a^2 - y^2)}} = \pm \int \omega \, \mathrm{d}t.$$

$$\therefore \sin^{-1}\left(\frac{y}{a}\right) = \pm \omega t + b \text{ where } b \text{ is constant}.$$

$$\therefore \ y = a \sin(\pm \omega t + b).$$

$$\therefore \ y = a \sin\omega t \cos b + a \cos\omega t \sin b$$

$$\text{or } y = a \cos\omega t \sin b - a \sin\omega t \cos b$$

and in either case since a and b are arbitrary constants we may write the solution

$$y = A \cos\omega t + B \sin\omega t$$

where A and B are arbitrary.

(ii) Solve the equation

$$\frac{\mathrm{d}^2 y}{\mathrm{d}x^2} = \mathrm{e}^{2y}.$$

Put $\dfrac{\mathrm{d}y}{\mathrm{d}x} = p$.　Then

$$\frac{\mathrm{d}^2 y}{\mathrm{d}x^2} = \frac{\mathrm{d}\left(\dfrac{\mathrm{d}y}{\mathrm{d}x}\right)}{\mathrm{d}y} \cdot \frac{\mathrm{d}y}{\mathrm{d}x} = p \frac{\mathrm{d}p}{\mathrm{d}y}.$$

The differential equation is transformed to

$$p \frac{\mathrm{d}p}{\mathrm{d}y} = \mathrm{e}^{2y}.$$

$$\therefore \int p \, \mathrm{d}p = \int \mathrm{e}^{2y} \, \mathrm{d}y.$$

$$\therefore \frac{p^2}{2} = \frac{\mathrm{e}^{2y}}{2} + \frac{a^2}{2}$$

where a is an arbitrary constant.

$$\therefore \frac{\mathrm{d}y}{\mathrm{d}x} = \pm \sqrt{(\mathrm{e}^{2y} + a^2)}.$$

$$\therefore \int \frac{\mathrm{d}y}{\sqrt{(\mathrm{e}^{2y} + a^2)}} = \pm \int \mathrm{d}x.$$

Put $\mathrm{e}^{2y} = t$ so that $\dfrac{\mathrm{d}t}{\mathrm{d}y} = 2\mathrm{e}^{2y} = 2t$. Then

$$\int \frac{\mathrm{d}t}{2t \sqrt{(t + a^2)}} = \pm \int \mathrm{d}x.$$

Put $(t + a^2) = u^2$ so that $\dfrac{\mathrm{d}t}{\mathrm{d}u} = 2u$. Then

$$\int \frac{2u\,\mathrm{d}u}{2\,(u^2 - a^2)\,u} = \pm \int \mathrm{d}x.$$

$$\therefore x = \pm \frac{1}{2a} \log_e \left(\frac{u - a}{u + a} \right) + b$$

where b is an arbitrary constant.

$$\therefore x = \frac{1}{2a} \log_e \left\{ \frac{\sqrt{(e^{2y} + a^2)} - a}{\sqrt{(e^{2y} + a^2)} + a} \right\} + b.$$

The ambiguous $\pm$ sign is omitted from the final statement of the solution because the range of possible values of the arbitrary a includes negative values as well as the corresponding positive values and therefore the solution as stated includes all possible solutions.

Exercises 17.8

1. Solve the equation $\dfrac{\mathrm{d}^2 y}{\mathrm{d}x^2} + 4y = 0$ given that $y = 1$ when $x = \dfrac{1}{4}\pi$ and $y = 2$ when $x = \frac{1}{2}\pi$.

2. Solve the equation $\dfrac{\mathrm{d}^2 y}{\mathrm{d}x^2} + 9y = 0$ given that when $x = 0$, $y = 2$, and $\dfrac{\mathrm{d}y}{\mathrm{d}x} = 0$.

3. Solve the equation $\dfrac{\mathrm{d}^2 y}{\mathrm{d}x^2} + 64y = 0$ given that $y = 0$ and $\dfrac{\mathrm{d}y}{\mathrm{d}x} = 1$ when $x = \frac{1}{2}\pi$.

4. Solve $\dfrac{\mathrm{d}^2 y}{\mathrm{d}x^2} + 9y = 0$ given that $y = \sqrt{2}$ when $x = \dfrac{\pi}{12}$ and $\dfrac{\mathrm{d}y}{\mathrm{d}x} = 1$ when $x = \dfrac{\pi}{6}$.

5. Solve the equation $\dfrac{\mathrm{d}^2 y}{\mathrm{d}x^2} = n^2 y$ $\quad (n \neq 0)$.

6. Solve the equation $\dfrac{\mathrm{d}^2 y}{\mathrm{d}x^2} - y = 0$ given that when $x = 0$, $y = 4$ and $\dfrac{\mathrm{d}y}{\mathrm{d}x} = 2$.

7. Solve the equation $\dfrac{\mathrm{d}^2 y}{\mathrm{d}x^2} + \sin x = 0$.

8. Solve the equation $y^3 \dfrac{\mathrm{d}^2 y}{\mathrm{d}x^2} = 1$ given that when $x = 0$, $y = 1$ and $\dfrac{\mathrm{d}y}{\mathrm{d}x} = 1$.

17.9 Linear equations of the second order with constant coefficients

We consider the equation

$$a \frac{\mathrm{d}^2 y}{\mathrm{d}x^2} + b \frac{\mathrm{d}y}{\mathrm{d}x} + cy = P \tag{17.4}$$

where a, b and c are constants and P is a function of x.

(i) Put $y = u + v$ where u and v are functions of x. Then

$$\frac{dy}{dx} = \frac{du}{dx} + \frac{dv}{dx}, \quad \frac{d^2y}{dx^2} = \frac{d^2u}{dx^2} + \frac{d^2v}{dx^2}.$$

The equation transforms to

$$\left(a\frac{d^2u}{dx^2} + b\frac{du}{dx} + cu\right) + \left(a\frac{d^2v}{dx^2} + b\frac{dv}{dx} + cv\right) = P$$

and it is therefore satisfied by the solution $y = u + v$ if

$$(1) \quad a\frac{d^2u}{dx^2} + b\frac{du}{dx} + cu = 0,$$

$$(2) \quad a\frac{d^2v}{dx^2} + b\frac{dv}{dx} + cv = P.$$

Thus, if $y = u$ is a solution of the original equation with the right hand side replaced by zero and $y = v$ is a solution of the original equation, then

$$y = u + v$$

is a solution of the original equation.

If, further, $y = u$ is the *general* solution of the equation

$$a\frac{d^2y}{dx^2} + b\frac{dy}{dx} + cy = 0 \qquad (17.5)$$

and therefore contains two arbitrary constants, and $y = v$ is a particular solution of

$$a\frac{d^2y}{dx^2} + b\frac{dy}{dx} + cy = P,$$

then $y = u + v$ is the general solution of the equation

$$a\frac{d^2y}{dx^2} + b\frac{dy}{dx} + cy = P.$$

Of these two parts of the solution, u is called the *complementary function* (C.F.) and v is called the *particular integral* (P.I.).

(ii) If $y = u_1$ and $y = u_2$ are particular solutions of

$a\dfrac{d^2y}{dx^2} + b\dfrac{dy}{dx} + cy = 0$, then $y = Au_1 + Bu_2$ is the general solution, for

(1) it contains two arbitrary constants,

(2) $a \dfrac{d^2}{dx^2} (A u_1 + B u_2) + b \dfrac{d}{dx} (A u_1 + B u_2) + c (A u_1 + B u_2)$

$$= A \left(a \dfrac{d^2 u_1}{dx^2} + b \dfrac{du_1}{dx} + c u_1 \right) + B \left(a \dfrac{d^2 u_2}{dx^2} + b \dfrac{du_2}{dx} + c u_2 \right) = 0.$$

We require therefore to find two particular solutions u_1 and u_2 of the equation

$$a \dfrac{d^2 y}{dx^2} + b \dfrac{dy}{dx} + cy = 0$$

in order to write down the complementary function as $A u_1 + B u_2$.

17.10 The complementary function

In the equation

$$a \dfrac{d^2 y}{dx^2} + b \dfrac{dy}{dx} + cy = 0,$$

we try as a solution $y = e^{mx}$, where m is a constant to be determined, so that $\dfrac{dy}{dx} = m e^{mx}$ and $\dfrac{d^2 y}{dx^2} = m^2 e^{mx}$. Then

$$a m^2 e^{mx} + b m e^{mx} + c e^{mx} = 0,$$

i.e., $\qquad\qquad e^{mx}(a m^2 + b m + c) = 0.$

Hence $y = e^{m_1 x}$ and $y = e^{m_2 x}$ where m_1, m_2 are the roots of $a m^2 + b m + c = 0$, are solutions of $a \dfrac{d^2 y}{dx^2} + b \dfrac{dy}{dx} + cy = 0$. (The equation in m is called the *auxiliary equation*.) The complementary function is therefore

$$A e^{m_1 x} + B e^{m_2 x} \quad \text{provided that } m_1 \neq m_2. \tag{17.6}$$

Special cases arise when $m_1 = m_2$, and when m_1 and m_2 are complex. These two cases are considered separately below.

(i) If $m_1 = m_2$, then

$$\dfrac{b}{a} = -2 m_1; \quad \dfrac{c}{a} = m_1^2,$$

and the differential Equation (17.5) can be written

$$\dfrac{d^2 y}{dx^2} - 2 m_1 \dfrac{dy}{dx} + m_1^2 y = 0.$$

Put $y = z\mathrm{e}^{m_1 x}$ where z is a function of x. Then

$$\frac{\mathrm{d}y}{\mathrm{d}x} = \mathrm{e}^{m_1 x}\frac{\mathrm{d}z}{\mathrm{d}x} + m_1 z\,\mathrm{e}^{m_1 x}, \quad \frac{\mathrm{d}^2 y}{\mathrm{d}x^2} = \mathrm{e}^{m_1 x}\frac{\mathrm{d}^2 z}{\mathrm{d}x^2} + 2\,m_1\,\mathrm{e}^{m_1 x}\frac{\mathrm{d}z}{\mathrm{d}x} + m_1^2\,\mathrm{e}^{m_1 x} z$$

and the equation transforms to

$$\mathrm{e}^{m_1 x}\left(\frac{\mathrm{d}^2 z}{\mathrm{d}x^2} + 2\,m_1\frac{\mathrm{d}z}{\mathrm{d}x} + m_1^2 z - 2\,m_1\frac{\mathrm{d}z}{\mathrm{d}x} - 2\,m_1^2 z + m_1^2 z\right) = 0,$$

$$\text{i.e., to} \quad \mathrm{e}^{m_1 x}\frac{\mathrm{d}^2 z}{\mathrm{d}x^2} = 0\,.$$

$$\therefore \frac{\mathrm{d}^2 z}{\mathrm{d}x^2} = 0\,.$$

$$\therefore z = A\,x + B \text{ where } A \text{ and } B \text{ are constants.}$$

Hence in this case the C.F. is

$$\mathrm{e}^{m_1 x} z = \mathrm{e}^{m_1 x}(A\,x + B)\,. \tag{17.7}$$

(ii) If m_1 and m_2 are complex, then, assuming that a, b, c are all real, we can write $m_1 = \alpha + i\beta$, $m_2 = \alpha - i\beta$. The C.F. is

$$A\,\mathrm{e}^{\alpha x + i\beta x} + B\,\mathrm{e}^{\alpha x - i\beta x} = \mathrm{e}^{\alpha x}(A\,\mathrm{e}^{i\beta x} + B\,\mathrm{e}^{-i\beta x})\,.$$

But $A\,\mathrm{e}^{i\beta x} = A(\cos\beta x + i\sin\beta x)$ from result **(16.14)**,
and $B\,\mathrm{e}^{-i\beta x} = B(\cos\beta x - i\sin\beta x)$.
Hence the C.F. is

$$\mathrm{e}^{\alpha x}(C\cos\beta x + D\sin\beta x) \tag{17.8}$$

where C and D are constants.

This result may also be obtained by the method used in (i), i.e. with the substitution $y = \mathrm{e}^{\alpha x} z$, reducing the equation to the form

$$\frac{\mathrm{d}^2 z}{\mathrm{d}x^2} + \beta^2 z = 0\,.$$

Examples. (i) *Roots of the auxiliary equation are real and different.*

$$2\frac{\mathrm{d}^2 y}{\mathrm{d}x^2} - 5\frac{\mathrm{d}y}{\mathrm{d}x} + 2y = 0\,.$$

The auxiliary equation is

$$2m^2 - 5m + 2 = 0,$$

$$\text{i.e.,} \quad (2m - 1)(m - 2) = 0\,.$$

$$\therefore m = \tfrac{1}{2} \text{ or } 2\,.$$

Hence the general solution is $y = A\,\mathrm{e}^{x/2} + B\,\mathrm{e}^{2x}$.

(ii) *Roots of the auxiliary equation are equal.*

$$\frac{d^2y}{dx^2} + 4\frac{dy}{dx} + 4y = 0.$$

The auxiliary equation is

$$m^2 + 4m + 4 = 0.$$

$$\therefore m = -2 \text{ (a repeated root).}$$

Hence the general solution is $y = (Ax + B)e^{-2x}$.

(iii) *Roots of the auxiliary equation are complex.*

$$\frac{d^2y}{dx^2} - 2\frac{dy}{dx} + 2y = 0.$$

The auxiliary equation is

$$m^2 - 2m + 2 = 0.$$

$$\therefore m = 1 \pm i.$$

Hence the general solution is $y = e^x(A\cos x + B\sin x)$.

(iv) *The linear equation is of higher order than the second.*

The methods used below are justified by similar reasoning to that used above for equations of the second order.

$$\frac{d^3y}{dx^3} - y = 0.$$

The auxiliary equation is

$$m^3 - 1 = 0.$$

The solutions of this equation are $m = 1, \omega, \omega^2$, i.e., $1, \frac{1}{2}(-1 \pm i\sqrt{3})$. The general solution of the differential equation is therefore

$$y = A\,e^x + e^{-x/2}\left\{B\cos\left(\frac{\sqrt{3}}{2}x\right) + C\sin\left(\frac{\sqrt{3}}{2}x\right)\right\}.$$

(v)

$$\frac{d^3y}{dx^3} + 2\frac{d^2y}{dx^2} + \frac{dy}{dx} = 0.$$

The auxiliary equation is

$$m^3 + 2m^2 + m = 0.$$

The solution of this equation is $m = 0, -1, -1$. The general solution of the differential equation is therefore

$$y = A + e^{-x}(Bx + C).$$

Summary. If $f(m) = 0$ is the auxiliary equation of the linear differential equation, then

(i) for every real root m_r which is not a repeated root, the C.F. contains a term $A_r e^{m_r x}$,

(ii) for every double root m_p the C.F. contains a term $e^{m_p x}(A_p x + B_p)$; [if there are n roots each equal to m_s the C.F. contains a term $e^{m_s x} f(x)$ where $f(x)$ is a polynomial involving n arbitrary constants and of degree $n - 1$ in x],

(iii) for every pair of conjugate complex roots $\alpha \pm i\beta$ the C.F. contains a term $e^{\alpha x} (A \cos\beta x + B \sin\beta x)$.

Exercises 17.10

Obtain the general solutions of the following equations:

1. $2\dfrac{d^2 y}{dx^2} - 5\dfrac{dy}{dx} - 3y = 0.$

2. $4\dfrac{d^2 y}{dx^2} - \dfrac{dy}{dx} - 5y = 0.$

3. $\dfrac{d^2 y}{dx^2} - 9y = 0.$

4. $\dfrac{d^2 y}{dx^2} + 5\dfrac{dy}{dx} = 0.$

5. $\dfrac{d^2 y}{dx^2} - 4\dfrac{dy}{dx} + 4y = 0.$

6. $2\dfrac{d^2 y}{dx^2} + 2\dfrac{dy}{dx} + y = 0.$

7. $\dfrac{d^2 y}{dx^2} + 3\dfrac{dy}{dx} + 4y = 0.$

8. $\dfrac{d^2 y}{dx^2} + 5\dfrac{dy}{dx} + y = 0.$

9. $4\dfrac{d^2 y}{dx^2} + 12\dfrac{dy}{dx} + 9y = 0.$

10. $\dfrac{d^3 y}{dx^3} + \dfrac{d^2 y}{dx^2} = 0.$

17.11 The particular integral

We have to find any particular solution of the equation

$$a\frac{d^2 y}{dx^2} + b\frac{dy}{dx} + cy = P \tag{17.4}$$

where P is a function of x. In special cases we can state intuitively the nature of the function which is a particular integral, and the intuitive statement can then be confirmed by substitution into the equation.

(i) If P is a polynomial function of x of degree n, there is a solution of the form $y = f(x)$ where $f(x)$ is a polynomial of degree n.

(ii) If P is a function of the form $g \cos nx + h \sin nx$, there is a solution of the form $y = k \cos nx + l \sin nx$.

(iii) If P is of the form $p e^{nx}$ there is a solution of the form $y = q e^{nx}$.

(iv) If P is a function of x which is the sum of some or all of the functions enumerated in (i) to (iii), there is a solution which is also the sum of such functions.

(v) If the C.F. already contains terms of the types suggested for the particular integral, it is possible to obtain particular integrals in case (i) of the form $y = kx^{p+1}$, where the term of highest degree in the C.F. involves x^p, and in case (ii) of the form $kx \cos nx + lx \sin nx$, and in case (iii) of the form $qx\, e^{nx}$ unless there is a similar term in the C.F. in which case the P.I. is of the form $qx^2\, e^{nx}$ etc.

Examples. (i) $\dfrac{d^2y}{dx^2} - 4\dfrac{dy}{dx} + 3y = x^3.$

The auxiliary equation is $m^2 - 4m + 3 = 0$ with roots $m = 3,\ 1$. Hence the C.F. is

$$A e^{3x} + B e^x.$$

Try $y = ax^3 + bx^2 + cx + d$ as the P.I. in the differential equation.

Then $\dfrac{dy}{dx} = 3ax^2 + 2bx + c, \qquad \dfrac{d^2y}{dx^2} = 6ax + 2b.$

$\therefore 6ax + 2b - 12ax^2 - 8bx - 4c + 3ax^3 + 3bx^2 + 3cx + 3d \equiv x^3.$

$\therefore 3ax^3 - (12a - 3b)x^2 + (6a - 8b + 3c)x + (2b - 4c + 3d) \equiv x^3.$

Equating coefficients of powers of x we find

$$a = \frac{1}{3},$$

$$12a - 3b = 0, \ \therefore b = \frac{4}{3},$$

$$6a - 8b + 3c = 0, \ \therefore c = \frac{26}{9},$$

$$2b - 4c + 3d = 0, \ \therefore d = \frac{80}{27}.$$

Hence the P.I. is $\dfrac{1}{3}x^3 + \dfrac{4}{3}x^2 + \dfrac{26x}{9} + \dfrac{80}{27}$ and the general solution of the equation is

$$y = A e^{3x} + B e^x + \frac{1}{3}x^3 + \frac{4}{3}x^2 + \frac{26x}{9} + \frac{80}{27}.$$

(ii) $\dfrac{d^2y}{dx^2} + 2\dfrac{dy}{dx} + 5y = \sin 2x.$

The auxiliary equation is $m^2 + 2m + 5 = 0$ with roots $m = -1 \pm 2i$. Hence the C.F. is
$$e^{-x}(A \cos 2x + B \sin 2x).$$

Try $y = a \sin 2x + b \cos 2x$ as the P.I. in the differential equation. Then
$$\frac{dy}{dx} = 2a \cos 2x - 2b \sin 2x, \quad \frac{d^2y}{dx^2} = -4a \sin 2x - 4b \cos 2x.$$

$\therefore \quad -4a \sin 2x - 4b \cos 2x + 4a \cos 2x - 4b \sin 2x + 5a \sin 2x +$
$+ 5b \cos 2x \equiv \sin 2x.$

$\therefore \quad (a - 4b) \sin 2x + (4a + b) \cos 2x \equiv \sin 2x.$

Equating coefficients of $\sin 2x$ and $\cos 2x$,
$$a - 4b = 1,$$
$$4a + b = 0.$$
$$\therefore \quad a = \frac{1}{17}, \quad b = -\frac{4}{17}.$$

Hence a particular integral is $\dfrac{1}{17} \sin 2x - \dfrac{4}{17} \cos 2x$, and the general solution is
$$y = e^{-x}(A \cos 2x + B \sin 2x) + \frac{1}{17} \sin 2x - \frac{4}{17} \cos 2x.$$

(iii) $\dfrac{d^2y}{dx^2} + 4\dfrac{dy}{dx} + 4y = e^x.$

The auxiliary equation is $m^2 + 4m + 4 = 0$ which has $m = -2$ as a repeated root. Hence the C.F. is
$$e^{-2x}(Ax + B).$$

Try $y = ae^x$ as the P.I. in the differential equation. Then
$$\frac{dy}{dx} = ae^x, \quad \frac{d^2y}{dx^2} = ae^x.$$
$$\therefore \quad ae^x + 4ae^x + 4ae^x \equiv e^x.$$
$$\therefore \quad a = \frac{1}{9}.$$

Hence a particular integral is $y = \dfrac{1}{9} e^x$ and the general solution is
$$y = e^{-2x}(Ax + B) + \frac{1}{9} e^x.$$

(iv) $\dfrac{d^2y}{dx^2} - 3\dfrac{dy}{dx} + 2y = 2\,e^x.$

The auxiliary equation is $m^2 - 3m + 2 = 0$ with roots $m = 2,\ 1$. Hence the C.F. is

$$A\,e^{2x} + B\,e^x.$$

If the substitution $y = a\,e^x$ is made, $\dfrac{d^2y}{dx^2} - 3\dfrac{dy}{dx} + 2y$ is identically zero. In this case we try $y = ax\,e^x$ as the P.I. Then

$$\frac{dy}{dx} = a\,e^x + a\,x\,e^x, \qquad \frac{d^2y}{dx^2} = 2\,a\,e^x + a\,x\,e^x.$$

$$\therefore\ 2\,a\,e^x + a\,x\,e^x - 3\,a\,e^x - 3\,a\,x\,e^x + 2\,a\,x\,e^x \equiv 2\,e^x.$$

$$\therefore\ -a\,e^x \equiv 2\,e^x.$$

$$\therefore\quad a = -2.$$

Hence a particular integral is $-2xe^x$, and the general solution is

$$y = A\,e^{2x} + B\,e^x - 2xe^x.$$

If $A\,x\,e^x$ had also been a term of the C.F., the trial substitution to make in order to find the P.I. would have been $y = ax^2e^x$.

(v) $\dfrac{d^2y}{dx^2} + n^2y = \sin nx.$

The C.F. is $A \sin nx + B \cos nx$.

Because $A \sin nx$ is a term of the C.F. we try $y = x(a \sin nx + b \cos nx)$ as the P.I. Then

$$\frac{dy}{dx} = (a \sin nx + b \cos nx) + x(an \cos nx - bn \sin nx),$$

$$\frac{d^2y}{dx^2} = 2n(a \cos nx - b \sin nx) + x(-an^2 \sin nx - bn^2 \cos nx).$$

$\therefore\ 2an \cos nx - 2bn \sin nx - an^2x \sin nx - bn^2x \cos nx + an^2x \sin nx +$
$+\ bn^2x \cos nx \equiv \sin nx.$

$$\therefore\ 2an \cos nx - 2bn \sin nx \equiv \sin nx.$$

$$\therefore\ a = 0, \quad b = -\frac{1}{2n}.$$

Hence a particular integral is $-\dfrac{x}{2n}\cos nx$ and the general solution is

$$y = A \sin nx + B \cos nx - \frac{x}{2n}\cos nx.$$

It must be emphasized that the analysis of methods for finding the P.I. of a linear equation of the second order with constant coefficients which we have just made is concerned only with special cases. In the equation

$$a\frac{d^2 y}{dx^2} + b\frac{dy}{dx} + cy = P,$$

P is any function of x. Those functions which we have considered here are the ones occurring most frequently.

Exercises 17.11

Obtain the general solution of each of the following equations:

1. $\dfrac{d^2 y}{dx^2} + 5\dfrac{dy}{dx} + 6y = 4.$

2. $\dfrac{d^2 y}{dx^2} + 9y = 1.$

3. $\dfrac{d^2 y}{dx^2} - \dfrac{dy}{dx} - 2y = \sin x.$

4. $\dfrac{d^2 y}{dx^2} - 2\dfrac{dy}{dx} + y = 2\,e^x.$

5. $\dfrac{d^2 y}{dx^2} + 3\dfrac{dy}{dx} + 2y = e^{-2x}.$

6. $3\dfrac{d^2 y}{dx^2} + 2\dfrac{dy}{dx} + y = \cos 2x.$

7. $\dfrac{d^2 y}{dx^2} - 3\dfrac{dy}{dx} + 3y = 2.$

8. $\dfrac{d^2 y}{dx^2} - 4\dfrac{dy}{dx} + 5y = x^2 + x + 1.$

9. $\dfrac{d^2 y}{dx^2} - \dfrac{dy}{dx} - 2y = 1 + x.$

10. $\dfrac{d^2 y}{dx^2} - 4\dfrac{dy}{dx} + 4y = x + e^x.$

11. $\dfrac{d^2 y}{dx^2} + 4\dfrac{dy}{dx} + 5y = \sin 2x.$

12. $\dfrac{d^2 y}{dx^2} - 5\dfrac{dy}{dx} = 4.$

13. $\dfrac{d^2 y}{dx^2} + 2\dfrac{dy}{dx} + 2y = \sin 2x + \cos x.$

14. $\dfrac{d^3 y}{dx^3} + \dfrac{d^2 y}{dx^2} = 1 + e^{-x}.$

15. $\dfrac{d^2 y}{dx^2} - 6\dfrac{dy}{dx} + 9y = e^{3x}.$

16. $\dfrac{d^2 y}{dx^2} - 3\dfrac{dy}{dx} + 2y = \sin 2x.$

17. $p^2\dfrac{d^2 y}{dx^2} - p\dfrac{dy}{dx} = a(1 + e^x), \quad p \ne 0 \text{ or } 1.$

18. $\dfrac{d^4 y}{dx^4} - 16y = x.$

19. $\dfrac{d^2 x}{dt^2} - 2\dfrac{dx}{dt} - 3x = \cos\left(2t + \dfrac{1}{4}\pi\right).$

20. $\dfrac{d^2 x}{dt^2} - 2\dfrac{dx}{dt} + 2x = \cos\left(2t + \dfrac{1}{3}\pi\right).$

21. The equation of motion for the small oscillations of a pendulum swinging in a resisting medium is

$$\frac{d^2\theta}{dt^2} + k\frac{d\theta}{dt} + n^2\theta = 0$$

where θ is the angular displacement of the pendulum from its mean position and k and n are constants. Show that if $k^2 > 4n$ and if the pendulum starts from rest at $\theta = \alpha$, then

$$\theta = \frac{\alpha}{m_2 - m_1}(m_2\,e^{m_1 t} - m_1\,e^{m_2 t})$$

where m_1 and m_2 are the roots of $m^2 + km + n^2 = 0$.

22. Solve the equation

$$\frac{d^2y}{dt^2} + 3\frac{dy}{dt} + 2y = 10\cos t$$

given that $y = 1$ and $\frac{dy}{dt} = 0$ when $t = 0$.

23. Solve the equation

$$\frac{d^2x}{dt^2} - 4x = 5e^{3t}$$

given that $x = -2$ and $\frac{dx}{dt} = -3$ when $t = 0$. Find the value of t for which $\frac{dx}{dt} = 0$. (O.C.)

24. In the usual notation for a particle moving in a straight line, the equation of motion of a particle is

$$\frac{d^2s}{dt^2} + 5\frac{ds}{dt} + 6s = ae^{-t}$$

where $a > 0$. If $s = 0$ and $v = 0$ when $t = 0$, prove that s is positive for all values of t and if $a = 100$ calculate each of s and v, correct to three significant figures, when $t = 1$.

25. If the equation of motion for a particle moving in a straight line is

$$\frac{d^2s}{dt^2} + 2\frac{ds}{dt} + 5s = 10a\sin t,$$

prove that for large values of t, and whatever the initial conditions,

$$s \doteqdot \sqrt{5}a\sin(t - \alpha) \text{ where } \tan\alpha = \tfrac{1}{2}.$$

26. If the equation of motion for a particle moving in a straight line is

$$\frac{d^2s}{dt^2} + 3\frac{ds}{dt} + 2s = 16e^{-3t},$$

and $s = 4$, $v = -15$ when $t = 0$, calculate the minimum value of s.

27. If $\frac{d^2y}{dx^2} + n^2y = kx$, where n and k are real constants and $y = 0$ when $x = 0$ and when $x = a$, find the value of y when $x = \frac{1}{2}a$. (L.)

28. Solve the equation

$$\frac{\mathrm{d}^2 y}{\mathrm{d} t^2} + 6 \frac{\mathrm{d} y}{\mathrm{d} t} + 25 y = 6 \sin t,$$

given that when $t = 0$, $y = 0$ and $\mathrm{d} y/\mathrm{d} t = 0$.   (L.)

29. Solve the equation

$$\frac{\mathrm{d}^2 y}{\mathrm{d} \theta^2} + 9 y = 8 \sin \theta$$

given that $y = \mathrm{d} y/\mathrm{d}\theta = 0$ when $\theta = \frac{1}{2}\pi$.   (L.)

30. A particle moves on a fixed straight line through a fixed point O and its equation of motion is

$$\frac{\mathrm{d}^2 x}{\mathrm{d} t^2} + 2 k \frac{\mathrm{d} x}{\mathrm{d} t} + n^2 x = \cos nt,$$

x being its displacement from O at time t, $(n > k)$. Find x in terms of t, given that $x = \dfrac{1}{2 k^2}$ and $\dfrac{\mathrm{d} x}{\mathrm{d} t} = 0$ when $t = 0$. Find the speed at time t if $n = 2 k$.   (L.)

Miscellaneous Exercises XVII

1. Assume that, when a thermometer is placed in a hot liquid at temperature T, the temperature θ indicated by the thermometer at time t rises at a rate proportional to $T - \theta$.

If T is kept constant, and if $\theta = \theta_0$ at $t = 0$, prove that

$$T - \theta = (T - \theta_0)\mathrm{e}^{-kt}$$

where k is constant.

If $\theta_0 = 15°$, and if $\theta = 35°$ when $t = 10$ sec, and $\theta = 50°$ when $t = 20$ sec prove that $T = 95°$.   (N.)

2. Integrate the equations

(i) $\dfrac{\mathrm{d} y}{\mathrm{d} x} = \dfrac{x + y - 1}{2 x}$,

(ii) $\dfrac{\mathrm{d}^2 y}{\mathrm{d} x^2} - \dfrac{\mathrm{d} y}{\mathrm{d} x} - 2 y = \sin x$.   (N.)

3. (a) Solve the differential equation

$$\frac{\mathrm{d} y}{\mathrm{d} x} - x \operatorname{cosec}^2 y = 3 \operatorname{cosec}^2 y.$$

(b) If

$$x \frac{\mathrm{d} y}{\mathrm{d} x} + (x - 1) y = x^2 \mathrm{e}^{2 x},$$

find the value of y in terms of x, given that $y = 1$ when $x = 3$.   (N.)

4. Gas is allowed to expand in a cylinder, maintained at a constant absolute temperature T, and does work on a piston sliding in the cylinder. For any small increase in the volume of the gas from V to $V + \delta V$, the work done by the gas

is approximately $P\delta V$, where P is the pressure of the gas at volume V. If P, V and T satisfy the equation

$$PV = 85{,}000\ T,$$

show that the work W done by the gas in expanding to volume V satisfies the differential equation

$$V\frac{\mathrm{d}W}{\mathrm{d}V} = 85{,}000\ T.$$

If $T = 300$ and P changes from 2,000 to 1,030, find the percentage increase in the volume of the gas, and calculate the work done by the gas during the change. (N.)

5. The rate of decay at any instant of a radioactive substance is proportional to the amount of substance remaining at that instant. If the initial amount of the substance is A and the amount remaining after time t is x, prove that

$$x = A\,\mathrm{e}^{-kt}$$

where k is constant.

If the amount remaining is reduced from $\frac{1}{2}A$ to $\frac{1}{3}A$ in 8 hours, prove that the initial amount of the substance was halved in about 13·7 hours. (N.)

6. Integrate the equation $\dfrac{\mathrm{d}^2 y}{\mathrm{d}x^2} + 2\dfrac{\mathrm{d}y}{\mathrm{d}x} + 5y = x^2$. (N.)

7. (a) If $x(x+1)\dfrac{\mathrm{d}y}{\mathrm{d}x} = y(y+1)$ and $y = 2$ when $x = 1$, find y when $x = 2$.

(b) Solve completely the differential equation

$$\frac{\mathrm{d}^2 x}{\mathrm{d}t^2} + 5\frac{\mathrm{d}x}{\mathrm{d}t} + 6x = 2\mathrm{e}^{-t}.$$

If $x = 0$ and $\dfrac{\mathrm{d}x}{\mathrm{d}t} = 1$ when $t = 1$, show that the maximum value of x is $\dfrac{1}{4}$. (N.)

8. Integrate the equation

$$\frac{\mathrm{d}^2 y}{\mathrm{d}x^2} - 3\frac{\mathrm{d}y}{\mathrm{d}x} + 2y = 1 + \cosh x. \qquad \text{(N.)}$$

9. During a fermentation process the rate of decomposition of substance at any instant is related to the amounts of y gm of substance and x gm of active ferment by the law

$$\frac{\mathrm{d}y}{\mathrm{d}t} = -0.25\,xy.$$

The value of x time t is $4/(1+t)^2$; when $t = 0$, $y = 10$. Express $\dfrac{\mathrm{d}y}{\mathrm{d}t}$ in terms of y and t, and hence determine y as a function of t.

Prove that the amount of substance ultimately remaining is approximately 3·7 gm. (N.)

10. (a) Solve the differential equation

$$x\frac{\mathrm{d}y}{\mathrm{d}x} + 3y = \sqrt{(1+3x^3)},$$

given that $y = 1$ when $x = 1$.

(b) If $e^{-x^2} \dfrac{dy}{dx} = x(y + 2)^2$, find the value of y in terms of x, given that $y = 0$ when $x = 0$.   (N.)

11. A vessel containing V cubic inches of liquid is in the form of a cone of semi-vertical angle 30° with vertex downwards and axis vertical. At time $t = 0$ a small plug is removed from the bottom of the vessel and liquid flows out at a rate $k\sqrt{y}$ cubic inches per minute, k being a constant and y inches the depth of the liquid in the vessel at time t minutes. Find t as a function of y.

When $t = t_1$ the vessel contains a volume $\frac{1}{2} V$ cubic inches of liquid; when $t = t_2$, it is empty. Calculate t_1/t_2.   (N.)

12. Solve the differential equation

$$\frac{d^2 y}{dx^2} + 2k \frac{dy}{dx} + (1 + k^2)\, y = \cos x, \quad 0 < k < 1.$$ (N.)

13. Integrate the equations

$$\text{(i)} \quad \frac{d^3 y}{dx^3} - y = x^4 + e^{2x},$$

$$\text{(ii)} \quad (1 + x^2) \frac{dy}{dx} + xy = 2x.$$ (N.)

14. (i) Find the general solution of the equation

$$(x^2 - 1) \frac{dy}{dx} + 2y = x + 1.$$

(ii) A point starts from rest at the origin and its abscissa at time t satisfies the equation

$$\frac{d^2 x}{dt^2} + 4 \frac{dx}{dt} + 3x = 6a e^{-2t}$$

where $a > 0$. Determine the greatest value of x.   (N.)

15. The horizontal cross-section of a tank has a constant area A sq. in. Water is poured into the tank at the rate of K cu. in. per sec. At the same time water flows out through a hole in the bottom at the rate of $H\sqrt{y}$ cu. in. per sec., where y in. is the depth of the water. Show that at time t sec.

$$A \frac{dy}{dt} = K - H\sqrt{y}.$$

By putting $y = x^2$ find the general solution of this equation when H, K are constants. If $y = 0$ at $t = 0$, deduce that

$$\frac{H^2 t}{2AK} = \log_e \left(\frac{K}{K - H\sqrt{y}} \right) - \frac{H\sqrt{y}}{K}.$$

If $A = 1000$, $H = 100$, $K = 600$, show that when $y = 25$ the value of t is 115 approximately. Take $\log_e 6$ as $1{\cdot}7918$.   (N.)

16. A particle moving along a straight line OX is at a distance x from O at time t, and its velocity is given by

$$8t^2 \frac{dx}{dt} = (1 - t^2)\, x^2.$$

If $x = 4$ when $t = 1$, prove that $x = 8t/(1 + t^2)$.

Find (i) the velocity when $t = 0$,

    (ii) the maximum distance from O,

    (iii) the maximum speed *towards* O for positive values of t.   (N.)

17. At time t after a battery has been switched on, the current x in an electric circuit satisfies the differential equation

$$L \frac{dx}{dt} + Rx = E,$$

where L, R and E are positive constants. Show that the substitution $x = y + (E/R)$ reduces this equation to

$$L \frac{dy}{dt} + Ry = 0.$$

Solve the latter equation and deduce x as a function of t, given that $x = 0$ when $t = 0$.

The total charge q that has crossed any section of the circuit in time t is given by

$$q = \int_0^t x \, dt.$$

By integrating the original differential equation with respect to t, or otherwise, show that

$$q = \frac{E}{R} t - \frac{L}{R} x. \qquad \text{(N.)}$$

18. (i) Find the general solution of the equation

$$\frac{dy}{dx} + y \cot x = x \cot x.$$

(ii) The abscissa x of a point moving along the x-axis satisfies the equation

$$\frac{d^2 x}{dt^2} + 2n \frac{dx}{dt} + 2n^2 x = n^2 a (\cos nt - 2 \sin nt).$$

Initially $x = 0$ and $\frac{dx}{dt} = na$. Show that when the point first returns to the origin, its speed is $na (1 - e^{-\pi/2})$.   (N.)

19. (a) Solve the equation

$$(1 + x)^2 \left(x \frac{dy}{dx} - 2y \right) = x^3.$$

If $y = 1$ when $x = 1$, find y when $x = 2$.

(b) The abscissa of a point P moving along the x-axis satisfies the equation

$$4 \frac{d^4 x}{dt^4} + \frac{d^2 x}{dt^2} = 9 \sin t + 12 \cos t.$$

When $t = 0$, $x = 9$, $\frac{dx}{dt} = 3$, $\frac{d^2 x}{dt^2} = -4$, $\frac{d^3 x}{dt^3} = -3$. Prove that P oscillates between the origin and the point $x = 10$.   (N.)

20. (a) If $(x^2 - 1)\dfrac{\mathrm{d}y}{\mathrm{d}x} + 2y = 0$, find the value of y in terms of x, given that $y = 3$ when $x = 2$.

(b) Solve the differential equation

$$x\frac{\mathrm{d}y}{\mathrm{d}x} + 2y = \frac{2\sin x}{x\cos^3 x}.$$ (N.)

21. In a certain chemical reaction the rate of decomposition of a substance at any time t is proportional to the amount m that remains. In any small change, the percentage increase in the pressure p inside the vessel in which the reaction takes place is proportional to the percentage decrease in the amount m of the substance.

Express the above statements in calculus notation using a and b as the respective constants of proportionality.

If m and p have the values m_0, p_0 when $t = 0$, find

    (i) m in terms of m_0, a and t,

    (ii) p in terms of p_0, m_0, b and m,

and hence obtain p in terms of p_0, a, b and t. (N.)

22. Integrate the equations

    (i) $x(1 + x)\dfrac{\mathrm{d}y}{\mathrm{d}x} + y^2(1 - y) = 0$,

    (ii) $\dfrac{\mathrm{d}^2 y}{\mathrm{d}x^2} + 2\dfrac{\mathrm{d}y}{\mathrm{d}x} + 5y = 3 - 2\mathrm{e}^{-x} + \cos 2x.$ (N.)

23. (a) If $y = Ax\sin x + Bx\cos x$, where A and B are constants, obtain a linear differential equation independent of A and B which is satisfied by y.

(b) Solve the differential equations

    (i) $\dfrac{\mathrm{d}^2 y}{\mathrm{d}x^2} + 3\dfrac{\mathrm{d}y}{\mathrm{d}x} + 2y = x$,

    (ii) $(x - x^2)\dfrac{\mathrm{d}y}{\mathrm{d}x} + y = x^2 + 2x$

giving the solution of (ii) for which $y = 0$ when $x = -1$. (N.)

24. (a) Solve the equation

$$\frac{\mathrm{d}y}{\mathrm{d}x} + \frac{y(x + 2y)}{x(y + 2x)} = 0.$$

(b) The coordinates of a point moving in the $x - y$ plane satisfy the equations

$$\frac{\mathrm{d}x}{\mathrm{d}t} - \omega y = \omega a, \qquad \frac{\mathrm{d}y}{\mathrm{d}t} + 4\omega x = 4\omega a.$$

Given that $x = 2a$ and $y = a$ when $t = 0$, show that the path is an ellipse of area $4\pi a^2$. (N.)

25. (i) Solve the differential equation

$$\frac{\mathrm{d}y}{\mathrm{d}x} = \frac{4x - 3y}{3x - 2y}.$$

(ii) Find the solution of the differential equation

$$\frac{d^2x}{dt^2} - 4\frac{dx}{dt} + 3x = 8e^{-t} - 2e^t \tag{N.}$$

if $x = 1$ and $\dfrac{dx}{dt} = 0$ when $t = 0$.

26. (a) The tangent to a curve at P meets the x-axis at T and NP is the ordinate at P. Find the equations of all curves for which TN is a given constant k.

(b) Solve the differential equation

$$\frac{d^2y}{dx^2} - 7\frac{dy}{dx} + 12y = e^{3x} + 12. \tag{N.}$$

27. (a) In the differential equation

$$(x+y)\frac{dy}{dx} = x^2 + xy + x + 1$$

change the dependent variable from y to z, where $z = x + y$. Deduce the general solution of the given equation.

(b) The normal at the point $P(x, y)$ on a curve meets the x-axis at Q, and N is the foot of the ordinate of P. If $NQ = \dfrac{x(1+y^2)}{1+x^2}$, find the equation of the curve, given that it passes through the point $(3, 1)$. (N.)

28. (i) Find the function y of x satisfying the equation

$$(x+1)\frac{dy}{dx} - y(y+1)(x+2) = 0$$

and such that $y = \frac{1}{2}$ when $x = 0$.

(ii) By means of the substitution $x = e$ find the function y of x satisfying the equation

$$\cdot x^2\frac{d^2y}{dx^2} + x\frac{dy}{dx} + y = 0$$

and such that $y = 1$ and $\dfrac{dy}{dx} = 0$ when $x = 1$. (L.)

29. The normal to a plane curve at a variable point P meets the axes of x and y at Q and R respectively. The orthogonal projection of PQ on the x-axis is equal to that of PR on the y-axis. Find the equation of the curve, given that it passes through $(1, 1)$ but not through $(0, 0)$. (L.)

30. Solve the differential equations

(i) $x(x-y)\,dy/dx + x^2 + y^2 = 0,$

(ii) $dy/dx + y\tan x = \sec x,$

(iii) $d^2y/dx^2 + 2\,dy/dx + 5y = x + 1.$ (L.)

31. A vessel has the shape formed by revolving that portion of the parabola $y^2 = 4ax$ lying in the positive quadrant about the y-axis, and it contains liquid to a depth h; find the volume of the liquid.

There is a small hole at the vertex so that the liquid escapes slowly at a rate proportional to its depth at any instant. Obtain a differential equation for the depth y of the liquid in the vessel at any instant. If it takes 1 hour for the liquid to fall to half its original depth, show that the remainder will all escape in a further 4 minutes.   (L.)

32. (a) Solve the equation

$$(1 - x^2)\frac{\mathrm{d}y}{\mathrm{d}x} = xy\,(1 + y^2),$$

given that $y = 1$ when $x = 0$.

(b) By the substitution $z = x + y$, transform the equation

$$\frac{\mathrm{d}y}{\mathrm{d}x} = \frac{1 + 2x + 2y}{1 - 2x - 2y}$$

into an equation involving z and x only. Hence, or otherwise, solve the given equation.   (L.)

33. Solve the differential equations

$$\text{(i)} \quad \frac{\mathrm{d}y}{\mathrm{d}x} = \frac{2x + 3y}{3y - 2x};$$

$$\text{(ii)} \quad \frac{\mathrm{d}y}{\mathrm{d}x} + y = xy^3;$$

$$\text{(iii)} \quad \frac{\mathrm{d}y}{\mathrm{d}x} + 2y = 3\mathrm{e}^{-x} \text{ given that } y = 0 \text{ when } x = 0. \qquad \text{(L.)}$$

34. Solve the differential equations

(i) $x\,\mathrm{d}y/\mathrm{d}x + 2y = \cos x$;

(ii) $\mathrm{d}^2y/\mathrm{d}t^2 + 10\,\mathrm{d}y/\mathrm{d}t + 29y = 0$, given that $y = 0$ and $\mathrm{d}y/\mathrm{d}t = 4$ when $t = 0$.   (L.)

35. (i) Find y in terms of x given that

$$x\frac{\mathrm{d}y}{\mathrm{d}x} + 3y = 3, \text{ and } y = 3 \text{ when } x = 1.$$

(ii)   Solve $\dfrac{\mathrm{d}^2y}{\mathrm{d}x^2} + 4\dfrac{\mathrm{d}y}{\mathrm{d}x} + 3y = \sin x + \mathrm{e}^x + 3.$ (L.)

36. Solve the differential equations

$$\text{(i)} \quad x\sin y\,\frac{\mathrm{d}y}{\mathrm{d}x} - \cos y = 1, \text{ given that } y = 0 \text{ when } x = 1;$$

$$\text{(ii)} \quad \frac{\mathrm{d}^2y}{\mathrm{d}x^2} + 4\frac{\mathrm{d}y}{\mathrm{d}x} + 8y = \cosh x. \qquad \text{(L.)}$$

37. (i) Obtain the solution of the equation

$$(1 - x^2)\frac{\mathrm{d}y}{\mathrm{d}x} = 1 + \cos 2y$$

for which $y = \dfrac{1}{4}\pi$ when $x = 0$.

(ii) By substituting $t = 3x - y - 3$, or otherwise, solve the equation

$$\frac{dy}{dx} = \frac{3x - y + 3}{3x - y - 1}.$$

(L.)

38. (i) Find y in terms of x if

$$\sin x \frac{dy}{dx} - 2y \cos x = 3 \sin x.$$

(ii) Solve

$$\frac{dy}{dx} = \frac{x + 3y}{3x + y}$$

given $y = 0$ when $x = 1$.

(L.)

39. (i) By means of the substitution $x - y = z$, solve the differential equation $\frac{dy}{dx} = \sin(x - y)$, given that $y = 0$ when $x = 0$.

(ii) In an electrical circuit the current i after t seconds is such that $\frac{di}{dt} = k(E - Ri)$ where k, E, R are constants. Initially i is zero. If after t_1 seconds the current is two-thirds of its maximum value, show that after $2t_1$ seconds it has risen to eight-ninths of the maximum.

(L.)

40. (i) Solve the equation $\cos^4 x \frac{dy}{dx} = \sin^3 x$ given that $y = \frac{5}{3}$ when $x = \frac{\pi}{3}$.

(ii) A curve whose equation is $y = f(x)$ passes through the points $(1, 1)$ and $(2, 16)$ and satisfies the equation $ny \log x = \frac{dy}{dx} - ny$. What is the ordinate when $x = 3$?

(L.)

41. If $\frac{d^2 y}{dx^2} + 4 \frac{dy}{dx} + 3y = 0$ and $y = 0$, $\frac{dy}{dx} = 2$ at $x = 0$, find the maximum value of y.

(L.)

42. Solve the equation

$$\frac{d^2 y}{dx^2} - 7 \frac{dy}{dx} + 6y = 36x,$$

given that $y = 0$ and $\frac{dy}{dx} = 4$ when $x = 0$.

Show that, for this solution, $\frac{d^2 y}{dx^2}$ is zero for $x = \frac{1}{5} \log \frac{2}{9}$.

(L.)

43. Solve the equation $\frac{d^2 y}{dx^2} + 4y = 12 \cos 4x$, given that $y = 1$ when $x = \pi/4$ and $y = -3$ when $x = \pi/2$. Hence, or otherwise, show that y is a maximum when $x = n\pi \pm \pi/6$, where n is an integer.

(L.)

44. Solve the equations

(i) $x \frac{dy}{dx} + \frac{y}{x - 1} = x(x - 1)$,

(ii) $x(x - 1) \cos y \frac{dy}{dx} + \sin y = x(x - 1)^2$.

(L.)

45. Solve the equation

$$x \frac{dy}{dx} = y + \sqrt{(x^2 - y^2)}, \text{ by writing } y = vx, \text{ or otherwise.}$$

(L.)

46. A heavy particle B is attached to one end of a light inextensible string AB of length c. The string rests on a smooth horizontal plane, with A at the origin and B at the point $(0, c)$, the axes of coordinates being taken to lie in the plane. The end A of the string is now moved along the x-axis, and the particle moves in such a way that the string remains taut and AB always touches at B the curve traced by B. Show that if (x, y) are the coordinates of B,

$$(c^2 - y^2)^{\frac{1}{2}}(dy/dx) + y = 0.$$

Deduce that the equation of the curve is

$$x + (c^2 - y^2)^{\frac{1}{2}} - c \log_e \left\{ \frac{c + (c^2 - y^2)^{\frac{1}{2}}}{y} \right\} = 0. \tag{N.}$$

47. A body of mass m moves in a straight line under a constant propelling force and a resistance mkv, where v is its velocity and k is a constant. The body starts from rest at time $t = 0$ and the velocity tends to a limiting value V as t increases. Write down differential equations connecting (i) v with t, (ii) v with the distance covered. Solve these equations.

Show that the average velocity during the interval from the start to the instant when $v = \frac{1}{2} V$ is

$$V\left(1 - \frac{1}{2 \log_e 2}\right). \tag{N.}$$

48. In a certain chemical reaction the amount x of one substance at time t is related to the velocity of the reaction dx/dt by the equation

$$\frac{dx}{dt} = k(a - x)(2a - x)$$

where a and k are constants, and $x = 0$ when $t = 0$. If $x = 2 \cdot 0$ when $t = 1$ and $x = 2 \cdot 8$ when $t = 3$, show that $a = 3$ and find (i) the value of k, (ii) the value of x when $t = 2$. (N.)

49. At time t a population of organisms has x members. The rate of increase due to reproduction is $a + bt$ per cent. and the rate of decrease due to mortality $a' + b't$ per cent., where a, a', b, b' are positive constants. Construct a differential equation connecting x with t, and solve it, given that initially $x = x_0$. If $a > a'$ and $b < b'$, show that x will at first increase but will ultimately tend to zero. Find also the maximum value of x. (N.)

50. (a) The normal at any point of a curve passes through the point $(1, 1)$. Express this condition in the form of a differential equation, and hence find the equation of the family of curves which satisfy the condition.

(b) Given that y satisfies the differential equation

$$\frac{dy}{dx} + 2y \tan x = \sin x,$$

and that $y = 1$ when $x = \pi/3$, express dy/dx in terms of x. (N.)

51. A flask contains a growing bacterial culture and food which is being consumed by the culture. Each unit of food consumed results in an increase of three-quarters of a unit in the culture. At time t the flask contains x units of culture and y units

of food, and the time-rate of increase of x is equal to xy; initially $x = 3$, $y = 100$. Construct a differential equation for dx/dt, and hence determine x as a function of the time. (N.)

52. A population of insects is placed in an experimental environment and allowed to grow for several days. Its net time-rate of increase is proportional to its size, x insects, at any time t days, and its increases during the fourth day and the fifth day are estimated (from counts of newly hatched and dead insects) as 3,566 and 4,143 insects respectively.

By constructing and solving a suitable differential equation, determine the initial size and initial time-rate of increase of the population. (N.)

53. (a) The curve $y = f(x)$ passes through the point $(3, 1)$, and its gradient at the point (x, y) is given by the differential equation

$$\frac{dy}{dx} = \tfrac{1}{2}\left(1 - \frac{y}{x}\right).$$

Find $f(x)$.

(b) By substituting $y = z^{-\frac{1}{2}}$, or otherwise, solve the differential equation

$$\frac{dy}{dx} = y - x e^{-2x} y^3. \tag{N.}$$

54. The output, N articles per day, of a machine slows down in such a way that the rate of decrease of N is proportional to the product of N and the total time t that the machine has been in use. Express this statement as a differential equation and solve it for N in terms of t and any necessary constants.

Initially the output was 1,000 articles per day but after 50 days it has dropped to 950 articles per day. Calculate how much longer the machine will be kept in use if it is to be discarded as soon as its output falls to 500 articles per day. (N.)

55. A radioactive substance disintegrates in accordance with the equation $dm/dt = -km$, where m is the mass remaining at time t and k is a constant. Prove that $m = m_0 e^{-kt}$, where m_0 is the initial mass.

A mixture of two radioactive substances is known to consist initially of 300 mgm of one substance and 100 mgm of the other. After one week the total mass of the mixture is found to be 200 mgm and after one further week the total mass is 112 mgm. It is required to deduce the values of the constants k_1 and k_2 which give the rates of disintegration of the two substances respectively in accordance with the above equation, the unit of time being one week. By using the substitutions $x_1 = e^{-k_1}$, $x_2 = e^{-k_2}$ show that the data allow of two different solutions. In order to select the correct solution a further measurement of the total mass is made three weeks after the start, and the total mass is then found to be 70·4 mgm. Deduce the values of k_1 and k_2. (N.)

CHAPTER XVIII

APPROXIMATE NUMERICAL SOLUTION OF EQUATIONS

18.1 Graphical methods

The real roots of an equation can usually be found, by graphical methods, to a degree of approximation which is determined by the nature of the tables which are used.

A first approximation to the roots is obtained by finding the intersections of appropriate graphs, and subsequent, closer, approximations are found by plotting the parts of the graphs near to the points of intersection to an increasingly large scale. The final stage of the approximation can frequently be concluded by reference to the tables and without further drawing.

Examples. (i) Find, approximately, the real root of the equation

$$x^3 + x + 1 = 0.$$

If $f(x) \equiv x^3 + x + 1$, then $f'(x) = 3x^2 + 1$.

$\therefore f'(x) > 0$ for all real values of x.

$\therefore f(x)$ increases continuously and there is only one real root of $f(x) = 0$.

But $f(-1) = -1$ and $f(0) = 1$.

Hence the real root α is such that $-1 < \alpha < 0$.

The root is given by the intersection of $y = -x - 1$ and $y = x^3$.

x	-1	-0.9	-0.8	-0.7	-0.6	-0.5	-0.4	-0.3	-0.2	-0.1
x^3	-1	-0.729	-0.512	-0.343	-0.216	-0.125	-0.064	-0.027	-0.008	-0.001

The graphs of $y = -x - 1$ and $y = x^3$ drawn on the same axes are shown in Fig. 161 and the intersection of these graphs gives $x = -0.68$ as the first approximation to the root of $x^3 + x + 1 = 0$.

Values of x^3 and of $-x - 1$ are now worked out for the range $-0.69 < x < -0.68$. *Without drawing the graphs,* we can at once see that the root lies in the range

$$-0.682 > x > -0.683.$$

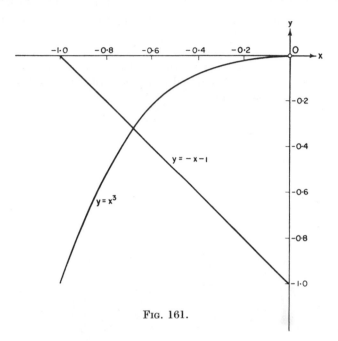

FIG. 161.

x	-0.680	-0.681	-0.682	-0.683
$\log(-x)$	$\overline{1}.8325$	$\overline{1}.8331$	$\overline{1}.8338$	$\overline{1}.8344$
$\log(-x)^3$	$\overline{1}.4975$	$\overline{1}.4993$	$\overline{1}.5014$	$\overline{1}.5032$
x^3	-0.3145	-0.3157	-0.3173	-0.3185
$-1-x$	-0.32	-0.319	-0.318	-0.317

A final calculation for $x = -0.6825$, shows that the intersection of the graphs occurs in the range $-0.682 > x > -0.6825$.

$$x \qquad -0.6825$$
$$\log(-x) \qquad \overline{1}.8341$$
$$\log(-x)^3 \qquad \overline{1}.5023$$
$$x^3 \qquad -0.3179$$
$$-1-x \qquad -0.3175.$$

Thus the real root of $x^3 + x + 1 = 0$ is -0.682 correct to three significant figures.

K

(ii) Find an approximate value of that real root of the equation $e^x = \tan x$ which is in the range $0 < x < \frac{1}{2}\pi$.

There is an intersection of the graphs of $y = e^x$ and $y = \tan x$ (Fig. 162), in the range $1\cdot3 < x < 1\cdot4$. Tabulated values of e^x and $\tan x$ for $1\cdot3 \leq x \leq 1\cdot4$ taken as far as necessary are

	1·30	1·31
x		
e^x	3·67	3·71
x in degrees and minutes	74° 29'	75° 3'
$\tan x$	3·60	3·75 .

(Values of e^x are obtained by the 'antilog' process from the tables of natural logarithms or directly from tables of e^x.) The real root is therefore in the range

$$1\cdot30 < x < 1\cdot31.$$

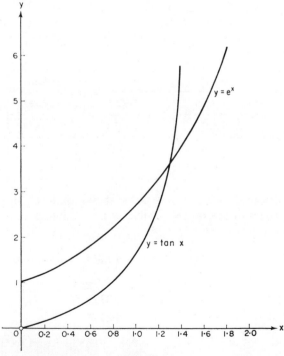

Fig. 162.

Further tabulation for a value of x in this range gives

x	1·305
e^x	3·688
x in degrees and minutes	74° 46′
$\tan x$	3·67.

The intersection of the graphs is therefore in the range

$$1·305 < x < 1·310.$$

Hence the real root of $e^x = \tan x$ in the range $0 < x < \frac{1}{2}\pi$ is $x = 1·31$ correct to 3 significant figures.

Exercises 18.1

Obtain approximate values of the real roots correct to two decimal places of the equations in the following questions:

1. $x^3 + 2x - 1 = 0$.   2. $x^3 + x^2 - 3 = 0$.

3. $\sin\theta = 1 - \frac{1}{2}\theta$.   4. $e^x - 1/x = 0$.

5. $\cosh x = 2x$.

18.2 The number of real roots of an equation

We shall assume that a polynomial equation of the nth degree has n roots, real or complex.

If $f(x)$ is a continuous function of x for $a \leqq x \leqq b$, and $f(a)$ and $f(b)$ have opposite signs, it follows from graphical considerations, illustrated by Fig. 163, that there is an odd number of roots of $f(x) = 0$ for

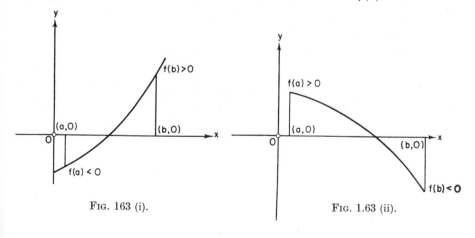

FIG. 163 (i).   FIG. 1.63 (ii).

$a < x < b$. In a case in which it is possible to deduce the fact that there is only one root within the chosen range we are able to use this idea to obtain the approximate location of a zero of the function $f(x)$.

The following additional rules, which we state without proof, provide help in the process of deducing the total number, and in particular cases the signs, of the real roots of an equation $f(x) = 0$ where $f(x)$ is a polynomial in x.

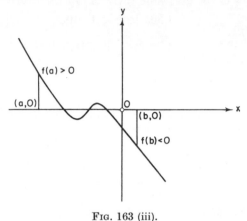

Fig. 163 (iii).

1. If $f(a)$ and $f(b)$ have like signs, there is an even number of roots of $f(x) = 0$ in the range $a < x < b$ or there are no roots.

2. If $f(x) \equiv x^n + a_1 x^{n-1} + a_2 x^{n-2} + \ldots + a_n = 0$, then (i) if n is odd, the equation has at least one real root, or (ii) if n is even and a_n is negative, the equation has at least one positive root and at least one negative root.

3. Each case in which *successive* coefficients of powers of x are of a different sign is called a *change of sign*. Thus in

$$x^3 - 3x^2 + x + 1 = 0$$

there are two changes of sign, from the coefficient of x^3 to the coefficient of x^2 and from the coefficient of x^2 to the coefficient of x. *Descartes' Rule of Signs* states that $f(x) = 0$ cannot have more positive roots than $f(x)$ has changes of sign, or more negative roots than $f(-x)$ has changes of sign.

It is important to notice that this rule does not find the actual number of real roots of an equation of this kind but finds only an upper bound to the number of roots.

Examples. (i) Show that, if q is positive, $x^3 + qx + r = 0$ has only one real root.

If r is positive, $x^3 + qx + r$ has no sign changes, and so there is no positive root. Also

$$(-x)^3 + q(-x) + r \equiv -x^3 - qx + r$$

has one sign change. Hence there is at most one negative root. The equation must have at least one real root because the highest power of x is odd and so the equation has one (negative) real root and two complex roots.

If r is negative, $x^3 + qx + r$ has one sign change, and $(-x)^3 + q(-x) + r$ has no sign changes. Hence, there are no negative roots and, at most, one positive root and so the equation has one real (positive) root and two complex roots.

(ii) Show that $x^5 - 3x^2 + x - 1 = 0$ has no negative roots and has at least two complex roots.

Show further that the equation has a real root between $x = 1$ and $x = 2$.

There are three changes of sign in the coefficients of

$$x^5 - 3x^2 + x - 1$$

and no changes of sign in the coefficients of

$$(-x)^5 - 3(-x)^2 + (-x) - 1 \equiv -x^5 - 3x^2 - x - 1.$$

Hence the equation has no more than three positive roots, no negative roots, and at least two complex roots. Also $f(1)$ is negative and $f(2)$ is positive, and therefore the equation has a real root between $x = 1$ and $x = 2$.

Exercises 18.2

1. Show that $x^3 + a = 0$ has two complex roots, whatever the sign of a.

2. Show that $x^5 + x^3 - 5 = 0$ has only one real root and find the two consecutive integers between which the root lies.

3. Show that $x^7 - 5x^6 + x^4 + x^2 + 1 = 0$ has at least four complex roots.

4. Show that $x^4 + 6x - 5 = 0$ has two and only two real roots and in each case find the consecutive integers between which the root lies.

5. Find the number of real roots of $x^3 - 3x^2 - 4 = 0$.

6. Find the number of real roots of $x^4 + 3x - 1 = 0$ and in each case find the consecutive integers between which the root lies.

18.3 The approximate value of a small root of a polynomial equation

A *small* root of a polynomial equation can be found by the method illustrated below.

Example. $x^3 + 5x^2 + 5x - 1 = 0$.

This equation has a small root which is given approximately by $5x - 1 = 0$, i.e., $x \doteqdot 0.2$, since for small values of x the terms x^3 and $5x^2$ are small compared with the other terms.

Writing the equation as

$$5x = 1 - 5x^2 - x^3 \tag{1}$$

we have, for small values of x, [putting $x = 0.2$ on the r.h. side of (1)]

$$5x \doteqdot 1 - 5(0.04) = 1 - 0.2 = 0.8 .$$

$$\therefore x \doteqdot 0.16 .$$

Using this approximation in the r.h. side of (1) gives

$$5x \doteqdot 1 - 5(0.0256) - (0.0041)$$

$$= 1 - 0.128 - 0.004 = 0.868 .$$

$$\therefore x \doteqdot 0.174 .$$

Similarly a further approximation gives

$$5x \doteqdot 1 - 5(0.0303) - 0.0053 = 0.843 .$$

$$\therefore x \doteqdot 0.169 .$$

The solution is $x = 0.17$ correct to 2 decimal places. (A check would show that $x^3 + 5x^2 + 5x - 1$ is negative when $x = 0.165$ and positive when $x = 0.175$.)

This process of successive approximation, whereby each approximation is used as the basis of the next approximation, is called the process of *iteration*. The method, for example, of obtaining square roots by the standard arithmetical process is an iterative one.

With this method a root which is not small could be found by transforming the equation, as shown in the example which follows.

Example. Find, correct to 3 significant figures, the real root of the equation

$$x^3 + x - 11 = 0 .$$

The equation has only one real root and this root is in the range $2 < x < 3$.

Put $x = y + 2$. Then

$$(y + 2)^3 + (y + 2) - 11 = 0,$$

i.e., $y^3 + 6y^2 + 13y - 1 = 0.$

A first approximation to the root of this equation is $y \doteqdot \dfrac{1}{13}$,

i.e., $y \doteqdot 0{\cdot}077.$

Using the same method as in the previous example,

$$y \doteqdot \frac{1}{13} \left\{1 - 6\,(0{\cdot}077)^2\right\} \doteqdot \frac{1}{13}\left\{1 - 6\,(0{\cdot}0059)\right\}$$

$$= \frac{1}{13}\,(1 - 0{\cdot}0354).$$

$$\therefore y \doteqdot 0{\cdot}074.$$

$$\therefore y \doteqdot \frac{1}{13}\left\{1 - 6\,(0{\cdot}074)^2 - (0{\cdot}074)^3\right\}$$

$$\doteqdot \frac{1}{13}\,(0{\cdot}967) \doteqdot 0{\cdot}074.$$

$$\therefore y \doteqdot 0{\cdot}07.$$

$$\therefore x \doteqdot 2{\cdot}07 \text{ is a root of } x^3 + x - 11 = 0.$$

Exercises 18.3

Use the method of this section to obtain the numerically small root of each of the equations in Questions 1–3 and in each case correct to 2 decimal places.

1. $x^3 + 4x^2 + 5x + 1 = 0.$ 2. $x^3 - x^2 + 6x - 1 = 0.$

3. $x^3 + x^2 + 10x - 2 = 0.$

4. Show that there is a real root of the equation

$$x^3 + 2x^2 + 5x - 27 = 0$$

between $x = 2$ and $x = 3$ and calculate its value correct to 3 significant figures.

5. Use the methods of this section to calculate $\sqrt{17}$ correct to 3 significant figures.

6. Use the methods of this section to calculate the cube root of 30 correct to 3 significant figures.

18.4 Newton's method for obtaining a closer approximation to a real root of an equation

If $x = a$ is an approximation to a root of the equation $f(x) = 0$, and if $f(a + h) = 0$ where h is small, then

$$f'(a) = \lim_{h \to 0} \left\{ \frac{f(a+h) - f(a)}{h} \right\}$$

$$\doteq \frac{f(a+h) - f(a)}{h}.$$

$$\therefore h \doteq \frac{f(a+h) - f(a)}{f'(a)}.$$

$$\therefore h \doteq \frac{-f(a)}{f'(a)}.$$

Hence, if $x = a$ is an approximation to the root of $f(x) = 0$, then, in general, a closer approximation to the root is

$$x = a - \frac{f(a)}{f'(a)}. \tag{18.1}$$

This is called *Newton's formula* for approximation to a root of the equation $f(x) = 0$. If $f'(a)$ is of the same order of smallness as $f(a)$, this generalization fails. Unless, however, $f'(a) = 0$ it is possible to choose h so that $f(a)$ is small compared with $f'(a)$ for a sufficiently small h.

Geometrical Interpretation. (Fig. 164). The point $P_1\{x_1, f(x_1)\}$ is on the curve $y = f(x)$. The curve cuts the x-axis at $B(b, 0)$ and $x_1 - b$ is small. The tangent to the curve at P_1 cuts the x-axis at T_2 where

$$OT_2 = x_2 \,(\text{say}) = OT_1 - T_1T_2 = x_1 - P_1T_1/\tan P_1T_2T_1$$

$$= x_1 - f(x_1)/f'(x_1).$$

Similarly, $OT_3 = x_3$ (say) $\quad = x_2 - f(x_2)/f'(x_2)$,

$\qquad OT_4 = x_4 \qquad\qquad = x_3 - f(x_3)/f'(x_3)$,

$$\begin{array}{ccc} \cdot & \cdot & \cdot \quad \cdot \quad \cdot \\ \cdot & \cdot & \cdot \quad \cdot \quad \cdot \\ \cdot & \cdot & \cdot \quad \cdot \quad \cdot \end{array}$$

$$OT_r = x_r \qquad\qquad = x_{r-1} - f(x_{r-1})/f'(x_{r-1}).$$

In the case illustrated in Fig. 164 the terms of the sequence x_1, x_2, x_3, ..., x_r approach the limit b as $r \to \infty$.

Fig. 165 shows the result of choosing P_1 so that $x_1 - b$ is not sufficiently small in a case in which $f'(a)$ is small.

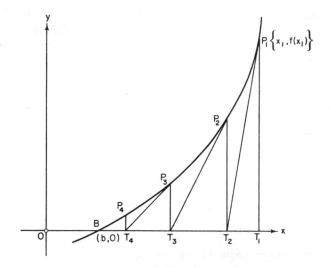

Fig. 164.

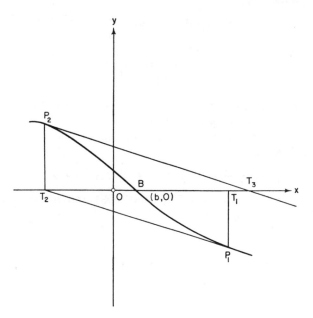

Fig. 165.

Examples. (i) Find the real roots of $x^4 - 12x + 7 = 0$.

(a) If $f(x) \equiv x^4 - 12x + 7$, there are two sign changes in $f(x)$ and no sign changes in $f(-x)$. There are at most two positive roots and no negative roots.

(b) $f(0) = 7, f(1) = -4, f(2) = -1, f(3) = 52$.
Hence there is a root between 0 and 1, and a root between 2 and 3 which is near to 2.

(c) $f'(x) = 4x^3 - 12$.

First approximation. Choose $x_1 = 0.6$ as a first approximation to the smaller root. This first approximation is obtained according to the principles discussed in § 18.3 by taking $12x \doteqdot 7$.

Second approximation.

$$f(0.6) = 0.1296 - 7.2 + 7 = -0.0704.$$

$$f'(0.6) = 4 \times 0.216 - 12 \doteqdot -11.$$

Hence a second approximation to the root is

$$0.6 - \frac{-0.0704}{-11} \doteqdot 0.594.$$

Third approximation.

$$f(0.594) \doteqdot 0.1246 - 7.128 + 7 = -0.0034.$$

Hence a third approximation is

$$0.594 - \frac{0.0034}{11} \doteqdot 0.594.$$

It is unnecessary when the variation in $f'(x)$ is small to make a new calculation at each approximation for $f'(x_r)$.

The final approximation is a check on the answer.

Final approximation.

$$f(0.5935) \doteqdot 0.1241 - 7.1220 + 7 > 0.$$

Hence the root lies between 0.5935 and 0.594 and so the root correct to 3 significant figures is 0.594.

Second root — first approximation.

We choose $x = 2$ as a first approximation to the second root.

Second approximation.

$$f(2) = -1, \quad f'(2) = 20.$$

Hence a second approximation is $2 - \dfrac{-1}{20} = 2.05$.

Third approximation.

$$f(2 \cdot 05) \fallingdotseq 17 \cdot 66 - 24 \cdot 6 + 7 = 0 \cdot 06.$$

Hence a better approximation is $2 \cdot 05 - \dfrac{0 \cdot 06}{20} \fallingdotseq 2 \cdot 047$.

Final approximation.

$$f(2 \cdot 045) \fallingdotseq 17 \cdot 48 - 24 \cdot 56 + 7 < 0$$

and since $f(2 \cdot 05) > 0$, the root is between $2 \cdot 045$ and $2 \cdot 05$. Thus the root correct to 3 significant figures is $2 \cdot 05$.

(ii) Show that the equation $x^3 + x^2 + 5 = 0$ has only one real root and determine this root correct to three significant figures. If the complex roots of this equation are $\alpha + i\beta$ and $\alpha - i\beta$, find α and β each correct to two places of decimals. (L.)

$$\text{Let } f(x) \equiv x^3 + x^2 + 5.$$

(a) There are no sign changes in $f(x)$ and therefore no positive roots of $f(x) = 0$ and only one sign change in $f(-x)$ and therefore at most one negative root. But the equation is a cubic and has at least one real root. Hence $f(x) = 0$ has one real (negative) root and two complex roots.

(b) *First approximation.*

$$f(-3) = -13, \quad f(-2) = 1.$$

Hence there is a root of $f(x) = 0$ which is near to $x = -2$. We therefore choose $x = -2$ as the first approximation to the real root.

Second approximation.

$$f'(x) = 3x^2 + 2x, \quad f'(-2) = 8.$$

Hence a second approximation to the root is $-2 - \dfrac{1}{8} = -2 \cdot 125$.

Third approximation.

$$f(-2 \cdot 125) \fallingdotseq -9 \cdot 592 + 4 \cdot 514 + 5 = -0 \cdot 078.$$

Hence a third approximation to the root is

$$-2 \cdot 125 + \frac{0 \cdot 078}{8} \fallingdotseq -2 \cdot 115.$$

Final approximation.

$$f(-2 \cdot 115) \fallingdotseq -9 \cdot 461 + 4 \cdot 473 + 5 = 0 \cdot 012.$$

Hence $f(x)$ changes sign between $x = -2 \cdot 115$ and $x = -2 \cdot 125$ and so the real root correct to 3 significant figures is $x = -2 \cdot 12$.

(c) The sum of the roots (real and complex)

$$= - \text{(the coefficient of } x^2) = 2\alpha - 2{\cdot}12 = -1.$$

$$\therefore \alpha \doteqdot 0{\cdot}56.$$

The product of the roots $= -$ (the absolute term)

$$= - 2{\cdot}12(\alpha^2 + \beta^2) = -5.$$

$$\therefore \alpha^2 + \beta^2 \doteqdot \frac{-5}{-2{\cdot}12} \doteqdot 2{\cdot}359.$$

$$\therefore \beta^2 \doteqdot 2{\cdot}359 - 0{\cdot}314 = 2{\cdot}045.$$

$$\therefore \beta \doteqdot 1{\cdot}43.$$

(iii) The acute angle θ satisfies the equation

$$\sin(2\theta + \alpha) = (\sqrt{3}) \cos(\theta - \alpha).$$

If α is zero, show that $\theta = \pi/3$. If α is so small that its square may be neglected and $\theta = (\pi/3) + \lambda$, prove that λ is approximately 4α. (N.)

If α is zero, $\sin 2\theta = \sqrt{3} \cos\theta$ so that

$$2 \sin\theta \cos\theta - \sqrt{3} \cos\theta = 0,$$

i.e., $$\cos\theta (2 \sin\theta - \sqrt{3}) = 0.$$

Rejecting the roots of this equation which correspond to non-acute angles θ we have

$$\sin\theta = \frac{\sqrt{3}}{2} \quad \text{whence} \quad \theta = \frac{1}{3}\pi.$$

$$\sin(2\theta + \alpha) = \sqrt{3} \cos(\theta - \alpha).$$

$$\therefore \sin 2\theta \cos\alpha + \cos 2\theta \sin\alpha = \sqrt{3} \cos\theta \cos\alpha + \sqrt{3} \sin\theta \sin\alpha.$$

If α is so small that its square may be neglected, $\sin\alpha \doteqdot \alpha$ and $\cos\alpha \doteqdot 1$ and hence

$$\sin 2\theta + \alpha \cos 2\theta \doteqdot \sqrt{3} \cos\theta + \sqrt{3}\alpha \sin\theta.$$

If $$f(\theta) \equiv \sin 2\theta + \alpha \cos 2\theta - \sqrt{3} \cos\theta - \sqrt{3}\alpha \sin\theta,$$

so that $f'(\theta) \equiv 2 \cos 2\theta - 2\alpha \sin 2\theta + \sqrt{3} \sin\theta - \sqrt{3}\alpha \cos\theta$, then $\theta = \frac{1}{3}\pi$ is near the root of $f(\theta) = 0$; But

$$f\left(\frac{1}{3}\pi\right) = \frac{1}{2}\sqrt{3} - \frac{1}{2}\alpha - \frac{1}{2}\sqrt{3} - \frac{3}{2}\alpha = -2\alpha,$$

$$f'\left(\frac{1}{3}\pi\right) = -1 - \alpha\sqrt{3} + \frac{3}{2} - \frac{1}{2}\alpha\sqrt{3} = \frac{1}{2} - \frac{3}{2}\sqrt{3}\alpha.$$

Hence $\theta = \dfrac{1}{3}\pi + \lambda$ is a better approximation to the root where

$$\lambda = \frac{-f\left(\dfrac{1}{3}\pi\right)}{f'\left(\dfrac{1}{3}\pi\right)}.$$

$$\therefore \lambda = \frac{+2\alpha}{+\dfrac{1}{2} - \dfrac{3}{2}\sqrt{3}\,\alpha} = 4\alpha\,(1 - 3\sqrt{3}\,\alpha)^{-1} \doteqdot 4\alpha$$

if terms involving powers of α higher than the first are neglected.

Exercises 18.4

In each of the Questions 1–8 calculate, correct to 2 decimal places, the specified root of the equation.

1. $x^3 + x - 3 = 0$; the real root.   2. $x^3 + 2x - 20 = 0$; the real root.

3. $x^4 - 2x - 1 = 0$; the positive root.   4. $x^3 - 3x + 1 = 0$; the smallest root.

5. $x^5 - 7 = 0$; the real root.   6. $x^5 - x - 1 = 0$; the root between 1 and 2.

7. $x^3 - 2x - 5 = 0$; the complex roots.   8. $x^3 + 2x^2 - 7x + 9 = 0$; the complex roots.

9. Show by means of a sketch graph that the equation $\tan x = 2x$ has three roots in the range $-\frac{1}{2}\pi < x < \frac{1}{2}\pi$ and find the non-zero roots correct to 3 significant figures.

10. Find the root of the equation $x e^x = 4$ correct to three significant figures.

11. Find the root of the equation $\log x + x = 2$ correct to three significant figures.

12. Find, by means of a sketch graph, the number of roots of the equation $10 \sin x = x$ and calculate the largest positive root.

13. Prove that if the cube and higher powers of α can be neglected, one root of $x^3 - \alpha x^2 + 1 = 0$ is approximately $-1 + \dfrac{1}{3}\alpha - \dfrac{1}{9}\alpha^2$.

14. If α is so small that its square and higher powers can be neglected, find approximate values for the roots of the equation

$$x(x - 1)(x - 2) - \alpha(x - 3) = 0.$$

15. If ε is so small that its square and higher powers can be neglected, find an approximation to the root of the equation $\tan x = 1 + \varepsilon x$ in the range $0 < x < \frac{1}{2}\pi$.

Miscellaneous Exercises XVIII

1. Draw in the same diagram the graphs of $x^2 - 3$ and of $-1/x$. Hence find the roots of the equation $x^3 - 3x + 1 = 0$ as accurately as your diagram permits. (O.C.)

2. Show that the equation

$$\log_e x + x^2 - 4x = 0$$

has a root lying between 3 and 4. By the use of Newton's method, or otherwise, calculate its value correct to two decimal places. (C.)

3. Draw in the same diagram the graphs of $\frac{1}{2}x^3$ and of $4 - 1/x$ from $x = -2$ to $x = 2$, taking 1 inch as the unit of length.

Hence find the roots of the equation

$$x^4 - 8x + 2 = 0$$

as accurately as your diagram permits. (O.C.)

4. Show that the equation

$$x^3 + 4x^2 - 8x + 2 = 0$$

has two positive roots and one negative root. State for each root the consecutive integers between which it lies and find the largest root correct to one decimal place. (N.)

5. Sketch the curve whose equation is

$$y = x^3 - 6x + 3.$$

Hence show that the equation $x^3 - 6x + 3 = 0$ has two positive roots and one negative root.

Use Newton's method to calculate the negative root correct to three significant figures. (N.)

6. Draw a sketch of the graph $y = e^x$, and find the equation of the tangent to the curve which passes through the origin.

Show by graphical considerations that the equation $e^x = cx$ has two real roots if $c > e$, but has no real roots if $0 < c < e$.

Taking the logarithm of each side of the equation $e^x = 4x$, verify that one root is approximately 2, and calculate this root correct to two decimal places. (N.)

7. Sketch the graph of the polynomial

$$(x - 1)^2(x - 4)$$

and calculate the coordinates of the turning points. Determine the number of real roots of the equation $(x - 1)^2(x - 4) = k$, when k has the following values: (i) 2, (ii) -2, (iii) -5. State in each case how many roots are positive.

Calculate the real root of the equation

$$(x - 1)^2(x - 4) = 60$$

correct to one decimal place. (N.)

8. Show that the roots of the quadratic equation in x,

$$(x - 2)(x - 6) - yx(x - 4) = 0$$

are real and unequal for all real values of y.

Draw the graph of

$$y = \frac{(x - 2)\,(x - 6)}{x\,(x - 4)},$$

and hence find approximately the three roots of the equation

$$(x - 2)(x - 6) + x^2(x - 4) = 0. \tag{N.}$$

9. Show, graphically or otherwise, that the equation $e^x = 2 - x$ has only one real root. Verify that the root is approximately equal to 0·4 and find its value correct to two decimal places. (L.)

10. Show that the equation $x^3 + 2x^2 + 5 = 0$ has only one real root, and that this root is negative.
Determine the root to three significant figures. (L.)

11. Show graphically or otherwise that the equation

$$x^3 + x^2 = 100$$

has only one real root.
Evaluate this root to two decimal places. (You must show some check on the accuracy of the figure in the second decimal place.)
Hence find the complex roots giving their real and imaginary parts each to one decimal place. (L.)

12. Find, graphically, an approximate root of the equation

$$\sin x = 1 - x,$$

and obtain the root correct to 3 decimal places. (L.)

13. Show that the equation $x^3 + 2x - 1 = 0$ has only one real root, and find it correct to two places of decimals. (L.)

14. Show that when λ is large and positive the equation $x^3 - \lambda x + 1 = 0$ has a small positive root given by $1/\lambda$ to a first approximation, and that a better approximation to this root is $\dfrac{1}{\lambda} + \dfrac{1}{\lambda^4}$.
Find the smaller positive root of $x^3 - 5x + 1 = 0$ correct to four decimal places. (L.)

15. Show, graphically, that the equation $\cosh x - 4 + 2x = 0$ has two real roots, one of which lies between 0 and 2. If this root is called a, verify that $a = 1·14$ approximately and obtain a better approximation. (L.)

16. In the same diagram sketch the graphs of the curve $y = \log(x - 1)$ and the line $y = x/m$. Hence show that there is a value of m for which the equation $x = m \log(x - 1)$ has two coincident real roots in x. Prove that this value of m satisfies the equation

$$1 + 1/m = \log m.$$

Show, further, that $m = 3·6$ is an approximation to the root of the equation for m, and obtain an improved approximation correct to three places of decimals. (L.)

17. Find the number of real roots of the equation

$$x^3 + 8x^2 - 32x + 14 = 0$$

and determine their signs.
Evaluate the smallest positive root *correct* to 3 decimal places. (L.)

18. By means of a sketch of the graph of $y = \tan x$, or otherwise, show that the equation $\tan x = kx$ (where k is a constant exceeding unity) has a root between 0 and $\pi/2$.

If $k = 1\cdot3$ find the value of this root correct to three significant figures, expressing the result in degrees.

When k is very large show, by substituting $x = \pi/2 - y$, or otherwise, that this root is approximately $\pi/2 - 2/(k\pi)$.   (L.)

19. If $4x^5 - 5x^4 + 1 = \alpha$ and α is small, show by Binomial expansion or otherwise, that the equation has two roots near unity or no real positive roots, according to whether α is positive or negative.

Determine the larger root correct to 3 significant figures when $\alpha = 0.1$.   (L.)

20. (a) Show that the equation

$$x^5 - 5x^3 + 15x - 4 = 0$$

has only one real root and that it is positive.

(b) Find the number of negative and of positive roots of the equation

$$x^5 - 5x^3 + 4 = 0,$$

and sketch the curve

$$y = x^5 - 5x^3 + 4.$$ (N.)

CHAPTER XIX

INEQUALITIES

19.1 Rules of manipulation

We have already in this book made use of the inequality signs

$$>, \text{ is greater than,}$$

and $\qquad <$, is less than.

We now consider the rules by which a statement of an inequality can be manipulated algebraically. Throughout this chapter letters used as symbols denote real numbers without other limitations unless it is stated otherwise.

(i) If $\qquad a > b,$

then $\qquad a + x > b + x.$

(ii) It follows from (i) that terms in an inequality can be transposed from one side of the inequality to the other according to the same rule that applies to equations, for if

$$a + b > c,$$

then $\qquad a + b - b > c - b,$

i.e., $\qquad a > c - b.$

This rule is implicit in the meaning of the symbol, $>$, for

$$a + b > c \quad \textit{means that} \quad a + b - c > 0$$

and $\qquad a + b - c > 0 \quad \textit{means that} \quad a > c - b.$

(iii) If an inequality is multiplied throughout by a positive number, the inequality is preserved. If an inequality is multiplied throughout by a negative number the inequality is reversed,

for, if x is positive and $a > b,$

$$x(a - b) > 0,$$

i.e., $\qquad xa - xb > 0,$

so that $\qquad xa > xb;$

if x is negative and $a > b$,

$$x(a - b) < 0,$$

i.e., $$xa - xb < 0,$$

so that $$xa < xb.$$

Important corollaries of this result are listed below

(a) If both sides of an inequality are positive and both sides are squared, the inequality is preserved. For example, $2 < 4 \to 4 < 16$.

(b) If both sides of an inequality are negative and both sides are squared, the inequality sign is reversed. For example,

$$-2 > -4 \to 4 < 16.$$

(c) If one side of an inequality is negative and one side is positive, *no further deduction can be drawn, with this information only, through the process of squaring both sides.* Thus, for example, if $a < b$, a is negative and b is positive,

then $a^2 > b^2$ if $|a| > |b|$,

but $a^2 < b^2$ if $|a| < |b|$,

and $a^2 = b^2$ if $|a| = |b|$.

Throughout this chapter the sign $\sqrt{}$ is to be taken to indicate *the positive square root.* With this definition, if a and b are positive and $a > b$,

then $$\sqrt{a} > \sqrt{b}.$$

It will be clear from the foregoing that very great care is needed if any of the processes of multiplication throughout by a factor, squaring or taking the square root are used with an inequality.

(iv) If corresponding sides of two inequalities are added, the inequality is preserved.

Thus $$a > b$$

and $$c > d,$$

imply $$a + c > b + d,$$

for $(a + c) - (b + d) = (a - b) + (c - d)$ which is positive. This is an example of the kind of irreversible statement which commonly occurs with inequalities. Thus

$$a + c > b + d$$

certainly does not imply $a > b$ and $c > d$. No valid deduction can be drawn by the process of subtracting corresponding sides of two inequalities.

(v) If corresponding sides of two inequalities are multiplied, deductions which can be validly drawn are limited by the same considerations as those discussed in (iii) above. Thus $a > b$ and $c > d$ only implies $ac > bd$ with certainty if all the quantities concerned are positive.

No valid deduction can be drawn by the process of dividing corresponding sides of two inequalities.

We have already discussed inequalities which are related to the sign of $Ax + By + C$ in Vol. I, § 5.6. The examples which follow illustrate some of the varied algebraic methods used for resolving problems concerning inequalities from first principles.

Examples. (i) Find the range of values of x for which

$$f(x) \equiv \frac{x(x-a)}{(x-b)(x-c)} > 0 \quad \text{where} \quad 0 < a < b < c.$$

Graphs of such functions as $f(x)$ were discussed in Volume I, Chapter VI. Sketch graphs of the functions concerned frequently provide good means of solving problems concerning inequalities. In this case the method which follows is also a useful one.

Consider the signs of the factors of the numerator and denominator in the ranges bounded by $x = 0, a, b, c$ respectively.

x	< 0	$0 < x < a$	$a < x < b$	$b < x < c$	$> c$
x	$-$	$+$	$+$	$+$	$+$
$x - a$	$-$	$-$	$+$	$+$	$+$
$x - b$	$-$	$-$	$-$	$+$	$+$
$x - c$	$-$	$-$	$-$	$-$	$+$
$f(x)$	$+$	$-$	$+$	$-$	$+$

As $x \to b_-$, $f(x) \to +\infty$ and as $x \to c_+$, $f(x) \to \infty$.

$$\therefore f(x) > 0 \quad \text{in the ranges} \quad x < 0, a < x < b, x > c.$$

[Note. $f(x)$ is undefined at the points $x = b$ and $x = c$.]

(ii) Find the range of values of x for which

$$\frac{x+1}{x-1} \geq 2.$$

This type of problem is best solved by transferring all the terms to one side of the inequality and resolving the expression on that side as far as possible into factors. Thus we require the range of values of x for which

$$\frac{x+1}{x-1} - 2 \geqq 0,$$

i.e. for which

$$\frac{3-x}{x-1} \geqq 0$$

and the required range is seen at once to be

$$3 \geqq x > 1.$$

[When $x = 1$, the given function is undefined.]

The student should note that multiplication of the original inequality throughout by $x - 1$ would have been multiplication by a quantity whose sign is ambiguous and *unless the two cases $x > 1$ and $x < 1$ were separately considered* no valid conclusions could have been drawn. The result could have been reached, however, if both sides of the inequality had been multiplied by $(x - 1)^2$ which, unless $x = 1$, is certainly positive.

We should then have $\qquad (x + 1)(x - 1) \geqq 2(x - 1)^2$

which reduces to $\qquad (x - 1)(x - 3) \leqq 0,$

giving the same result as before (after special consideration of the point $x = 1$).

(iii) Find the range of values of x for which

$$|x - 2| < 2x.$$

Fig. 166 shows that the graphs of $y = |x - 2|$ and $y = 2x$ intersect at the point of intersection of $y = 2x$ with $y = -x + 2$,

i.e., where $\qquad x = \dfrac{2}{3}.$ $\qquad$ Then, from the graph,

$$|x - 2| < 2x \quad \text{for} \quad x > \frac{2}{3}.$$

Otherwise. If $x < 0$, then $|x - 2|$ is certainly not less than $2x$. If $x > 0$, then

$$|x - 2| < 2x$$

implies $\qquad (x - 2)^2 < 4x^2,$

i.e., $\qquad x^2 - 4x + 4 < 4x^2,$

i.e., $\qquad 3x^2 + 4x - 4 > 0,$

i.e., $\qquad (3x - 2)(x + 2) > 0.$

$$\therefore x > \frac{2}{3} \text{ since } x \text{ must be positive.}$$

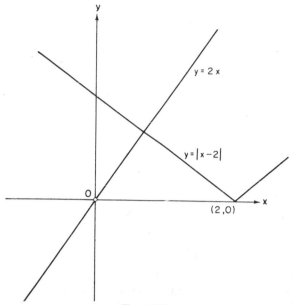

FIG. 166.

(iv) Find the range of values of x for which

$$|x - 4| + |2x - 1| > 4.$$

In cases such as this it is useful to consider, as follows, the separate cases which arise according as each of $x - 4$ and $2x - 1$ is positive or negative:

(a) If $x \geq 4$, the inequality requires

$$(x - 4) + (2x - 1) > 4 \quad \text{and also} \quad x \geq 4,$$

i.e.,
$$3x - 5 > 4 \quad \text{and also} \quad x \geq 4,$$

i.e.,
$$x > 3 \quad \text{and also} \quad x \geq 4.$$

Hence, if $x \geq 4$,
$$|x - 4| + |2x - 1| > 4.$$

(b) If $\dfrac{1}{2} \leq x < 4$, the inequality requires

$$4 - x + 2x - 1 > 4 \quad \text{and also} \quad \frac{1}{2} \leq x < 4,$$

i.e.,
$$x > 1 \quad \text{and also} \quad \frac{1}{2} \leq x < 4.$$

Hence, if $1 < x < 4$, $|x - 4| + |2x - 1| > 4$.

(c) If $x < \dfrac{1}{2}$, the inequality requires

$$4 - x + 1 - 2x > 4 \quad \text{and also} \quad x < \frac{1}{2},$$

i.e., $-3x > -1$ and also $x < \dfrac{1}{2}$,

i.e., $x < \dfrac{1}{3}$ and also $x < \dfrac{1}{2}$.

Hence, if $x < \dfrac{1}{3}$, $|x - 4| + |2x - 1| > 4$.

$\therefore |x - 4| + |2x - 1| > 4$ for the ranges of values $x < \dfrac{1}{3}$ and $x > 1$.

(v) Prove that $x^2 + y^2 - 4x \geqq 6y - 13$.

(a) In general, it is easier to prove that an expression is greater than zero than it is to prove directly that one expression is greater than another.

(b) If an algebraic expression, involving real numbers only, can be put into the form of the sum of squares it is necessarily positive or zero.

Here, $(x - 2)^2 + (y - 3)^2 \geqq 0$.

$$\therefore x^2 + y^2 - 4x - 6y + 13 \geqq 0.$$

$$\therefore x^2 + y^2 - 4x \geqq 6y - 13.$$

(vi) For what values of x is

$$0 \leqq \frac{x}{x - 1} \leqq 2 \, ?$$

If $f(x) \equiv \dfrac{x}{x - 1} \equiv 1 + \dfrac{1}{x - 1}$, the graph of $y = f(x)$, a rectangular hyperbola, of centre $(1, 1)$ and with asymptotes parallel to the coordinate axes, is shown in Fig. 167.

The line $y = 2$ intersects $y = f(x)$ at $(2, 2)$ and $y = 0$ intersects $y = f(x)$ at $(0, 0)$. Hence (from the graph) $0 \leqq \dfrac{x}{x - 1} \leqq 2$ in the ranges $x \leqq 0$ and $x \geqq 2$.

Otherwise, the example may be worked analytically thus:

Clearly $\dfrac{x}{x - 1} \geqq 0$

when $x > 1 \quad \text{or} \quad x \leqq 0$.   (1)

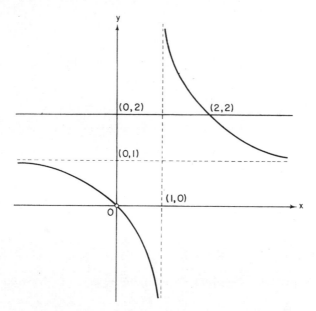

$(0,2)$ $(2,2)$

$(0,1)$

$(1,0)$

FIG. 167.

Also
$$\frac{x}{x-1} \leqq 2$$

when
$$\frac{x}{x-1} - 2 \leqq 0,$$

i.e., when
$$\frac{2-x}{x-1} \leqq 0,$$

i.e., when $x \geqq 2$ or $x < 1$.   (2)

The ranges of x common to (1) and (2) are $x \leqq 0$ and $x \geqq 2$.

(vii) (a) If $0 < x < 1$ and $0 < y < 1$, prove that
$$0 < x + y - xy < 1.$$

(b) Prove that, if $a < b < c$, there is no real value of x for which $\dfrac{1}{x-a} - \dfrac{1}{x-b} + \dfrac{1}{x-c}$ is zero.   (N.)

(a) Since $0 < x < 1$ and $0 < y < 1$,
$$\therefore x > xy.$$

Similarly $\qquad\qquad\qquad\qquad y > xy.$

$$\therefore x + y > 2xy.$$

$$\therefore x + y > xy \quad \text{since} \quad xy > 0.$$

$$\therefore x + y - xy > 0. \tag{1}$$

Also $\qquad\qquad x + y - xy - 1 = (x - 1)(1 - y).$

But $(x - 1)$ is negative and $(1 - y)$ is positive.

$$\therefore x + y - xy - 1 < 0.$$

$$\therefore x + y - xy < 1. \tag{2}$$

Combining results (1) and (2) gives

$$0 < x + y - xy < 1.$$

(b) $\dfrac{1}{x - a} - \dfrac{1}{x - b} + \dfrac{1}{x - c} = \dfrac{x^2 - 2bx + (bc + ab - ac)}{(x - a)(x - b)(x - c)}.$

But the equation $x^2 - 2bx + (bc + ab - ac) = 0$ has real roots only if

$$4b^2 - 4(bc + ab - ac) = 4(b - a)(b - c) \geqq 0.$$

And, since $a < b < c$, $4(b - a)(b - c) < 0$. Hence there are no real values of x for which $x^2 - 2bx + (bc + ab - ac)$ is zero, and so there are no real values of x for which

$$\frac{1}{x - a} - \frac{1}{x - b} + \frac{1}{x - c} \quad \text{is zero.}$$

Exercises 19.1

In each of the following questions, find the ranges of values of x within which the stated inequality obtains.

1. $\dfrac{2x - 4}{x - 1} < 1.$

2. $x - \dfrac{6}{x} < 1.$

3. $\dfrac{(x - 2)(x + 5)}{(x - 3)(x + 1)} < 0.$

4. $\dfrac{x - 1}{x^2 + x + 1} \leqq 0.$

5. $\dfrac{x - 1}{x + 1} \leqq x.$

6. $\dfrac{1}{x + 1} - \dfrac{1}{x + 2} < \dfrac{1}{x}.$

7. $\dfrac{x + a}{b} > \dfrac{a}{x + b}$

   (i) where a and b are each positive,

   (ii) when $a < b < 0$.

8. $|x + 1| > 1.$

9. $|2x + 3| - |x + 4| < 2.$

10. $\dfrac{|x|}{|x - 1|} \leqq 1.$

11. $|\sin x| \leqq \frac{1}{2}.$

12. $2 \geqq \dfrac{x - 1}{x + 1} \geqq 0.$

13. $\dfrac{2}{x - 1} < x < \dfrac{3}{x - 2}.$

14. $|\sin x + \cos x| \leqq 1.$

15. $x^2 < |x - 2|.$

In each of Questions 16–20, use sketch-graphs to find the ranges of values of x for which the stated inequality obtains.

16. $\left|\dfrac{1}{x}\right| \geqq x^2.$

17. $\sqrt{(8 - x^2)} \leqq |x|.$

18. $7 - x^2 \geqq \dfrac{6}{x}.$

19. $\dfrac{(x + 1)(x - 4)}{x(x - 7)} > x.$

20. $|\sec x| < 2 \tan x$, for values of x between 0 and 2π.

21. Prove that $x^2 + y^2 - 10x + 25 \geqq 0.$

22. Show that $x^2 + y^2 - xy \geqq 0$ for all real values of x and y. State the necessary condition for $x^2 + y^2 - xy$ to be equal to zero.

23. Show that $x^2 + y^2 + z^2 - xy - yz - zx \geqq 0$ for all real values of x, y and z.

24. Show that the roots of the quadratic equation in x,

$$(x - 2)(x - 6) - yx(x - 4) = 0,$$

are real and unequal for all real values of y.   (N.)

25. Show that if $x > 0$,

$$x + \frac{1}{x} \geqq 2.$$

26. Prove that if $a < x < y$ and $a > 0$, then $\dfrac{y}{y - a} < \dfrac{x}{x - a}.$

27. Prove that if $x + y - 3z = 0$, then $x^2 + y^2 - 3z^2 \geqq 0.$

28. Prove that if $x + y + z = a$ and $xy + xz + yz = 0$, then each of x, y and z lies between $-\dfrac{1}{3}a$ and a.

29. By considering the graphs of $x^2 + y^2 - x = 0$ and $x^2 + y^2 - 1 = 0$, or otherwise, prove that if $x^2 + y^2 < x$, then $x^2 + y^2 < 1$.

30. (a) Find the ranges of x for which

$$\left|\frac{x - 3}{x + 1}\right| < 2.$$

(b) If x, y and z are positive, show that

$$2(x^3 + y^3) \geqq (x^2 + y^2)(x + y)$$

and that

$$3(x^3 + y^3 + z^3) \geqq (x^2 + y^2 + z^2)(x + y + z).$$ (N.)

19.2 Fundamental inequalities

$$\because (a - b)^2 \geq 0,$$

$$\therefore a^2 - 2ab + b^2 \geq 0.$$

$$\therefore a^2 + b^2 \geq 2ab. \tag{19.1}$$

From (19.1) $$a^2 + b^2 + 2ab \geq 4ab.$$

$$\therefore (a + b)^2 \geq 4ab. \tag{19.2}$$

From (19.2), *if a and b are positive,*

$$a + b \geq 2 \sqrt{(ab)}.$$

$$\therefore \tfrac{1}{2}(a + b) \geq \sqrt{(ab)}, \tag{19.3}$$

the equality occurring when $a = b$.

The arithmetic mean of two positive numbers is greater than their geometric mean, unless the two numbers are equal, in which case the means are equal.

This last result (19.3) can be generalised in the form:—*The arithmetic mean of n positive numbers is greater than or equal to their geometric mean.*

There are several proofs of this important theorem of inequalities. We give here a proof by the methods of the calculus, methods which we shall discuss further in the next section.

If $f(x) = \log x - x + 1$, then $f'(x) = \dfrac{1}{x} - 1$. Hence $f'(x)$ is zero when $x = 1$, is positive for $0 < x < 1$, and is negative for $x > 1$ i.e., $f(x)$ has a maximum at $(1, 0)$.

$$\therefore \log x \leq x - 1 \text{ for all positive values of } x.$$

Then, if $A = \dfrac{1}{n} \sum_{r=1}^{n} a_r,$

$$\sum_{r=1}^{n} \log \left(\frac{a_r}{A} \right) \leq \sum_{r=1}^{n} \left(\frac{a_r}{A} \right) - n.$$

$$\therefore \log \left(\frac{a_1 a_2 a_3 \cdots a_n}{A^n} \right) \leq n - n = 0.$$

$$\therefore A^n \geq a_1 a_2 a_3 \cdots a_n.$$

$$\therefore A \geq \sqrt[n]{(a_1 a_2 a_3 \cdots a_n)},$$

i.e., $$\frac{a_1 + a_2 + a_3 + \cdots + a_n}{n} \geq \sqrt[n]{(a_1 a_2 a_3 \cdots a_n)}. \tag{19.4}$$

At this stage we make little use of the fundamental inequalities we have discussed here. We approach similar problems from first principles and by the same methods.

Examples. (i) Prove that $(a^2 + b^2)(a^4 + b^4) \geqq (a^3 + b^3)^2$.

$$(a^2 + b^2)(a^4 + b^4) = a^6 + b^6 + a^2 b^4 + b^2 a^4$$
$$= (a^3 + b^3)^2 - 2a^3 b^3 + a^2 b^4 + b^2 a^4$$
$$= (a^3 + b^3)^2 + a^2 b^2 (a - b)^2.$$

But $$a^2 b^2 (a - b)^2 \geqq 0.$$

$$\therefore \; (a^2 + b^2)(a^4 + b^4) \geqq (a^3 + b^3)^2.$$

(ii) If x and y are positive, prove that

$$\frac{1}{x^2} + \frac{1}{y^2} \geqq \frac{8}{(x + y)^2}.$$

$$\frac{1}{x^2} + \frac{1}{y^2} - \frac{8}{(x + y)^2} = \frac{x^4 + x^2 y^2 + 2 x^3 y + x^2 y^2 + y^4 + 2 x y^3 - 8 x^2 y^2}{(x + y)^2 x^2 y^2}$$

$$= \frac{x^4 + y^4 - 6 x^2 y^2 + 2 x^3 y + 2 x y^3}{(x + y)^2 x^2 y^2}$$

$$= \frac{(x^2 - y^2)^2 + 2 x y (x - y)^2}{(x + y)^2 x^2 y^2}$$

Each term of the numerator is positive or zero and the denominator is positive.

$$\therefore \; \frac{1}{x^2} + \frac{1}{y^2} - \frac{8}{(x + y)^2} \geqq 0.$$

$$\therefore \; \frac{1}{x^2} + \frac{1}{y^2} \geqq \frac{8}{(x + y)^2}.$$

(iii) Prove that

$$a^4 + b^4 + c^4 \geqq a^2 b^2 + b^2 c^2 + c^2 a^2 \geqq abc(a + b + c).$$

From (19.1)

$$a^4 + b^4 \geqq 2 a^2 b^2,$$
$$b^4 + c^4 \geqq 2 b^2 c^2,$$
$$c^4 + a^4 \geqq 2 c^2 a^2.$$

Adding

$$2(a^4 + b^4 + c^4) \geqq 2(a^2b^2 + b^2c^2 + c^2a^2).$$

$$\therefore\ a^4 + b^4 + c^4 \geqq a^2b^2 + b^2c^2 + c^2a^2.$$

Also from (19.1)

$$a^2(b^2 + c^2) \geqq 2a^2bc,$$

$$b^2(c^2 + a^2) \geqq 2b^2ac,$$

$$c^2(a^2 + b^2) \geqq 2c^2ab.$$

Adding

$$2\,\Sigma a^2b^2 \geqq 2abc\,\Sigma a.$$

$$\therefore\ \Sigma a^2b^2 \geqq abc\Sigma a.$$

Exercises 19.2

1. Prove that $2(a^2 + b^2) \geqq (a + b)^2$.

2. Prove that $3(a^2 + b^2 + c^2) \geqq (a + b + c)^2$.

3. Prove that $a^2 + b^2 + c^2 \geqq bc + ca + ab$.

4. Prove that, if a, b and c are positive,

$$(a + b + c)(a^2 + b^2 + c^2) \geqq 9abc.$$

5. Prove that $\left(\dfrac{a + b + c}{3}\right)^2 \leqq \dfrac{a^2 + b^2 + c^2}{3}$.

6. Prove that $a^3b + ab^3 \leqq a^4 + b^4$.

7. Prove that, if a, b and c are positive,

$$ab^2 + ac^2 + bc^2 + ba^2 + ca^2 + cb^2 \geqq 6abc.$$

8. Prove that

$$(x + y)(y + z)(z + x) \geqq 8xyz.$$

9. If x, y are positive, show that

$$\left(\frac{x^3 + y^3}{2}\right)^2 \geqq \left(\frac{x^2 + y^2}{2}\right)^3,$$

and that

$$\left(\frac{2x^3 + y^3}{3}\right)^2 \geqq \left(\frac{2x^2 + y^2}{3}\right)^3.$$

By the substitution $x^2 + y^2 = 2w^2$, or otherwise, show that, if x, y, z are positive,

$$\left(\frac{x^3 + y^3 + z^3}{3}\right)^2 \geqq \left(\frac{x^2 + y^2 + z^2}{3}\right)^3. \tag{N.}$$

10. If x and y are positive and unequal, prove that

$$x^4 + y^4 > x^3y + xy^3 > 2x^2y^2. \tag{N.}$$

19.3 The calculus applied to inequalities

We have already discussed in § 19.2 one case in which the methods of the calculus are used to obtain an inequality. The examples which follow are further illustrations of the use of these methods.

Examples. (i) Prove that $x > \dfrac{3 \sin x}{2 + \cos x}$ for all values of $x > 0$.

If $f(x) \equiv x - \dfrac{3 \sin x}{2 + \cos x}$, then

$$f'(x) \equiv 1 - \frac{3 \cos x (2 + \cos x) + \sin x (3 \sin x)}{(2 + \cos x)^2}$$

$$\equiv \frac{1 - 2 \cos x + \cos^2 x}{(2 + \cos x)^2}$$

$$\equiv \left(\frac{1 - \cos x}{2 + \cos x} \right)^2.$$

Hence $f'(x)$ is finite and ≥ 0 for all values of x. Thus, except at the points where $x = 2n\pi$, where $f'(x) = 0$, and which are points of inflexion, the function increases throughout the range.

But when $x = 0$, $f(x) = 0$.

$$\therefore \ f(x) > 0 \text{ for all values of } x > 0.$$

(ii) Without evaluating the integrals, prove that

$$\int_0^{\frac{1}{2}\pi} \sin^n x \, dx > \int_0^{\frac{1}{2}\pi} \sin^{n+1} x \, dx,$$

if n is positive.

In the range $0 \leq x \leq \tfrac{1}{2}\pi$

$$\sin^{n+1} x = \sin^n x \cdot \sin x \leq \sin^n x$$

because $\sin x \leq 1$. If n is positive $\sin^n(0) = \sin^{n+1}(0) = 0$ and $\sin^n(\tfrac{1}{2}\pi) = \sin^{n+1}(\tfrac{1}{2}\pi) = 1$. The graphs of $\sin^n x$ and $\sin^{n+1} x$ are therefore as shown in Fig. 168 and the area enclosed by the curve $y = \sin^n x$, the x-axis and the ordinate $x = \tfrac{1}{2}\pi$ is greater than the area enclosed by the curve $y = \sin^{n+1} x$, the x-axis and the ordinate $x = \tfrac{1}{2}\pi$.

$$\therefore \int_0^{\frac{1}{2}\pi} \sin^n x \, dx > \int_0^{\frac{1}{2}\pi} \sin^{n+1} x \, dx,$$

if n is positive.

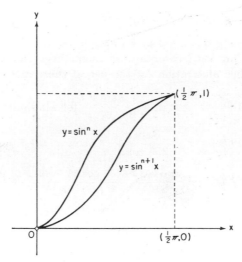

F<small>IG</small>. 168.

Exercises 19.3

1. If $y = \dfrac{e^{\frac{1}{2}x}}{3 + x^2}$, prove that $y > \dfrac{1}{3}$ for all positive values of x.

2. Prove that for $-2 < x < 0$, $x^2 e^x < 4/e^2$.

3. Prove that for $x > 0$, $x^3 - 2x^2 + x \geqq 0$.

4. Prove that if $x^2 + \dfrac{2}{x} \geqq 3$, $x \leqq -2$ or $x > 0$.

5. If a and b are positive, prove that $\dfrac{a + b \sinh x}{a + b \cosh x} > \dfrac{a}{a + b}$ for all positive values of x and also prove that $\dfrac{a + b \sinh x}{a + b \cosh x} > -1$ for all values of x.

6. Prove that $4 \left(\theta - 2 \sin \frac{1}{2}\theta\right) - \theta + \sin \theta$ is positive when $\theta > 0$.   (O.C.)

7. If $y = x e^{x - x^2}$, prove that y increases as x increases from $-\frac{1}{2}$ to 1.   (N.)

8. If $x > 0$, show that $\sin x < x < \sinh x$.

If $V = \sin^2 x \cosh^2 y + \cos^2 x \sinh^2 y$ and $x = a \cos t$, $y = a \sin t$, show that

$$\frac{dV}{dt} = x \sinh 2y - y \sin 2x,$$

and deduce that, as t increases from 0 to $\frac{1}{2}\pi$, V increases steadily from $\sin^2 a$ to $\sinh^2 a$.   (N.)

9. Prove that $\displaystyle\int_0^1 \frac{\sqrt{(1 - x^2)}}{1 + x^2} \, dx = \frac{1}{2}\pi \left(\sqrt{2} - 1\right)$.

Hence prove that

$$\tfrac{1}{2}\pi\,(\sqrt{2}-1) < \int_0^1 \frac{\sqrt{(1-x^2)}}{1+x^4}\,dx < \frac{1}{4}\pi.\qquad\text{(N.)}$$

10. Find the equation of the tangent to the curve $y = 1/x$ at the point $(1, 1)$ and the equation of the tangent to the curve $y = \cos x$ at the point $(\tfrac{1}{2}\pi, 0)$. Deduce that $1/x > \cos x$ for $0 \leq x \leq \tfrac{1}{2}\pi$. (N.)

Miscellaneous Exercises XIX

1. If $x + y < 2$, $x - y < 4$ and $2x + y > 2$, show that $0 < x < 3$ and find a similar inequality for y. (N.)
(See Volume I § 5.6.)

2. (a) Show that for real values of α, β and x the value of the function

$$\frac{x^2 - \alpha\beta}{2x - \alpha - \beta}$$

cannot lie between α and β.

(b) If x is positive, prove that

$$x \geq 1 + \log_e x.\qquad\text{(N.)}$$

3. Show that $\sinh x > \operatorname{sech} x$ if $x > \tfrac{1}{2}\log_e (2 + \sqrt{5})$. (N.)

4. Prove that if a, b, c, d are positive constants such that $bc > ad$, the function

$$\frac{a\cos x + b\sin x}{c\cos x + d\sin x}$$

increases throughout the range $x = 0$ to $x = \tfrac{1}{2}\pi$. (L.)

5. Prove that, for all values of a and b,

$$\sin a \sin b \leq \sin^2 \tfrac{1}{2}(a + b).$$

Show further that, if a, b, c and d all lie between 0 and π, then

$$\sin a \sin b \sin c \sin d \leq \left\{\sin\frac{1}{4}(a + b + c + d)\right\}^4$$

and, by writing $d = \dfrac{1}{3}(a + b + c)$, deduce that

$$\sin a \sin b \sin c \leq \left\{\sin\frac{1}{3}(a + b + c)\right\}^3.$$

Show precisely where the restrictions on a, b, c and d are used in the proof of the last two inequalities. (C.)

6. (a) If $pq > 0$, prove that $p/q > 0$. Find

(i) the range of values of x in which $\dfrac{1 - 4x}{2x - 3} > 0$,

(ii) the ranges of values of x in which

$$\frac{2x}{x - 1} + \frac{x - 5}{x - 2} > 3.$$

(b) In one diagram sketch the three straight lines

$$x + y - 3 = 0, \; y - \tfrac{1}{2}x - 5 = 0, \; y - 3x + 10 = 0,$$

and shade the region in which the following three inequalities are all satisfied:

$$x + y - 3 > 0, \; y - \tfrac{1}{2}x - 5 > 0, \; y - 3x + 10 > 0.$$

In a second diagram shade the region in which none of them is satisfied.   (N.)

7. Find which term is greatest in the expansion of $(x + y)^n$ when $n = 16$, $x = 2$, $y = \tfrac{1}{2}$; calculate the value of this term.   (N.)

8. Prove, by differentiation or otherwise, that if x is positive

(i) $\log_e (1 + x) > x - \tfrac{1}{2}x^2$,

(ii) $x > \dfrac{5 \sin x}{4 + \cos x}$.   (N.)

9. If a, b are positive numbers, prove that

$$(1 - a)(1 - b) > 1 - a - b.$$

Deduce, or prove otherwise, that if a, b, c are positive, then

$$(1 - a)(1 - b)(1 - c) > 1 - a - b - c$$

when one at least of a, b, c is less than unity.   (O.C.)

10. (a) Show that $x(x - 1)(x - 2)(x - 3)$ differs by a constant from a perfect square. Determine the range of values of x for which

$$64x(x - 1)(x - 2)(x - 3) < 17.$$

(b) Sketch the graph of the function $|x|/(1 + x)$.   (N.)

11. Draw the graph

$$y = x^3 - 3x^2 + 5$$

as accurately as you can between the limits $x = -2$, $x = 4$, taking 1 in. (or 2 cm.) as the unit for x and $\dfrac{1}{5}$ in. (or $\tfrac{1}{2}$ cm.) as the unit for y. Indicate clearly the turning points and the point of inflexion.

By means of your graph, or otherwise, determine the values of x for which

$$x^3 - 3x^2 - x + 3 \geqq 0.$$

12. Find the gradient of the curve described by the point

$$x = 1 + t^2, \; y = t(3 + t^2)$$

as t varies.

Also find the value of $\dfrac{d^2 y}{dx^2}$ for this curve, and deduce that the product $\dfrac{dy}{dx} \cdot \dfrac{d^2 y}{dx^2}$ is positive at any point of the curve at which $x > 2$.   (N.)

13. (a) If x, y are real numbers satisfying the condition $x^2 + y^2 \leqq 2$, prove that $|x + y| \leqq 2$.

(b) Indicate in a sketch the shape of the graph of

$$y = \frac{2x - 1}{x + 1}.$$

Calculate the range or ranges of values of x for which

$$\left| \frac{2x - 1}{x + 1} - 2 \right| < 1,$$

and show that the result is confirmed by the graph.   (N.)

14. (a) Prove that

(i) $a(1 - a) \leq \dfrac{1}{4}$ for all values of a,

(ii) if $b \leq c \leq 1$, then $b(1 - c) \leq \dfrac{1}{4}$.

(b) Prove that $xe^{1-x} \leq 1$ for all values of x.   (N.)

15. (a) By considering separately the three ranges

$$x < 0, \ 0 < x < 1, \ 1 < x,$$

sketch the graph of $y = |2x| - |1 - x|$.

(b) By considering separately the signs of $x - 1$ and $x - y$ in the four regions into which the plane is divided by the straight lines $x - 1 = 0$ and $x - y = 0$, sketch the graph of

$$2y = |x - 1| + |x - y|.$$ (N.)

16. Show that the function

$$px + \cos^{-1}(p \sin x), \ 0 < p < 1, \ 0 \leq x \leq \tfrac{1}{2}\pi$$

increases as x increases throughout its range, p being constant.

Find the greatest value of the function if x and p can each assume any value in their respective ranges.   (L.)

17. If a, b, c are three different real numbers, show that the roots of the equation

$$(x - b)(x - c) + (x - c)(x - a) + (x - a)(x - b) = 0$$

are real and unequal.   (L.)

18. Prove that the value of

$$3 \sin x - \sin 3x$$

steadily increases as x increases from $-\tfrac{1}{2}\pi$ to $\tfrac{1}{2}\pi$.

Deduce that $\tan 3x \cot x$ steadily increases as x increases from 0 to $\pi/6$.

19. Prove that, for $x > 0$, each of the functions

$$\log\left(1 + \frac{1}{x}\right) - \frac{2}{2x + 1}$$

and

$$\frac{1}{\sqrt{\{x(x + 1)\}}} - \log\left(1 + \frac{1}{x}\right)$$

decreases as x increases. To what limit do these functions tend as $x \to \infty$?

L

Deduce that $\dfrac{2}{2x+1} < \log\left(1 + \dfrac{1}{x}\right) < \dfrac{1}{\sqrt{\{x\,(x+1)\}}}$ for $x > 0$.   (N.)

20. (a) Find the ranges of values of θ between $0°$ and $360°$ for which

$$\cos 2\theta > \cos 4\theta.$$

(b) If α, β, γ, δ are angles, each of which lies between $-\tfrac{1}{2}\pi$ and $+\tfrac{1}{2}\pi$, prove that,

$$\text{(i)}\quad \cos\alpha + \cos\beta \leqq 2\cos\tfrac{1}{2}(\alpha + \beta),$$

$$\text{(ii)}\quad \cos\alpha + \cos\beta + \cos\gamma + \cos\delta \leqq 4\cos\left(\frac{\alpha + \beta + \gamma + \delta}{4}\right).$$

21. (a) Find what restrictions must be imposed on the values of x and y in order to satisfy both the inequalities

$$x > y, \quad \frac{x}{x+1} > \frac{y}{y+1}.$$

(b) Show that $b^2 + c^2 \geqq \tfrac{1}{2}(b + c)^2$, and hence find the range of values of a for which the simultaneous equations

$$a + b + c = 1,$$
$$a^2 + b^2 + c^2 = 3$$

may be satisfied by real values of b and c.   (N.)

22. (a) If $la + mb + n > 0$, show that the point (a, b) lies on the same side of the straight line

$$lx + my + n = 0$$

as the point $(x_0 + l, y_0 + m)$, where (x_0, y_0) is any point on the line.

(b) Show that the condition that the quadratic function

$$at^2 + bt + 1$$

shall be positive when $t = 1$, negative when $t = 2$, and positive when $t = 3$, is that the point (a, b), referred to a pair of coordinate axes, should belong to a certain region in the plane of the axes. Draw a diagram showing the region.   (N.)

23. (a) Find the least positive integer n such that

$$\frac{54 + n}{77} > \frac{85}{52 + n}.$$

(b) If the value of the fraction

$$\frac{x\,(2 + x)}{2x + 1}$$

lies within the range from 0 to 1, find the ranges within one of which x must lie.   (N.)

24. If A, B, C, D are positive angles not greater than two right angles, prove that
$$\operatorname{Sin} A + \operatorname{Sin} B < 2 \operatorname{Sin} \frac{1}{2}(A + B),$$

$$\operatorname{Sin} A + \operatorname{Sin} B + \operatorname{Sin} C + \operatorname{Sin} D < 4 \operatorname{Sin} \frac{1}{4}(A + B + C + D).$$

By taking D to be $\frac{1}{3}(A + B + C)$, prove that

$$\operatorname{Sin} A + \operatorname{Sin} B + \operatorname{Sin} C < 3 \operatorname{Sin} \frac{1}{3}(A + B + C). \tag{O.C.}$$

25. If h and k are positive constants and x is a real variable, prove that the expression

$$\frac{x - h}{x^2 - k^2}$$

can assume any real value if $k > h$, and that there will be two numbers between which it cannot lie if $h > k$; find the difference between these two numbers.

Prove also that if a line parallel to the x-axis cuts the graph of

$$y = \frac{x - h}{x^2 - k^2}$$

where $x = a$ and $x = b$, then

$$(a - h)(b - h) = h^2 - k^2. \tag{N.}$$

CHAPTER XX

COORDINATE GEOMETRY

Throughout the two volumes of this book we have discussed topics of coordinate geometry in relation to other branches of mathematics. In this chapter we summarize the results that have so far been obtained, add other results, and discuss other methods which are necessary to complete an elementary treatment of the subject.

20.1 The straight line

1. The length of the line joining the points (x_1, y_1), (x_2, y_2) is $\sqrt{\{(x_1 - x_2)^2 + (y_1 - y_2)^2\}}$.

2. The coordinates of the point which divides the line joining (x_1, y_1), (x_2, y_2) in the ratio $l : m$ are $\left(\dfrac{lx_2 + mx_1}{l + m}, \dfrac{ly_2 + my_1}{l + m} \right)$. By convention, division in the ratio $a : -b$, where a and b are positive numbers, indicates external division in the ratio $a : b$.

3. Standard forms of the equation of a straight line are as follow:

(i) $Ax + By + C = 0$, the general equation.

(ii) $y = mx + c$, the equation of the line of gradient m which intercepts the y-axis at $(0, c)$. This excludes the case $x = $ constant.

(iii) $(y - y_1) = m(x - x_1)$, the line through (x_1, y_1) with gradient m.

(iv)
$$\begin{vmatrix} x & y & 1 \\ x_1 & y_1 & 1 \\ x_2 & y_2 & 1 \end{vmatrix} = 0 \quad \text{or} \quad \frac{x - x_1}{x_2 - x_1} = \frac{y - y_1}{y_2 - y_1},$$

the line through (x_1, y_1), (x_2, y_2).

(v) $x/a + y/b = 1$, the line which makes intercepts a and b on Ox, Oy respectively.

(vi) $x \cos\alpha + y \sin\alpha = p$, the *perpendicular* form of the equation of a straight line; the length of the perpendicular from the origin on to the line is p, defined as positive, and the anti-clockwise angle which this perpendicular makes with the x-axis is α.

4. The acute angle between the lines $y = m_1 x + c_1$ and $y = m_2 x + c_2$ is $\tan^{-1} \left| \dfrac{m_1 - m_2}{1 + m_1 m_2} \right|$ unless $m_1 = m_2$, in which case the lines are parallel, or $m_1 m_2 = -1$, in which case they are perpendicular.

5. The function $Ax + By + C$ is positive for all points on one side of the line $Ax + By + C = 0$ and negative for all points on the other side of the line. The equation of the line can be written in the form in which $C > 0$ (unless $C = 0$). If this is done, the positive side of the line is defined as that side on which the origin lies.

6. The length of the perpendicular from (x_1, y_1) to the line $Ax + By + C = 0$ is $\dfrac{\pm (A x_1 + B y_1 + C)}{\sqrt{(A^2 + B^2)}}$. By convention the positive square root of $A^2 + B^2$ is taken and the perpendicular is defined as positive if it is drawn from a point on the positive side of the line and negative if it is drawn from a point on the negative side of the line.

7. The equations of the bisectors of the angles between

$$a_1 x + b_1 y + c_1 = 0 \quad \text{and} \quad a_2 x + b_2 y + c_2 = 0 \quad \text{are}$$

$$\frac{a_1 x + b_1 y + c_1}{\sqrt{(a_1^2 + b_1^2)}} = \pm \frac{a_2 x + b_2 y + c_2}{\sqrt{(a_2^2 + b_2^2)}}$$

8. The area of the triangle whose vertices are (x_1, y_1), (x_2, y_2), (x_3, y_3) is given by the modulus of

$$\tfrac{1}{2} \begin{vmatrix} x_1 & y_1 & 1 \\ x_2 & y_2 & 1 \\ x_3 & y_3 & 1 \end{vmatrix}$$

9. The lines
$$a_1 x + b_1 y + c_1 = 0,$$
$$a_2 x + b_2 y + c_2 = 0,$$
$$a_3 x + b_3 y + c_3 = 0$$

are concurrent if
$$\begin{vmatrix} a_1 & b_1 & c_1 \\ a_2 & b_2 & c_2 \\ a_3 & b_3 & c_3 \end{vmatrix} = 0$$

provided that no two of the lines are parallel.

20.2 Line pairs

1. $ax^2 + 2hxy + by^2 = 0$, $h^2 \geqq ab$, represents a pair of lines through the origin.

2. The lines are at right angles if $a + b = 0$ and coincident if $h^2 = ab$.

3. In other cases the acute angle between the lines is given by

$$\tan^{-1} \left| \frac{2\sqrt{(h^2 - ab)}}{a + b} \right|.$$

4. The equation

$$ax^2 + 2hxy + by^2 + 2gx + 2fy + c = 0$$

represents a pair of straight lines if

$$\begin{vmatrix} a & h & g \\ h & b & f \\ g & f & c \end{vmatrix} = 0.$$

5. The equation of the lines joining the origin to the common points of $lx + my = 1$ and $ax^2 + 2hxy + by^2 + 2gx + 2fy + c = 0$ is $ax^2 + 2hxy + by^2 + (2gx + 2fy)(lx + my) + c(lx + my)^2 = 0$.

Example. The equation of the pair of lines which bisect the angles between the lines represented by $ax^2 + 2hxy + by^2 = 0$ is

$$hx^2 - (a - b)xy - hy^2 = 0.$$

For (Fig. 169) if the lines represented by $ax^2 + 2hxy + by^2 = 0$ are $y = x\tan\theta_1$ and $y = x\tan\theta_2$, if $P(X, Y)$ is a point on one of the bisectors of the angles between them, and if $\angle POx = \theta$, then

$$\theta = \tfrac{1}{2}(\theta_1 + \theta_2) \quad \text{or} \quad \theta = \tfrac{1}{2}(\pi + \theta_1 + \theta_2)$$

and in either case $\qquad \tan 2\theta = \tan(\theta_1 + \theta_2)$,

i.e., $\qquad \dfrac{2\tan\theta}{1 - \tan^2\theta} = \dfrac{\tan\theta_1 + \tan\theta_2}{1 - \tan\theta_1 \tan\theta_2}.$

But $\tan\theta_1 + \tan\theta_2 = -\dfrac{2h}{b}$, $\tan\theta_1 \tan\theta_2 = \dfrac{a}{b}$ and $\tan\theta = \dfrac{Y}{X}$.

$$\therefore \quad \frac{\dfrac{2Y}{X}}{1 - \dfrac{Y^2}{X^2}} = \frac{-\dfrac{2h}{b}}{1 - \dfrac{a}{b}},$$

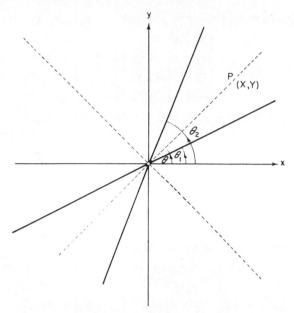

Fig. 169.

i.e.,
$$\frac{XY}{X^2 - Y^2} = \frac{h}{a - b}$$

or
$$h X^2 - (a - b) X Y - h Y^2 = 0.$$

It follows that the equation of the line pair which bisects the angles between the given lines is $h x^2 - (a - b) xy - h y^2 = 0$, as stated above.

20.3 The circle

1. The equation of the circle of radius a and centre at the origin is $x^2 + y^2 = a^2$.

2. The equation of the circle of radius a and centre (α, β) is $(x - \alpha)^2 + (y - \beta)^2 = a^2$.

3. The general equation of the circle is
$$x^2 + y^2 + 2gx + 2fy + c = 0;$$
the centre of this circle is $(-g, -f)$ and the radius is $\sqrt{(g^2 + f^2 - c)}$.

4. The equation of the circle through (x_1, y_1), (x_2, y_2), (x_3, y_3) is

$$\begin{vmatrix} x^2 + y^2 & x & y & 1 \\ x_1^2 + y_1^2 & x_1 & y_1 & 1 \\ x_2^2 + y_2^2 & x_2 & y_2 & 1 \\ x_3^2 + y_3^2 & x_3 & y_3 & 1 \end{vmatrix} = 0.$$

5. The equation of the circle with the line joining $A(x_1, y_1)$ to $B(x_2, y_2)$ as diameter is

$$(x - x_1)(x - x_2) + (y - y_1)(y - y_2) = 0,$$

for, if $P(x, y)$ is a point on the circle, PA and PB are perpendicular.

$$\therefore \text{ gradient } PA \times \text{ gradient } PB = -1.$$

$$\therefore \frac{y - y_1}{x - x_1} \times \frac{y - y_2}{x - x_2} = -1.$$

giving the stated equation of the circle.

6. The equation of the tangent at the point (x_1, y_1), on the circle $x^2 + y^2 + 2gx + 2fy + c = 0$, is

$$xx_1 + yy_1 + g(x + x_1) + f(y + y_1) + c = 0.$$

7. The equation $y = mx \pm a\sqrt{(1 + m^2)}$ is the equation of a tangent to the circle $x^2 + y^2 = a^2$ for all values of m. There is no simple form of the corresponding equation for the general circle. Usually, the condition that a line is a tangent to a circle is best obtained from the property which states that the perpendicular distance from the centre to the tangent is equal to the radius of the circle.

8. The length of the tangent from (x_1, y_1) to the circle $x^2 + y^2 + 2gx + 2fy + c = 0$ is $\sqrt{(x_1^2 + y_1^2 + 2gx_1 + 2fy_1 + c)}$.

9. The equation of the common chord of two intersecting circles

$$S_1 \equiv x^2 + y^2 + 2g_1x + 2f_1y + c_1 = 0,$$
$$S_2 \equiv x^2 + y^2 + 2g_2x + 2f_2y + c_2 = 0$$

is $S_1 - S_2 = 0$, and the equation of any circle through the intersection of $S_1 = 0$ and $S_2 = 0$ can be expressed in the form

$$S_1 + \lambda S_2 = 0.$$

10. $$x^2 + y^2 + 2g_1x + 2f_1y + c_1 = 0,$$
$$x^2 + y^2 + 2g_2x + 2f_2y + c_2 = 0$$

are orthogonal if

$$2g_1g_2 + 2f_1f_2 = c_1 + c_2.$$

20.4 The radical axis

The Power of a point. If a straight line drawn through $T(x_1, y_1)$ cuts the circle

$$S \equiv x^2 + y^2 + 2gx + 2fy + c = 0 \text{ at } A, B,$$

then $\quad TA \cdot TB = TC^2 - r^2 = x_1^2 + y_1^2 + 2gx_1 + 2fy_1 + c,$

where C is the centre and r is the radius of $S = 0$ and where TA, TB are the *directed* lengths of the displacements from T to A and from T to B respectively.

In Figs. 170 (i), (ii), if $\overrightarrow{AB}$ makes an angle θ with Ox, $TA = r_1$ and $TB = r_2$ where r_1 and r_2 are directed lengths and the direction AB is positive, then A is

$$(x_1 + r_1 \cos\theta, \, y_1 + r_1 \sin\theta) \text{ and } B \text{ is } (x_1 + r_2 \cos\theta, \, y_1 + r_2 \sin\theta).$$

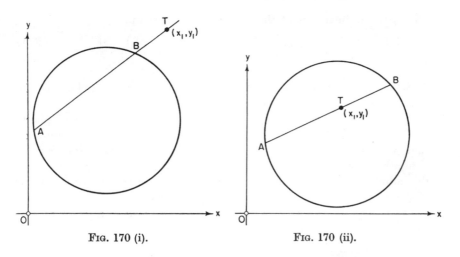

FIG. 170 (i).   FIG. 170 (ii).

Because A and B are points on the circle $S = 0$, r_1 and r_2 are the roots of

$$(x_1 + r \cos\theta)^2 + (y_1 + r \sin\theta)^2 + 2g(x_1 + r \cos\theta)$$
$$+ 2f(y_1 + r \sin\theta) + c = 0.$$

$$\therefore \; r_1 r_2 = (x_1^2 + y_1^2 + 2gx_1 + 2fy_1 + c)/(\cos^2\theta + \sin^2\theta),$$

i.e., $\qquad\qquad r_1 r_2 = x_1^2 + y_1^2 + 2gx_1 + 2fy_1 + c.$

The quantity $TC^2 - r^2$ is defined as the *power* of T with respect to the circle. If T is outside the circle, Fig. 170 (i), the power of T is equal to the square on the tangent drawn from T to the circle. If T is on the circle the power of T is zero. If T is inside the circle, Fig. 170 (ii), the power of T is negative and is numerically equal to the product of the segments of the chords of the circle drawn through T.

The *Radical Axis* of two circles is defined as the locus of points the powers of which, with respect to the two circles, are equal.

If

$$S_1 \equiv x^2 + y^2 + 2g_1x + 2f_1y + c_1 = 0,$$

$$S_2 \equiv x^2 + y^2 + 2g_2x + 2f_2y + c_2 = 0$$

and if $P(h, k)$ is a point at which the powers with respect to S_1 and S_2 are equal, then

$$h^2 + k^2 + 2g_1h + 2f_1k + c_1 = h^2 + k^2 + 2g_2h + 2f_2k + c_2,$$

i.e., $$2(g_1 - g_2)h + 2(f_1 - f_2)k + (c_1 - c_2) = 0.$$

Hence the radical axis of the two circles is the line

$$2(g_1 - g_2)x + 2(f_1 - f_2)y + (c_1 - c_2) = 0 \qquad (20.1)$$

or $$S_1 - S_2 = 0.$$

If S_1 and S_2 intersect, this is the equation of the common chord of the two circles.

If S_1 and S_2 do not intersect, the radical axis is the locus of points from which tangents to the two circles are equal. If the circles intersect, that part of the radical axis which is outside the circles is the locus of points from which tangents to the two circles are equal. In both cases the radical axis is a straight line perpendicular to the line of the centres of the circles.

The following are useful properties of circles related to their radical axes.

1. *The Radical Centre.*

The radical axes of three circles taken in pairs are concurrent at a point which is called the radical centre of the circles. This property follows directly from the definition of radical axis as a locus. The power of the point of intersection of any two of the radical axes is the same with respect to all three circles and this point must therefore be on the third radical axis.

2. *The difference between the powers of any point with respect to two given circles is proportional to the distance of the point from the radical axis of the circles.*

With the notation used previously, if the circles are $S_1 = 0$ and $S_2 = 0$, and P is the point (x_1, y_1), then the difference between the powers of P with respect to $S_1 = 0$ and $S_2 = 0$ is $2x_1(g_1 - g_2) + 2y_1(f_1 - f_2) + (c_1 - c_2)$. But the equation of the radical axis is

$$2(g_1 - g_2)x + 2(f_1 - f_2)y + (c_1 - c_2) = 0.$$

Hence the distance of P from the radical axis is

$$\left| \frac{2(g_1 - g_2)\, x_1 + 2(f_1 - f_2)\, y_1 + (c_1 - c_2)}{2\sqrt{\{(g_1 - g_2)^2 + (f_1 - f_2)^2\}}} \right|. \tag{20.2}$$

Thus, if PN is the perpendicular from P to the radical axis, and A and B are the centres of $S_1 = 0$ and $S_2 = 0$ respectively, the difference between the powers of P with respect to $S_1 = 0$ and $S_2 = 0$ is $2PN \cdot AB$.

3. It is a corollary to (20.2) that the power of any point on one circle with respect to another is proportional to the distance of the point from the radical axis of the two circles.

20.5 Coaxal circles

The radical axis of the circles $S_1 = 0$ and $S_2 = 0$ is $S_1 - S_2 = 0$. Similarly the radical axis of $S_1 + \lambda(S_1 - S_2) = 0$ (where λ is any constant), and $S_1 = 0$ is $\lambda(S_1 - S_2) = 0$, i.e. $S_1 - S_2 = 0$. It follows that there exists an infinite system of circles all of which have a common radical axis with $S_1 = 0$ and $S_2 = 0$. Such a system is called a *coaxal system of circles* and, in general, the equation of a coaxal system is

$$S_1 + \lambda L = 0 \tag{20.3}$$

where $S_1 = 0$ is the equation of any circle of the system and $L = 0$ is the equation of the radical axis of the system.

The line joining any pair of centres of a coaxal system of circles is at right angles to the radical axis. All the centres of the circles of a coaxal system are therefore collinear. It is convenient, in order to discuss the properties of a coaxal system of circles, to choose the line of centres as the x-axis and the radical axis as the y-axis. In this case the equation of any circle of the system is expressible in the form

$$x^2 + y^2 + 2gx + c = 0 \tag{20.4}$$

and since the power of the origin (a point on the radical axis) is the same with respect to every circle of the system, c is constant for the coaxal system.

Orthogonal coaxal systems and limiting points.

If c is negative, the coaxal circles $x^2 + y^2 + 2gx + c = 0$ cut the y-axis at the points $(0, \pm \sqrt{(-c)}$ and the circles constitute an inter-secting set of coaxal circles with the y-axis as common chord.

If c is positive, none of the circles intersects the y-axis. Therefore in this case the circles constitute a non-intersecting set of coaxal circles. The radius of $x^2 + y^2 + 2gx + c = 0$ is equal to $\sqrt{(g^2 - c)}$ and as $g \to \pm \sqrt{c}$ the radius of the circle of the coaxal system tends to zero. The points $(\pm \sqrt{c}, 0)$ are called the *point circles* or the *limiting points* of the system. If P is any point on the radical axis and L_1 and L_2 are the limiting points of the system, then $PL_1^2 = PL_2^2 = $ the power of P with respect to the system.

The equation of the intersecting coaxal system of circles through $(\pm\sqrt{c}, 0)$, where c is positive, is

$$x^2 + y^2 + 2fy - c = 0$$

where f varies. Any circle of this system is orthogonal with any circle of the system

$$x^2 + y^2 + 2gx + c = 0$$

because the square on the distance between the centres is $g^2 + f^2$ and the sum of the squares on the radii is $g^2 - c + f^2 + c = g^2 + f^2$. It follows that *a non-intersecting system of coaxal circles is orthogonal to the intersecting system which passes through its limiting points.*

Examples. (i) Prove that the squares of the tangents drawn from any point on the circle $x^2 + y^2 + 2x - 16y = 0$ to the two circles

$$x^2 + y^2 - 10y + 2\lambda(x - 3y) = 0,$$

$$x^2 + y^2 - 10y + 2\lambda'(x - 3y) = 0$$

are in the constant ratio $\lambda - 1 : \lambda' - 1$. $[\lambda > 1, \lambda' > 1.]$ (O.C.)

The ratio of the squares of the tangents to the given circles is

$$p = \frac{x^2 + y^2 - 10y + 2\lambda(x - 3y)}{x^2 + y^2 - 10y + 2\lambda'(x - 3y)} \tag{1}$$

where (x, y) lies on the circle $x^2 + y^2 + 2x - 16y = 0$. Substitution of $-2x + 16y$ for $x^2 + y^2$ in (1) gives

$$p = \frac{-2x + 16y - 10y + 2\lambda(x - 3y)}{-2x + 16y - 10y + 2\lambda'(x - 3y)} = \frac{2(\lambda - 1)(x - 3y)}{2(\lambda' - 1)(x - 3y)} = \frac{(\lambda - 1)}{(\lambda' - 1)}.$$

This example illustrates the property spoken of in § 20.4, that the power of a point on one circle with respect to another is proportional to the distance of the point from the radical axis of the two circles. The three circles of this example are coaxal with radical axis $x - 3y = 0$, and the ratio of the distance between the centres of $x^2 + y^2 + 2x - 16y = 0$ and $x^2 + y^2 - 10y + 2\lambda(x - 3y) = 0$ to the distance between the centres of $x^2 + y^2 + 2x - 16y = 0$ and $x^2 + y^2 - 10y + 2\lambda'(x - 3y) = 0$ is $(\lambda - 1) : (\lambda' - 1)$.

(ii) Find the coordinates of the common points of the coaxal system of circles given by the equation

$$x^2 + y^2 + 2gx - 16 = 0,$$

where g varies. Show that only one circle of the system touches the line $x + 2y = 8$, and find its equation.

If S is any member of the system and S' is the circle

$$x^2 + y^2 + 6x + 8y + 24 = 0,$$

prove that the radical axis of S and S' passes through a fixed point on the axis of y. (N.)

The circles of the system $x^2 + y^2 + 2gx - 16 = 0$ cut the y-axis at $(0, \pm 4)$, and these are the common points of the system. The condition that $x + 2y = 8$ is a tangent to $x^2 + y^2 + 2gx - 16 = 0$ is that the distance of $(-g, 0)$ from $x + 2y = 8$ should be numerically equal to the radius of the circle,

i.e.,
$$\left| \frac{-8 - g}{\sqrt 5} \right| = \sqrt{(g^2 + 16)},$$

i.e.,
$$g^2 + 16g + 64 = 5g^2 + 80.$$

This equation has the repeated root $g = 2$, and therefore the only circle of the system which touches $x + 2y = 8$ is $x^2 + y^2 + 4x - 16 = 0$.

The radical axis of S and S' is

$$(x^2 + y^2 + 2gx - 16) - (x^2 + y^2 + 6x + 8y + 24) = 0,$$

i.e.,
$$(g - 3)x - 4y - 20 = 0,$$

and this passes through the fixed point $(0, -5)$ for all values of g.

(iii) Find the equation of the circle passing through the point $(1, -1)$ and cutting orthogonally all circles of the coaxal system

$$x^2 + y^2 + 2gx + 2gy + 2 = 0. \qquad \text{(N.)}$$

The radius of any circle of the system is $\sqrt{(2g^2 - 2)}$. Hence the limiting points of the system are $(1, 1)$ and $(-1, -1)$. Therefore any circle through $(1, 1)$, $(-1, -1)$ cuts all the circles of the system orthogonally. The circle which passes through these points and the point $(1, -1)$ has centre at the origin, radius $\sqrt{2}$, and equation

$$x^2 + y^2 = 2.$$

(iv) Find the coordinates of the limiting points of the system of circles which is coaxal with

$$S_1 \equiv x^2 + y^2 - 2x - 2y + 1 = 0,$$
$$S_2 \equiv x^2 + y^2 + 4x + 4y + 4 = 0$$

and find the equation of that circle of the system which passes through $(3, 1)$.

The radical axis of S_1 and S_2 is $2x + 2y + 1 = 0$ and so from (20.3) the coaxal system is

$$x^2 + y^2 - 2x - 2y + 1 + \lambda(2x + 2y + 1) = 0.$$

The radius of a circle of this system is

$$\sqrt{\{(1 - \lambda)^2 + (1 - \lambda)^2 - (1 + \lambda)\}} = \sqrt{(2\lambda^2 - 5\lambda + 1)}.$$

Therefore the limiting points are given by $\lambda = \alpha$ and $\lambda = \beta$, where α and β are the roots of $2\lambda^2 - 5\lambda + 1 = 0$; so the limiting points are $(1 - \alpha, 1 - \alpha)$ and $(1 - \beta, 1 - \beta)$, i.e., $\left\{\frac{1}{4}(\sqrt{17} - 1), \frac{1}{4}(\sqrt{17} - 1)\right\}$ and $\left\{-\frac{1}{4}(\sqrt{17} + 1), -\frac{1}{4}(\sqrt{17} + 1)\right\}$.

The circle $x^2 + y^2 - 2x - 2y + 1 + \lambda(2x + 2y + 1) = 0$ goes through $(3, 1)$ if $\lambda = -\frac{1}{3}$, and so the circle of the system which goes through $(3, 1)$ is

$$x^2 + y^2 - 2x - 2y + 1 - \frac{1}{3}(2x + 2y + 1) = 0,$$

i.e., $\quad 3x^2 + 3y^2 - 8x - 8y + 2 = 0.$

Exercises 20.5

1. Write down the equation of the radical axis of each of the following pairs of circles. State, in each case, whether the circles are intersecting or non-intersecting circles and write down the equation of the coaxal system to which the circles belong.

(i) $x^2 + y^2 + 4x - 6y + 1 = 0$, $x^2 + y^2 - 5x + y - 4 = 0$,

(ii) $2x^2 + 2y^2 - 5x + 4y = 0$, $4x^2 + 4y^2 - x + 6y = 0$,

(iii) $x^2 + y^2 + 5x - 2y - 1 = 0$, $3x^2 + 3y^2 - x + y - 1 = 0$,

(iv) $x^2 + y^2 + 2ax + 2by + c = 0$, $x^2 + y^2 - 2ax - 2by + c = 0$.

2. Find the coordinates of the points, the powers of which, with respect to each of the circles

$$x^2 + y^2 + 10x + 8y + 16 = 0,$$

$$x^2 + y^2 - 4x + 2y - 4 = 0$$

are each $+40$.

3. Find the coordinates of the points, the powers of which, with respect to each of the circles

$$x^2 + y^2 - 2x + y - 24 = 0,$$

$$x^2 + y^2 - 4x - y - 18 = 0$$

are each -1.

4. Find the radical centre of the circles

$$x^2 + y^2 - 4x + 1 = 0,$$

$$x^2 + y^2 + 2x + 1 = 0,$$

$$x^2 + y^2 - 5x + y - 5 = 0$$

and hence find the equation of the circle which cuts the three circles orthogonally.

5. Prove that the origin is one of the limiting points of the coaxal system to which the circles

$$x^2 + y^2 - 2x + 6y + 4 = 0,$$

$$5x^2 + 5y^2 - 4x + 12y + 8 = 0$$

belong. Find the other limiting point and the equation of the orthogonal system of circles.

6. Find the coordinates of the limiting points of the system of circles to which the circles

$$x^2 + y^2 - 9x - 5y + 1 = 0,$$

$$x^2 + y^2 + 11x + 7y + 17 = 0$$

belong, and find the equation of the circle which cuts these two circles orthogonally and passes through the origin.

7. Find the coordinates of the limiting points of the coaxal system to which the circles

$$x^2 + y^2 - 3x + 4y + 4 = 0,$$

$$x^2 + y^2 - 6x + 8y + 24 = 0$$

belong. Give a geometrical explanation of the fact that in this case the limiting points coincide.

8. Find the equations of the circles touching the line $x + y + 10 = 0$ and belonging to the coaxal system with limiting points $(1, -2)$ and $(-3, 1)$.

9. The limiting points of a coaxal system are $(\pm k, 0)$, where k is a positive constant. Find the coordinates of the centres of the circles of the system which have a radius $k \sqrt{3}$ units.

10. $S \equiv x^2 + y^2 + 2gx + 2fy + c = 0$ is a fixed circle and $A(a, b)$ is a fixed point. Find the equation of the locus of a point P such that the ratio of PA to the length of the tangent from P to $S = 0$ is k, where k is constant. Show that the locus is a circle which is coaxal with $S = 0$ and the limiting point A.

20.6 Conic sections

We summarize those properties of these curves which we have already obtained together with a few properties which can be obtained by the methods we have already discussed.

	Parabola $y^2 = 4ax$	Ellipse $b^2x^2 + a^2y^2 = a^2b^2$	Hyperbola $b^2x^2 - a^2y^2 = a^2b^2$
Eccentricity	$e = 1$	$e < 1$, $b^2 = a^2(1 - e^2)$	$e > 1$, $b^2 = a^2(e^2 - 1)$
Foci	$(a, 0)$	$(\pm ae, 0)$	$(\pm ae, 0)$
Directrices	$x + a = 0$	$x \pm a/e = 0$	$x \pm a/e = 0$
Tangent at (x_1, y_1) on the curve	$yy_1 = 2a(x + x_1)$	$b^2xx_1 + a^2yy_1 = a^2b^2$	$b^2xx_1 - a^2yy_1 = a^2b^2$
General equation of tangent	$y = mx + a/m$	$y = mx \pm \sqrt{(a^2m^2 + b^2)}$	$y = mx \pm \sqrt{(a^2m^2 - b^2)}$
Parametric equations	$x = at^2$, $y = 2at$	$x = a\cos\theta$, $y = b\sin\theta$	$x = a\sec\theta$, $y = b\tan\theta$ ($x = a\cosh t$, $y = b\sinh t$ for x – positive branch only)
Tangent at a point on the curve in parametric coordinates	$x - ty + at^2 = 0$	$bx\cos\theta + ay\sin\theta = ab$	$bx\sec\theta - ay\tan\theta = ab$
Normal	$tx + y - 2at - at^3 = 0$	$\dfrac{ax}{\cos\theta} - \dfrac{by}{\sin\theta} = a^2 - b^2$	$\dfrac{ax}{\sec\theta} + \dfrac{by}{\tan\theta} = a^2 + b^2$
Chord	$2x - (t_1 + t_2)y + 2at_1t_2 = 0$	$bx\cos\frac{1}{2}(\theta_1 + \theta_2) + ay\sin\frac{1}{2}(\theta_1 + \theta_2) = ab\cos\frac{1}{2}(\theta_1 - \theta_2)$	$bx\cos\frac{1}{2}(\theta_1 - \theta_2) - ay\sin\frac{1}{2}(\theta_1 + \theta_2) = ab\cos\frac{1}{2}(\theta_1 + \theta_2)$
Director Circle	Perpendicular tangents intersect on the directrix $x + a = 0$	$x^2 + y^2 = a^2 + b^2$	$x^2 + y^2 = a^2 - b^2$
The locus of the feet of perpendiculars from the focus on to tangents	The tangent at the vertex $x = 0$	$x^2 + y^2 = a^2$	$x^2 + y^2 = a^2$
A diameter. (The locus of the mid-points of chords parallel to $y = mx$)	$y = 2a/m$	$y = -b^2x/a^2m$	$y = b^2x/a^2m$
Asymptotes	—	—	$y = \pm\dfrac{b}{a}x$

20.7 Note on the general equation of the second degree

The general equation of the second degree is

$$S \equiv ax^2 + 2hxy + by^2 + 2gx + 2fy + c = 0$$

and it can be shown (but it will not be shown here) that this equation, in all cases, represents a conic section and is reducible by suitable change of origin and rotation of axes to one of the special cases we have considered in this book. In some of the work which follows, results are obtained for $S = 0$ so that they may be applied to particular cases. $S = 0$ certainly includes these particular cases so that the only assumption we have made, that $S = 0$ is in all cases a conic, does not weaken the validity of our arguments.

20.8 The chord of contact of tangents drawn from an external point to a conic. Pole and polar

The chord of contact. If two tangents are drawn to a conic from a point outside it, the straight line joining the points of contact is called a chord of contact.

It was shown in Volume I, Chapter III (§ 3.6) that the equation of the tangent at the point $A(x_1, y_1)$ on the conic

$$ax^2 + 2hxy + by^2 + 2gx + 2fy + c = 0$$

is $L_1 \equiv axx_1 + h(xy_1 + x_1y) + byy_1 + g(x + x_1) + f(y + y_1) + c = 0$
and the equation of the tangent at the point $B(x_2, y_2)$ is

$$L_2 \equiv axx_2 + h(xy_2 + x_2y) + byy_2 + g(x + x_2) + f(y + y_2) + c = 0.$$

If $L_1 = 0$ passes through $P(l, m)$, then

(1) $ax_1l + h(y_1l + x_1m) + bmy_1 + g(x_1 + l) + f(y_1 + m) + c = 0$

and, if $L_2 = 0$ passes through P, then

(2) $ax_2l + h(y_2l + x_2m) + bmy_2 + g(x_2 + l) + f(y_2 + m) + c = 0.$

These two purely algebraical relationships (1) and (2) are the conditions that A and B lie on the line

$$axl + h(xm + yl) + bym + g(x + l) + f(y + m) + c = 0 \qquad (20.5)$$

and this line is therefore the equation of AB.

In particular cases the equations of the chords of contact of tangents drawn from an external point (x_1, y_1) to each of the conics we have considered separately are:

	Curve	Chord of contact
Circle	$x^2 + y^2 + 2gx + 2fy + c = 0$	$xx_1 + yy_1 + g(x + x_1)$ $+ f(y + y_1) + c = 0$
Parabola	$y^2 = 4ax$	$yy_1 = 2a(x + x_1)$
Ellipse	$a^2y^2 + b^2x^2 = a^2b^2$	$a^2yy_1 + b^2xx_1 = a^2b^2$
Hyperbola	$a^2y^2 - b^2x^2 = a^2b^2$	$a^2yy_1 - b^2xx_1 = a^2b^2$
Rectangular Hyperbola	$xy = c^2$	$xy_1 + yx_1 = 2c^2$

Pole and Polar. The point $P(x_1, y_1)$ is defined as the *pole* of the line
$$L \equiv axx_1 + h(xy_1 + yx_1) + byy_1 + g(x + x_1) + f(y + y_1) + c = 0$$
whatever the position of P in the coordinate plane, and $L = 0$ is defined as the *polar* of P.

If the polar of $P_1(x_1, y_1)$ with respect to the conic S passes through $P_2(x_2, y_2)$, then the polar of P_2 with respect to S passes through P_1. The polar of P_1 is
$$axx_1 + h(xy_1 + x_1y) + byy_1 + g(x + x_1) + f(y + y_1) + c = 0,$$
and if P_2 lies on this line
$$ax_2x_1 + h(x_2y_1 + x_1y_2) + by_2y_1 + g(x_2 + x_1) + f(y_2 + y_1) + c = 0 \cdot$$
This purely algebraic relationship is the condition that P_1 lies on the line
$$axx_2 + h(xy_2 + x_2y) + byy_2 + g(x + x_2) + f(y + y_2) + c = 0.$$
This line is the polar of P_2 with respect to S, and therefore the polar of P_2 passes through P_1.

20.9 The equation of a pair of tangents drawn from an external point to a conic

If $S = 0$ meets the line joining $P_1(x_1, y_1)$ and $P_2(x_2, y_2)$ at A so that $P_1A : AP_2 = t_2 : t_1$, the coordinates of A are
$$\left(\frac{t_1 x_1 + t_2 x_2}{t_1 + t_2}, \ \frac{t_1 y_1 + t_2 y_2}{t_1 + t_2} \right), \quad \text{and, since } A \text{ is on the conic,}$$
$$a\left(\frac{t_1 x_1 + t_2 x_2}{t_1 + t_2} \right)^2 + 2h\left(\frac{t_1 x_1 + t_2 x_2}{t_1 + t_2} \right)\left(\frac{t_1 y_1 + t_2 y_2}{t_1 + t_2} \right) +$$
$$b\left(\frac{t_1 y_1 + t_2 y_2}{t_1 + t_2} \right)^2 + 2g\left(\frac{t_1 x_1 + t_2 x_2}{t_1 + t_2} \right) + 2f\left(\frac{t_1 y_1 + t_2 y_2}{t_1 + t_2} \right) + c = 0.$$
This equation can be written as an equation for $t_2 : t_1$ in the form
$$(ax_1^2 + 2hx_1y_1 + by_1^2 + 2gx_1 + 2fy_1 + c)t_1^2$$
$$+ 2\{ax_1x_2 + h(x_1y_2 + x_2y_1) + by_1y_2 + g(x_1 + x_2) +$$
$$f(y_1 + y_2) + c\}t_1t_2 + (ax_2^2 + 2hx_2y_2 + by_2^2 + 2gx_2 + 2fy_2 + c)t_2^2 = 0.$$
$$(20.6)$$

This is a quadratic equation for the ratio $t_2 : t_1$ and therefore a straight line cuts a conic in two distinct points, two coincident points or not at all. The equation (20.6) is called *Joachimsthal's ratio equation.*

If P_1 is a fixed point outside the conic and if P_2 lies on any tangent from P_1 to the conic, the equation (20.6) has equal roots. The condition for this is

$$\{ax_1x_2 + h(x_1y_2 + x_2y_1) + by_1y_2 + g(x_1 + x_2) + f(y_1 + y_2) + c\}^2 =$$
$$(ax_1^2 + 2hx_1y_1 + by_1^2 + 2gx_1 + 2fy_1 + c)(ax_2^2 + 2hx_2y_2 + by_2^2$$
$$+ 2gx_2 + 2fy_2 + c).$$

Hence the equation of the tangent pair from (x_1, y_1) to the conic is

$$\{ax_1x + h(x_1y + y_1x) + by_1y + g(x + x_1) + f(y + y_1) + c\}^2 =$$
$$(ax_1^2 + 2hx_1y_1 + by_1^2 + 2gx_1 + 2fy_1 + c)(ax^2 + 2hxy + by^2$$
$$+ 2gx + 2fy + c). \tag{20.7}$$

This equation (20.7) can conveniently be abbreviated in the form

$$P_1^2 = S_1 S,$$

where P_1 denotes the left hand side of the equation $P_1 = 0$ of the polar of (x_1, y_1), and S_1 the result of substituting (x_1, y_1) for (x, y) in S.

In particular cases the equations of the tangent-pairs are as follows:

Curve	Tangent Pair
$x^2 + y^2 = a^2$	$(xx_1 + yy_1 - a^2)^2 = (x_1^2 + y_1^2 - a^2)(x^2 + y^2 - a^2)$
$y^2 = 4ax$	$\{yy_1 - 2a(x + x_1)\}^2 = (y_1^2 - 4ax_1)(y^2 - 4ax)$
$b^2x^2 + a^2y^2 = a^2b^2$	$(b^2xx_1 + a^2yy_1 - a^2b^2)^2 =$
	$\quad (b^2x_1^2 + a^2y_1^2 - a^2b^2)(b^2x^2 + a^2y^2 - a^2b^2)$
$xy = c^2$	$(xy_1 + yx_1 - 2c^2)^2 = 4(x_1y_1 - c^2)(xy - c^2)$

Examples. (i) Show that the poles of an arbitrary fixed line through the point $(a, 0)$ with respect to all circles through the points $(a, 0)$ and $(-a, 0)$ lie on a conic which also passes through $(a, 0)$.

Circles through $(a, 0)$, $(-a, 0)$ are of the form

$$x^2 + y^2 + 2fy - a^2 = 0$$

where f varies. An arbitrary line through $(a,0)$ is of the form $kx - y - ka = 0$ where k is constant. The polar of (x_1, y_1) with respect to the circle is

$$xx_1 + yy_1 + f(y + y_1) - a^2 = 0,$$

i.e., $\quad x_1 x + (y_1 + f)y + (fy_1 - a^2) = 0,$

and this represents the same line as

$$kx - y - ka = 0$$

if $$\frac{x_1}{k} = -(y_1 + f) = \frac{a^2 - f y_1}{ka},$$

i.e., if $$f = -y_1 - \frac{x_1}{k} \quad \text{and} \quad x_1 = \frac{a^2 - f y_1}{a},$$

i.e., if $$a x_1 = a^2 + y_1^2 + \frac{x_1 y_1}{k}$$

or $$k y_1^2 + x_1 y_1 - k a x_1 - k a^2 = 0.$$

The poles of the line through $(a, 0)$ with respect to all the circles through $(a, 0)$, $(-a, 0)$ therefore lie on the curve

$$ky^2 + xy - kax - ka^2 = 0$$

which passes through $(a, 0)$.

(ii) Prove that, if tangents from the point (x_1, y_1) to the ellipse $b^2 x^2 + a^2 y^2 = a^2 b^2$ intersect at $45°$, then

$$2 \sqrt{(b^2 x_1^2 + a^2 y_1^2 - a^2 b^2)} = x_1^2 + y_1^2 - a^2 - b^2.$$

The equation of the tangent pair from (x_1, y_1) to the ellipse is

$$(b^2 x_1 x + a^2 y_1 y - a^2 b^2)^2 = (b^2 x_1^2 + a^2 y_1^2 - a^2 b^2)(b^2 x^2 + a^2 y^2 - a^2 b^2),$$

i.e., $b^2 x^2 (a^2 y_1^2 - a^2 b^2) - 2 a^2 b^2 x_1 y_1 xy + a^2 y^2 (b^2 x_1^2 - a^2 b^2) + \text{terms of lower degree} = 0$.

The acute angle between these lines is given by the formula

$$\tan^{-1} \left| \frac{2\sqrt{(H^2 - A B)}}{A + B} \right|, \text{ whence, if the tangents intersect at } 45°,$$

$$\frac{2 \sqrt{\{a^4 b^4 x_1^2 y_1^2 - b^2 (a^2 y_1^2 - a^2 b^2) a^2 (b^2 x_1^2 - a^2 b^2)\}}}{b^2 (a^2 y_1^2 - a^2 b^2) + a^2 (b^2 x_1^2 - a^2 b^2)} = 1.$$

$$\therefore \frac{2 a^2 b^2 \sqrt{\{x_1^2 y_1^2 - (y_1^2 - b^2)(x_1^2 - a^2)\}}}{a^2 b^2 \{(y_1^2 - b^2) + (x_1^2 - a^2)\}} = 1.$$

$$\therefore 2\sqrt{(b^2 x_1^2 + a^2 y_1^2 - a^2 b^2)} = x_1^2 + y_1^2 - a^2 - b^2.$$

Exercises 20.9

1. Write down the equation of the polar of each of the following points with respect to the conic whose equation is given.

(i) $(0, 0)$, $x^2 + y^2 - 6x + 2y - 4 = 0$,
(ii) $(5, 6)$, $5(x - 1)^2 + 2(y - 4)^2 = 10$,

(iii) $(2, 1)$, $2x^2 + xy + y^2 + 5x + y - 1 = 0$,

(iv) (x_1, y_1), $x^2 + xy + y^2 + 2x + y + 4 = 0$,

(v) (a, b), $axy + by = c^2$.

2. Find the coordinates of the pole of each of the following lines with respect to the conic whose equation is given.

(i) $3x + 5y - 15 = 0$, $x^2 + y^2 = 9$,

(ii) $ax + y - 4a^2 = 0$, $y^2 = 4a(x + 2a)$,

(iii) $4x + y - 7 = 0$, $x^2 + y^2 - 2x + 4y = 0$,

(iv) $x + y - 1 = 0$, $(x + 2)(y - 1) = 16$,

(v) $13x + 24y - 3 = 0$, $3x^2 + 4xy - 2y^2 + x - 8y + 7 = 0$.

3. In each of the following cases write down and simplify the equations of the pair of tangents from the point to the conic.

(i) $(1, 1)$, $3x^2 + 4y^2 - 1 = 0$,

(ii) $(0, 0)$, $xy = 4$,

(iii) $(0, -1)$, $3x^2 + 4xy + y^2 = 10$,

(iv) $(1, -1)$, $4x^2 - y^2 = 10$,

(v) $(0, 4)$, $(y + 1)^2 = 4(x - 1)$.

4. Prove that the equation of the chord of contact of the tangents drawn from the origin of coordinates to the circle

$$x^2 + y^2 + 2gx + 2fy + c = 0$$

is $\qquad$ $gx + fy + c = 0$.

Prove that the area of the triangle formed by the two tangents from the origin of coordinates and the chord of contact is

$$\{c^{3/2} \sqrt{(g^2 + f^2 - c)}\}/(g^2 + f^2).$$

5. Find the locus of the poles of normal chords of the parabola

$$x = at^2, \ y = 2at.$$

6. A point P moves so that the chord of contact of the tangents from P to the ellipse $b^2x^2 + a^2y^2 = a^2b^2$ touches the ellipse $4(b^2x^2 + a^2y^2) = a^2b^2$. Find the locus of P. (L.)

7. Show that the line $lx + my + n = 0$ touches the parabola $y^2 = 4ax$, if $am^2 = nl$.

If the polar of a point P with respect to the ellipse $(x + a)^2/a^2 + y^2/b^2 = 1$ touches the parabola $y^2 = 4ax$, show that the locus of P is the hyperbola $x(x + a)/a^4 - y^2/b^4 = 0$. (L.)

8. Show that the equation of the pair of tangents from the point $P(\alpha, \beta)$ to the circle $x^2 + y^2 = r^2$ can be expressed in the form

$$(\beta x - \alpha y)^2 = r^2[(x - \alpha)^2 + (y - \beta)^2].$$

These tangents cut the x-axis at the points A and B.

(i) If the mid-point of AB is the fixed point $(k, 0)$, show that the locus of P is the parabola $ky^2 = r^2(k - x)$.

(ii) If AB is of length $4r$, show that P lies on the quartic curve

$$x^2 y^2 = (y^2 - r^2)(3y^2 - 4r^2).$$ (L.)

9. Show that if m_1, m_2 are the gradients of the two tangents from the point (α, β) to the hyperbola $b^2 x^2 - a^2 y^2 = a^2 b^2$, then m_1, m_2 are the roots of the equation

$$m^2(\alpha^2 - a^2) - 2\alpha\beta m + b^2 + \beta^2 = 0.$$

Show that the locus of points from which the two tangents to the curve are equally inclined to a line of gradient $\tan\theta$ is the rectangular hyperbola

$$x^2 - y^2 - 2xy \cot 2\theta = a^2 + b^2.$$ (L.)

10. Show that the equation of the chord joining the points $P(2t_1, 2/t_1)$ and $Q(2t_2, 2/t_2)$ of the hyperbola $xy = 4$, is

$$x + y t_1 t_2 = 2(t_1 + t_2).$$

If the chord varies in such a way as to be always a tangent to the circle

$$x^2 + y^2 = 1,$$

prove that the point of intersection of the tangents at P and Q to the hyperbola lies always on a circle with centre at the origin and radius 8. (L.)

20.10 Conjugate diameters

The ellipse. Chords of the ellipse $b^2 x^2 + a^2 y^2 = a^2 b^2$ which are parallel to $y = mx$ are all bisected by the line $y = -b^2 x/a^2 m$ which is called a diameter of the ellipse. The symmetry of this result shows that chords of the ellipse which are parallel to $y = m_1 x$ where $m_1 = -b^2/a^2 m$ are all bisected by the diameter $y = mx$. The two diameters $y = mx$ and $y = m_1 x$ are called *conjugate* diameters.

Two of the properties of conjugate diameters of the ellipse are worked out below as examples. Others are included for the student to work out in Exercises 20.10.

Examples. (i) Prove that the eccentric angles of the ends of a pair of conjugate diameters of an ellipse differ by a right angle.

The equation of the diameter through the point $(a \cos\theta, b \sin\theta)$ on the ellipse $b^2 x^2 + a^2 y^2 = a^2 b^2$ is $y = \dfrac{b \sin\theta}{a \cos\theta} x$ and the equation of the diameter conjugate to this is

$$y = -\frac{b^2}{a^2} \frac{a \cos\theta}{b \sin\theta} x,$$

i.e., $\quad y = -\dfrac{b \cos\theta}{a \sin\theta} x,$ or $y = \dfrac{b}{a} \dfrac{\sin\left(\theta + \frac{1}{2}\pi\right)}{\cos\left(\theta + \frac{1}{2}\pi\right)} x.$

This is the diameter through the point of the curve

$$\{a \cos(\theta + \tfrac{1}{2}\pi),\ b \sin(\theta + \tfrac{1}{2}\pi)\}$$

and hence the required result follows. Note that the coordinates of corresponding extremities of conjugate diameters can be expressed as $(a \cos\theta,\ b \sin\theta)$, $(-a \sin\theta,\ b \cos\theta)$.

(ii) Find the equations of the *equiconjugate* diameters of the ellipse $b^2 x^2 + a^2 y^2 = a^2 b^2$. (Equiconjugate diameters are conjugate diameters of equal length.)

If the equiconjugate diameters are POP' and QOQ', where O is the centre of the ellipse, and if P is $(a \cos\theta,\ b \sin\theta)$ and Q is $(-a \sin\theta, b \cos\theta)$, then, since $OP^2 = OQ^2$,

$$a^2 \cos^2\theta + b^2 \sin^2\theta = a^2 \sin^2\theta + b^2 \cos^2\theta.$$

$$\therefore \tan^2\theta = 1.$$

$$\therefore \theta = \frac{1}{4}\pi \quad \text{or} \quad \frac{3}{4}\pi.$$

Hence OP is $y = b\,x/a$ and OQ is $y = -b\,x/a$

or OP is $y = -b\,x/a$ and OQ is $y = b\,x/a$.

(iii) Prove that the poles of all chords parallel to a diameter of an ellipse lie on the conjugate diameter.

The polar of (x_1, y_1) with respect to $b^2 x^2 + a^2 y^2 = a^2 b^2$ is

$$b^2 x_1 x + a^2 y_1 y - a^2 b^2 = 0$$

and this is the same line as

$$y = mx + c$$

if

$$\frac{b^2 x_1}{m} = -\frac{a^2 y_1}{1} = -\frac{a^2 b^2}{c},$$

i.e., if

$$x_1 = -\frac{a^2 m}{c}, \quad y_1 = \frac{b^2}{c}.$$

The pole of $y = mx + c$ with respect to the ellipse is therefore $\left(-\dfrac{a^2 m}{c},\ \dfrac{b^2}{c}\right)$. This point lies on the line $y = \dfrac{-b^2}{a^2 m} x$ and therefore the poles of all chords parallel to the diameter $y = mx$ lie on the conjugate diameter $y = \dfrac{-b^2}{a^2 m} x$.

The Hyperbola. The locus of the mid-points of chords of a hyperbola parallel to the line $y = mx$ is the line $y = m_1 x$ where $m m_1 = b^2/a^2$.

If $y = mx$ intersects the hyperbola $b^2x^2 - a^2y^2 = a^2b^2$, it does so at points whose x-coordinates are given by $x^2(b^2 - a^2m^2) = a^2b^2$. The line $y = mx$ therefore intersects the hyperbola if $a^2m^2 < b^2$ but does not intersect the hyperbola if $a^2m^2 > b^2$. But, since $mm_1 = b^2/a^2$, one of $y = mx$ and $y = m_1x$ intersects the hyperbola and the other one does not. The two lines $y = mx$ and $y = m_1x$ are *defined* as conjugate diameters of the hyperbola.

The Conjugate Hyperbola. The hyperbolas $b^2x^2 - a^2y^2 = a^2b^2$ and $a^2x^2 - b^2y^2 = a^2b^2$ are called conjugate hyperbolas. (See Exercises 20.10 Nos. 8 − 10).

Exercises 20.10

1. Prove that the tangents drawn at the ends of a diameter of an ellipse are parallel to the chords bisected by that diameter.

2. If OP and OQ are semi-conjugate diameters of the ellipse $b^2x^2 + a^2y^2 = a^2b^2$, prove that $OP^2 + OQ^2 = a^2 + b^2$.

3. If OP and OQ are semi-conjugate diameters of $b^2x^2 + a^2y^2 = a^2b^2$, prove that the area of triangle OPQ is constant and find its value.

4. Use the result of Exercise 3 to prove that the area of the parallelogram formed by the tangents drawn at the extremities of conjugate diameters of an ellipse is constant and state its value in terms of the lengths of the axes of the ellipse.

5. Prove that a pair of conjugate diameters of the ellipse $b^2x^2 + a^2y^2 = a^2b^2$ can be represented by the equation

$$b^2x^2 + 2hxy - a^2y^2 = 0.$$

Find the equation of the conjugate diameters of the ellipse

$$9x^2 + 4y^2 = 36$$

which intersect at an angle of $\tan^{-1}3$.

6. Prove that if POP' is one of the equiconjugate diameters of an ellipse of centre O and TS is a variable chord of the ellipse parallel to POP', the locus of the intersection of normals drawn from T and S is the other equiconjugate diameter of the ellipse.

7. If POP' and QOQ', conjugate diameters of the ellipse $b^2x^2 + a^2y^2 = a^2b^2$, intersect the tangent at a point L on the ellipse at T and T' respectively, prove that $TL.\,LT' = OM^2$ where OM is the semi-conjugate diameter to OL.

8. Prove that if a pair of diameters are conjugate with respect to a hyperbola they are conjugate with respect to the conjugate hyperbola.

9. If POP' and QOQ' are conjugate diameters of $b^2x^2 - a^2y^2 = a^2b^2$, P and P' are points on this hyperbola, and Q and Q' points on the conjugate hyperbola, prove that $OP^2 - OQ^2 = a^2 - b^2$.

10. With the data of Exercise 9 prove that the tangents at P, P', Q, Q' form a parallelogram whose vertices lie on the asymptotes and whose area is constant.

11. POP', DOD' are conjugate diameters of an ellipse, PD is produced to Q so that DQ is twice PD; find the locus of Q.   (O.C.)

12. POP', QOQ' are conjugate diameters of an ellipse whose centre is O. Find the locus of the intersection of the tangents at P and Q.

20.11 The polar equation of a conic

We have already discussed polar equations in general, and the polar equations of the straight line and circle in particular, in Chapter XV. In the exercises of that chapter reference was made to the polar equation of a conic with the focus as the pole. We shall discuss this equation more fully here.

Figs. 171 (i), (ii). Consider the conic as defined by the focus-directrix property, so that $SP = ePN$ where P is a point on the locus, S the focus and PN the perpendicular from P to the directrix. S is the pole and SX the initial line.

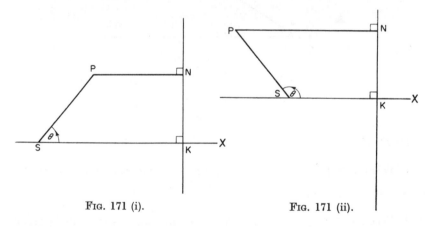

FIG. 171 (i).   FIG. 171 (ii).

Then, if P is (r, θ),

$$r = SP = ePN = e(SK - r\cos\theta).$$

But $eSK = l$ is the semi-latus rectum of the conic;

$$\therefore r = l - er\cos\theta.$$

$$\therefore \frac{l}{r} = 1 + e\cos\theta.$$ (20.8)

and since $e \leqq 1$ for the parabola and the ellipse all the values of r given by this equation are, in these cases, positive.

Fig. 171 (iii). Here S and P are on opposite sides of the directrix, a case which can arise when the figure is a hyperbola. If P is $(-r, \theta)$,

$$r = -SP = -ePN = -e\{-r\cos(\theta - \pi) - SK\} = -er\cos\theta + l.$$

$$\therefore \frac{l}{r} = 1 + e\cos\theta.$$

Thus, if negative values of r are not excluded, the hyperbola can be represented by a similar polar equation to that for the parabola and the ellipse.

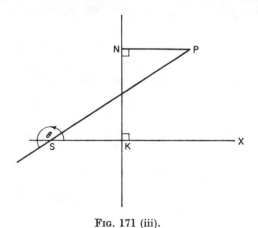

Fig. 171 (iii).

Particular cases.

1. When $e = 1$, $\dfrac{l}{r} = 1 + \cos\theta$ represents a parabola and is given by $-\pi < \theta < \pi$, $r > 0$.

2. When $e < 1$, $\dfrac{l}{r} = 1 + e\cos\theta$ represents an ellipse and is given completely by $-\pi < \theta \leqq \pi$, $r > 0$.

3. When $e > 1$, $\dfrac{l}{r} = 1 + e\cos\theta$ represents a hyperbola. The branch $KA'K'$ shown in Fig. 172 is given by $0 < \theta < \pi - \sec^{-1}e$ and $\pi + \sec^{-1}e < \theta \leqq 2\pi$ and $r > 0$. The branch LAL' is given by $\pi - \sec^{-1}e < \theta < \pi + \sec^{-1}e$ and $r < 0$.
In each case the equation of the directrix is $\dfrac{l}{r} = e\cos\theta$.

The equation of a chord. Let the vectorial angles of two points P and Q on the conic $\dfrac{l}{r} = 1 + e\cos\theta$ be $(\alpha - \beta)$ and $(\alpha + \beta)$ respectively $(\beta \neq 90°)$. The equation:

$$\frac{l}{r} = A\cos\theta + B\cos(\theta - \alpha),$$

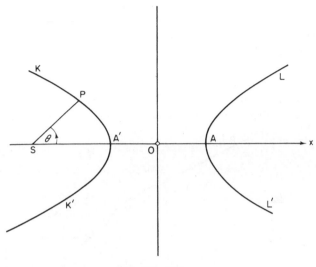

Fig. 172.

where A and B are constants, represents a straight line which does not pass through the pole. [It can be put into the form $r = k\sec(\theta - \gamma)$ which is one of the forms of the equation of a straight line obtained in Chapter XV, §15.3.] If this equation represents the straight line PQ, then

at P, $\qquad \dfrac{l}{r} = 1 + e\cos(\alpha - \beta) = A\cos(\alpha - \beta) + B\cos\beta,$

at Q, $\qquad \dfrac{l}{r} = 1 + e\cos(\alpha + \beta) = A\cos(\alpha + \beta) + B\cos\beta,$

and therefore by inspection,

$$A = e \quad \text{and} \quad B\cos\beta = 1.$$

The equation of PQ is therefore

$$\frac{l}{r} = e\cos\theta + \sec\beta\cos(\theta - \alpha). \tag{20.9}$$

Cases in which $\beta = 90°$, and the chord passes through the pole, should be considered from first principles.

The equation of a tangent. If α is the vectorial angle of a point P on the conic $l/r = 1 + e\cos\theta$, the equation of the tangent at P is obtained by letting $\beta \to 0$ in equation (20.9).

The equation of the tangent is therefore

$$l/r = e\cos\theta + \cos(\theta - \alpha). \tag{20.10}$$

Example. Points P, Q on the parabola $l/r = 1 + \cos\theta$ and on the same side of the axis are such that

$$1/SP + 1/SQ = 2/c,$$

where S is the focus and c is a constant. The bisector of the angle PSQ meets PQ at R. Prove that R lies on the straight line $r\cos\theta = l - c$.

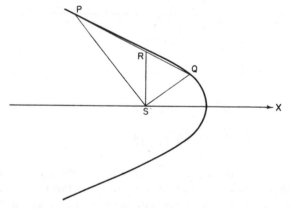

Fig. 173.

Fig. 173. Let the vectorial angles of P and Q be $(\alpha + \beta)$, $(\alpha - \beta)$ respectively. Then the vectorial angle of R is α. Let R be (r_1, α).

P and Q are on the parabola, and therefore

$$\frac{1}{SP} + \frac{1}{SQ} = \frac{1 + \cos(\alpha + \beta)}{l} + \frac{1 + \cos(\alpha - \beta)}{l} = \frac{2}{c}.$$

$$\therefore 2 + 2\cos\alpha\cos\beta = \frac{2l}{c}.$$

$$\therefore \sec\beta = \frac{c\cos\alpha}{l - c}.$$

But R is on the line $\dfrac{l}{r} = \cos\theta + \sec\beta\,\cos(\theta - \alpha)$.

$$\therefore \frac{l}{r_1} = \cos\alpha + \sec\beta.$$

$$\therefore \frac{l}{r_1} = \cos\alpha + \frac{c\cos\alpha}{l - c} = \frac{l\cos\alpha}{l - c}.$$

$$\therefore r_1\cos\alpha = l - c.$$

Hence R lies on the straight line $r\cos\theta = l - c$.

Exercises 20.11

1. Write down the cartesian equations referred to the pole as origin and the initial line as the positive x-axis of each of the following:

   (i) $2a/r = 1 - \cos\theta$,   (ii) $2a/r = 1 + \tfrac12\cos\theta$,

   (iii) $2a/r = 1 + 2\cos\theta$,   (iv) $2a/r = 2\cos\theta - 1$.

2. Write down the equations of the chords which join the points named in each of the following:

   (i) $2a/r = 1 + \cos\theta$, $\left(\dfrac{4a}{3},\ \dfrac13\pi\right)$, $(2a,\ \tfrac12\pi)$,

   (ii) $l/r = 1 + \dfrac{1}{\sqrt{3}}\cos\theta$, $\left(\dfrac{2l}{3}, \dfrac{\pi}{6}\right)$, $\left(2l, \dfrac{5\pi}{6}\right)$.

3. Write down the equation of the tangent to each of the following curves at the point named.

   (i) $2a/r = 1 + \cos\theta$, $\left(4a, \dfrac{4\pi}{3}\right)$,

   (ii) $l/r = 1 + 2\cos\theta$, $(l, \tfrac12\pi)$,

   (iii) $l/r = 1 + \tfrac12\cos\theta$, $\left(\dfrac{4l}{5}, \dfrac13\pi\right)$.

4. Prove that, if PSP_1 and QSQ_1 are two focal chords of a conic which are at right angles, then $\dfrac{1}{PS\cdot SP_1} + \dfrac{1}{QS\cdot SQ_1} = $ constant.

5. Use the polar equation of a conic to prove that if the tangents at two points A and B of a conic whose focus is S intersect at P, then PS bisects the angle APB.

6. Prove that if P is the point of intersection of the tangents at A and B of a parabola, then $SP^2 = SA\cdot SB$.

7. P is any point on the ellipse $l/r = 1 + e\cos\theta$. The focus O is the pole. A square is drawn with OP as diagonal. Prove that the other two vertices of the square each lie on an ellipse with the same eccentricity and the same focus. State the length of the semi-latus rectum and the direction of the major axis in each case.

8. Obtain the condition to be satisfied in order that the tangents at the points of vectorial angles $(\alpha \pm \beta)$ on the conic $l/r = 1 + e\cos\theta$ should be perpendicular, and deduce that the polar equation of the director-circle of the conic is

$$(1 - e^2)\,r^2 + 2ler\cos\theta - 2l^2 = 0.$$

Miscellaneous Exercises XX

1. If the radii of two circles are 7 in. and 1 in. and the distance between their centres is 10 in., show that the distance between the limiting points is 4·8 in. (N.)

2. Find the coordinates of the centre and the radius of each of the circles

$$x^2 + y^2 - 2x - 4y - 4 = 0,$$
$$x^2 + y^2 - 10x + 2y + 10 = 0$$

and verify that the circles are orthogonal. (N.)

3. Show that the x-axis is the radical axis of the system of circles given by the equation

$$x^2 + y^2 - 4x - 2ky + 3 = 0,$$

where k varies. Find the coordinates of the two common points of the system. Find the equations of
   (i) the circle of the system which passes through the point (4, 3),
   (ii) the two circles of the system which touch the y-axis,
   (iii) the two circles of the system which touch the line

$$x + y = 5. \tag{N.}$$

4. Define the radical axis of two circles, and show that the y-axis is the radical axis of the circles

$$x^2 + y^2 - 4x - 9 = 0,$$
$$x^2 + y^2 + 6x - 9 = 0.$$

Find the equation of the smallest circle through the common points A and B of these two circles. Find also the equations of the circles through A and B which have radius 5.

Show that any circle which cuts orthogonally all circles through A and B has its centre on the y-axis. If such a circle also cuts orthogonally the circle

$$x^2 + y^2 - 2x - 4y - 25 = 0$$

find its equation. (N.)

5. A diameter AB of the hyperbola $\dfrac{x^2}{16} - \dfrac{y^2}{8} = 1$ subtends right angles at the foci F, F'. Find the area and the perimeter of the rectangle $AFBF'$. (N.)

6. Show that the equation of the chord joining the points $P(cp, c/p)$, $Q(cq, c/q)$ on the curve $xy = c^2$ is

$$pqy + x = c(p + q)$$

Hence or otherwise find the equation of the tangent at P.
Find the coordinates of the point of intersection T of the tangents at P, Q. If p and q vary so that the chord PQ passes through the fixed point $(a, 0)$, find the equation of the locus of T. (N.)

7. The line $x \cos \alpha + y \sin \alpha = p$ meets the circle $x^2 + y^2 = a^2$ at the points P, Q. Prove that the equation of the circle S of which PQ is a diameter is

$$x^2 + y^2 - a^2 = 2p(x \cos \alpha + y \sin \alpha - p).$$

The line PQ varies in such a manner that S is always orthogonal to the fixed circle $x^2 + y^2 + 2fy + a^2 = 0$. Prove that the equation of the locus of the centre of S is $x^2 + y^2 + fy = 0$. (N.)

8. One circle of a coaxal system is

$$x^2 + y^2 - 4x - 8y + 10 = 0$$

and the radical axis is

$$2x + 4y - 5 = 0.$$

Show that the equation of any circle of the system may be written in the form

$$x^2 + y^2 + k(2x + 4y - 5) = 0.$$

Find the coordinates of the limiting points of the system.
Verify that the circle

$$x^2 + y^2 - x - 2y = 0$$

is orthogonal to each circle of the system. (N.)

9. The vertices of a triangle ABC lie on the rectangular hyperbola $xy = c^2$ and the sides AB, AC touch the parabola $y^2 = 4ax$. Prove that BC also touches the parabola.

Deduce that, if E is the point of intersection of the two conics and if the tangent at E to the parabola meets the hyperbola again at F, then the tangent at F to the parabola will touch the hyperbola. (N.)

10. Write down the equation of the radical axis of the circles

$$x^2 + y^2 + 6x + 4 = 0,$$

$$x^2 + y^2 - 8x + 4 = 0,$$

and also the equation of the system of circles coaxal with these circles. Show that the system has real limiting points, and state their coordinates.
Find the equations of

(i) the members of the system which touch the line

$$2x + y = 1,$$

(ii) the circle which is orthogonal to all members of the system and passes through the point $(3, 1)$. . (N.)

11. Tangents are drawn from the point $\left(\dfrac{10}{3}, \dfrac{5}{3} \right)$ to the ellipse

$$\frac{x^2}{20} + \frac{y^2}{5} = 1.$$

Find their equations. (N.)

12. A circle of radius R is drawn through a focus of a conic of eccentricity e and latus rectum $2l$; show that the sum of the reciprocals of the focal distances of the four points at which the circle cuts the conic is $2/l$. Show also that the product of these reciprocals is $e^2/(4l^2R^2)$. (N.)

13. From a fixed point P on the parabola $y^2 = 4ax$ chords PQ, PQ' are drawn making equal angles θ with the tangent at P; show that, as θ varies, QQ' passes through a fixed point R.

Show also that, if P moves along the parabola, the locus of R has the equation

$$(x + 2a) y^2 + 4a^3 = 0. \tag{N.}$$

14. Four points P_1, P_2, P_3, P_4 on the rectangular hyperbola given by $x = ct$, $y = c/t$, are determined respectively by the values t_1, t_2, t_3, t_4 of t. Show that, if these points lie on a circle through the origin O, then

$$t_1 t_2 t_3 t_4 = 1, \ (t_1 + t_2)(t_3 + t_4) + t_1 t_2 + t_2 t_4 = 0.$$

A variable straight line through the fixed point (x_0, y_0) meets the hyperbola at A, B and the circumcircle of the triangle OAB meets the hyperbola again at C, D. Show that the locus of the pole of CD with respect to the hyperbola is a circle through the origin. (N.)

15. Points P, Q, R on the rectangular hyperbola $xy = c^2$ are such that PQ, PR are perpendicular. Prove that QR is parallel to the normal to the hyperbola at P. Deduce the result that the circle on QR as diameter meets the hyperbola at the ends of a diameter of the hyperbola. (C.)

16. Find the coordinates of the point of intersection T of the tangents to the ellipse $b^2 x^2 + a^2 y^2 = a^2 b^2$ at the points P and Q, whose coordinates are

$$\{a \cos(\alpha - \beta), \ b \sin(\alpha - \beta)\} \quad \text{and} \quad \{a \cos(\alpha + \beta), \ b \sin(\alpha + \beta)\}$$

respectively.

If p_1, p_2, p_3 are respectively the lengths of the perpendiculars from P, Q and T on a variable tangent to the ellipse, prove that $p_1 p_2 / p_3^2$ is constant. (C.)

17. Two normals to the parabola $y^2 = 4ax$ make angles $\tan^{-1} m$ and $\tan^{-1}\left(\dfrac{2}{m}\right)$ with the axis of x; show that the locus of the point of intersection of the normals is the given parabola, and that the locus of the point of intersection of the corresponding tangents is a straight line parallel to the directrix. (O.C.)

18. Find the equation of the polar of the point (x', y') with respect to the ellipse $x^2/a^2 + y^2/b^2 = 1$. Show also that, if the point Q lies on the polar of P, then P lies on the polar of Q.

Show that the polar of a point on the circle $x^2 + y^2 = c^2$ with respect to the ellipse $x^2/a^2 + y^2/b^2 = 1$ touches a second ellipse, and find its equation. (O.C.)

19. A and B are two fixed points on a given circle, P and Q are the extremities of any diameter of the circle. Prove that the locus of the intersections of AP and BQ and of AQ and BP is the circle which cuts the given circle orthogonally at A and B. (O.C.)

20. Prove that the equation of a circle, which touches the straight lines $x^2 \sin^2\alpha - y^2 \cos^2\alpha = 0$ and lies in the angle 2α between them, may be written in the form

$$y^2 \cos^2\alpha - x^2 \sin^2\alpha + (x - a)^2 = 0,$$

and that the equation of a circle, which touches these two straight lines and lies in the other angle between them, may be written in the form

$$x^2 \sin^2\alpha - y^2 \cos^2\alpha + (y - b)^2 = 0,$$

where a and b may have any values.

Prove that the locus of the poles of the straight line $lx + my = 0$ with respect to the circles of either system is the straight line

$$mx \sin^2\alpha + ly \cos^2\alpha = 0.$$ (O.C.)

21. Find the equation of the tangent to the circle,

$$x^2 + y^2 - 2hx + \delta = 0,$$

at the point (x', y'); and deduce the equation of the polar of the point (p, q) with regard to the circle.

The pole with regard to the circle A of the radical axis of two circles A and B is the centre of B; prove that the centre of B is the pole of the radical axis with regard to A.   (O.C.)

22. Prove that the equation

$$8x^2 + 38xy - 33y^2 + 26x + 68y + 21 = 0$$

represents two straight lines.

Show also that the angle between them is $\tan^{-1}2$.   (O.C.)

23. Obtain the equation of the pair of lines from the origin to the points of intersection of the circle

$$x^2 + y^2 + 2gx + 2fy + c = 0$$

and the straight line $lx + my + n = 0$.

Prove that the locus of the middle points of chords of the circle

$$x^2 + y^2 - 2ax + 2b^2 = 0$$

which subtend a right angle at the origin, is the circle

$$x^2 + y^2 - ax + b^2 = 0.$$ (O.C.)

24. Define the polar of a point with respect to a conic, and find the equation of the polar of (x', y') with respect to the hyperbola $x^2/a^2 - y^2/b^2 = 1$.

Prove that the pole with respect to the hyperbola of the tangent to the ellipse $x^2/a^2 + y^2/b^2 = 1$ at a point P is a point Q on the ellipse; and that the pole of the tangent at Q is P.   (O.C.)

25. Find the coordinates of the pole of the chord joining the points P, Q on the conic $x^2/a^2 + y^2/b^2 = 1$ whose eccentric angles are $\alpha - \beta, \alpha + \beta$.

If T is this pole and S is a focus of the conic, prove that

$$ST^2 = \sec^2\beta.\ SP..SQ.$$ (O.C.)

26. If the origin be at one of the limiting points of a system of coaxal circles of which
$$x^2 + y^2 + 2gx + 2fy + c = 0$$

is a member, the equation of the system of circles cutting them all orthogonally is

$$(x^2 + y^2)(g + \mu f) + c(x + \mu y) = 0.$$ (L.)

27. Find the coordinates of T, the point of intersection of the tangents to the ellipse $\dfrac{x^2}{a^2} + \dfrac{y^2}{b^2} = 1$, at the points

$$P(a \cos\theta, b \sin\theta), \quad Q(a \cos\varphi, b \sin\varphi).$$

M

Find the area of the triangle OPQ (where O is the origin), and prove that the locus of the points T for which this area is as great as possible is

$$\frac{x^2}{a^2} + \frac{y^2}{b^2} = 2.$$

(O.C.)

28. P is a fixed point and Σ a given circle. Show that the locus of points Q, such that the circle on PQ as diameter is orthogonal to Σ, is the polar of P with respect to Σ. Show also that these circles belong to a coaxal system.   (L.)

29. The tangents at P_1 and P_2 to the ellipse

$$b^2 x^2 + a^2 y^2 = a^2 b^2$$

meet at the point T with coordinates (X, Y). Prove that the equation of the chord $P_1 P_2$ is

$$\frac{xX}{a^2} + \frac{yY}{b^2} = 1.$$

Prove that, when TP_1 and TP_2 are perpendicular, the chord $P_1 P_2$ touches the ellipse

$$\frac{x^2}{a^4} + \frac{y^2}{b^4} = \frac{1}{a^2 + b^2}.$$

(O.C.)

30. Prove that the polar equation of a circle passing through the pole is of the form

$$r = d \cos(\theta - \beta),$$

and give the geometrical interpretation of the constants d, β.

The parabola $l = 2r \sin^2 \frac{1}{2}\theta$ has its focus at S, and P is the point of vectorial angle α on the curve. Find the equation of the circle SPR which touches the parabola at P.

If Q is the other extremity of the focal chord through P, and the circle SQR touches the parabola at Q, prove that the equation of SR is

$$\tan\left(\theta - \frac{3}{2}\alpha\right) = \tan^3 \frac{1}{2}\alpha.$$

(N.)

31. Establish the equation of the tangent to the parabola

$$\frac{l}{r} = 1 + \cos\theta$$

at the point P of vectorial angle α in the form

$$\frac{l}{r} = \cos\theta + \cos(\theta - \alpha).$$

Show that the tangents at the points P, Q whose vectorial angles are α, β intersect at T whose vectorial angle is $\theta = \frac{1}{2}(\alpha + \beta)$.

Prove that if PSQ is a right angle the locus of T is a rectangular hyperbola with the same focus S.   (N.)

32. A fixed ellipse has a focus at S, and K is a fixed point on its major axis, and XL a fixed straight line perpendicular to its major axis. A variable chord through K cuts the ellipse at points P, Q, and the internal bisector of the angle PSQ cuts XL at L. If M is the foot of the perpendicular from L to SP, prove that SM is of constant length. (N.)

33. Prove that the equation of the circle having the points (x_1, y_1) and (x_2, y_2) as extremities of a diameter is

$$(x - x_1)(x - x_2) + (y - y_1)(y - y_2) = 0.$$

The line $x \sin\alpha + y \cos\alpha = 1$ intersects the conic

$$ax^2 + 2hxy + by^2 = 1$$

at points P, Q which lie at finite distances from the origin. Find the equation of the circle on PQ as diameter and prove that, if this circle passes through the origin and has its centre on the x-axis, then

$$a + b = 1 \quad \text{and} \quad h = a \cot\alpha. \tag{L.}$$

34. (i) Find the polar equation of a parabola, referred to its focus and axis. [You should reduce your answer to one of the forms

$$r = a \operatorname{cosec}^2 \tfrac{1}{2}\theta, \quad r = a \sec^2 \tfrac{1}{2}\theta,$$

where $2a$ is the semi-latus rectum.]

(ii) Prove that the semi-latus rectum is the harmonic mean of the segments into which the focus divides any focal chord. (O.C.)

35. A chord through the focus S of the conic

$$l/r = 1 + e \cos\theta$$

meets the conic at P and Q. Points P', Q' on this line are such that $SP \cdot SP' = SQ \cdot SQ' = c^2$, where c is a constant. Find the equation of the locus of the mid-point of $P'Q'$. (O.C.)

36. The lines OP and OQ, where O is the origin, are conjugate semi-diameters of the ellipse $b^2x^2 + a^2y^2 = a^2b^2$. Prove that the eccentric angles of P and Q differ by a right angle.

The circles on OP, OQ as diameters meet again at R. Prove that, as P and Q vary, the locus of R is the curve

$$2(x^2 + y^2)^2 = a^2x^2 + b^2y^2.$$

Prove also that the area enclosed by this curve is one-quarter of the area of the director circle of the ellipse. (O.C.)

37. Prove that a variable circle which passes through the origin and is orthogonal to the circle

$$x^2 + y^2 - 4x - 6 = 0.$$

also passes through another fixed point, and find the coordinates of this point. (O.C.)

38. Show that the equation of the tangent to the conic $l/r = 1 + e\cos\theta$ at the point whose vectorial angle is α is

$$l/r = e\cos\theta + \cos(\theta - \alpha).$$

Prove that if p be the length of the perpendicular from the pole of coordinates upon the tangent at the point whose radius vector is r,

$$l^2/p^2 = 2l/r + e^2 - 1.$$

39. The equation of a coaxal system of circles is given in the form $x^2 + y^2 + 2\lambda x + c = 0$, where λ is a parameter. Show that the circle with parameter λ_1 lies entirely inside the circle with parameter λ_2 if and only if $c > 0$ and λ_1 lies between 0 and λ_2.

Show also that the circles with parameters λ', λ'' are orthogonal if and only if $c < 0$ and $\lambda'\lambda'' = c$. In this case, if the centres of the two circles are O', O'', prove that the circle on $O'O''$ as diameter belongs to the given coaxal system.   (L.)

40. Find the equation of the tangent to the ellipse $x^2/a^2 + y^2/b^2 = 1$ at the point whose eccentric angle is φ. If (h, k) denote the coordinates of the point of intersection of the tangents at the points whose eccentric angles are φ and φ', prove that

$$\tan^2\frac{\varphi - \varphi'}{2} = \frac{h^2}{a^2} + \frac{k^2}{b^2} - 1.$$

41. Prove that the equation of the director circle of the conic

$$ax^2 + 2hxy + by^2 = 1$$

is $$(ab - h^2)(x^2 + y^2) = a + b.$$

Show that the locus of poles, with respect to this circle, of tangents drawn to the conic is

$$(ab - h^2)(bx^2 - 2hxy + ay^2) = (a + b)^2.$$

42. Find the equation of the pair of straight lines joining the origin of coordinates to the intersection of the circle $x^2 + y^2 + 2gx + 2fy + c = 0$ and the straight line

$$lx + my = 1.$$

Hence, or otherwise, find the coordinates of the circumcentre of the triangle formed by the lines

$$ax^2 + 2hxy + by^2 = 0,\ lx + my = 1.$$

If the lines $ax^2 + 2hxy + by^2 = 0$ vary, but are equally inclined to the axes, show that the circumcentre moves on a line through the origin.   (L.)

43. P_1, P_2, P_3, P_4 are four points on the rectangular hyperbola

$$x = ct,\ y = c/t.$$

Find the equation of the chord joining P_1 to P_2 and the condition that this chord shall be perpendicular to that joining P_3 to P_4. Show that if this condition is satisfied P_1P_4 is perpendicular to P_2P_3 and P_1P_3 is perpendicular to P_2P_4 also. (L.)

44. A circle passes through the vertex of the parabola $y^2 = 4ax$ and cuts the parabola again in the three points $(at_1^2, 2at_1)$, $(at_2^2, 2at_2)$, $(at_3^2, 2at_3)$. Prove that the circle intercepts on the tangent to the parabola at the vertex a length equal to $at_1t_2t_3/2$.

45. Find the equation of the tangent to the parabola $y^2 = 4ax$ at the point $P(at^2, 2at)$.

If this tangent meets the x-axis at T, and $S(a, 0)$ is the focus, find the coordinates of the orthocentre H of the triangle SPT and show that as t varies the locus of H is the curve $ay^2 = x(a - x)^2$. (L.)

46. A circle of variable radius r moves so as always to touch each of the two fixed circles

$$(x - 1)^2 + y^2 = 4; \ (x + 1)^2 + y^2 = 9.$$

If both the fixed circles lie inside the variable circle, prove that its centre lies on the curve whose equation is

$$12x^2 - 4y^2 = 3. \tag{L.}$$

47. Prove that the focal distances from any point P on an ellipse are equally inclined to the tangent at the point.

The foci of an ellipse are S, S' and P is any point on the curve. If the normal at P intersects the line SS' at G, prove that $PG^2 = SP . S'P . (1 - e^2)$, where e is the eccentricity of the ellipse. (L.)

48. Find the equation of the pair of lines joining the origin to the points in which the pair of lines

$$4x^2 - 15xy - 4y^2 + 39x + 65y - 169 = 0$$

are met by the line $x + 2y - 5 = 0$.

Show that the quadrilateral having the first pair, and also the second pair, as adjacent sides is cyclic, and find the equation of its circumcircle. (L.)

49. Find the equation of the normal to the parabola $y^2 = 4ax$ at the point $(at^2, 2at)$.

The tangents at the points P and Q on the parabola meet at R, and the normals at P and Q meet at S. If S lies on the line $y = -a$, prove that R lies on the rectangular hyperbola $xy = a^2$. (L.)

50. Find the equation of the chord joining the two points $(ct_1, c/t_1)$ and $(ct_2, c/t_2)$ on the rectangular hyperbola $xy = c^2$, and deduce that the equation of the tangent to the curve at the point $(ct, c/t)$ is $t^2y + x = 2ct$.

The points $P(ct_1, c/t_1)$, $Q(ct_2, c/t_2)$ and $R(ct_3, c/t_3)$ on the hyperbola form a triangle right-angled at Q; prove that $t_1t_2^2t_3 = -1$, and deduce that the tangent to the hyperbola at Q is perpendicular to the straight line PR. (L.)

51. Prove that any two conjugate diameters of an ellipse form with either directrix a triangle whose orthocentre is at the corresponding focus. (L.)

52. Show that the equation $ax^2 + 2hxy + by^2 = 0$ represents a pair of straight lines. Show also that the above equation and the equation

$$a(x - p)^2 + 2h(x - p)(y - q) + b(y - q)^2 = 0$$

represent the sides of a parallelogram and find the equations of its diagonals.

A variable straight line through the fixed point (x_0, y_0) cuts the fixed lines $x^2 + 2kxy - y^2 = 0$ at the points A and B. O is the origin and P the point such that the parallelogram $OAPB$ is completed. Show that the locus of P is a rectangular hyperbola. (L.)

53. Define the radical axis of two circles and show that the family of circles given by the equation

$$x^2 + y^2 + 2\lambda x + c = 0,$$

where λ is a parameter, form a co-axial system.

Ox and Oy are two perpendicular lines. P is a variable point on Ox, and the polar of P with respect to a given fixed circle in the plane Oxy intersects Oy at the point Q. Prove that, for different positions of P on Ox, the circles on PQ as diameter form a co-axial system whose radical axis passes through the centre of the given fixed circle. (L.)

54. Prove that, for all values of the constants λ and μ, the circle whose equation is

$$(x - \alpha)(x - \alpha + \lambda) + (y - \beta)(y - \beta + \mu) = r^2$$

bisects the circumference of the circle,

$$(x - \alpha)^2 + (y - \beta)^2 = r^2.$$

Find the equation of the circle which bisects the circumference of the circle $x^2 + y^2 + 2y - 3 = 0$ and touches the line $x - y = 0$ at the origin. (L.)

5. Find the equation of the tangent to the parabola $y^2 = 4ax$ at the point $(at^2, 2at)$.

Find the coordinates of the centre of the circle which touches the parabola at this point and passes through the focus, and show that as t varies the locus of the centres of these circles is the curve

$$27ay^2 = (2x - a)(5a - x)^2. \tag{L.}$$

56. Show that

$$x^2 + 4xy - 2y^2 + 6x - 12y - 15 = 0$$

represents a pair of straight lines, and that these lines together with the pair of lines $x^2 + 4xy - 2y^2 = 0$ form a rhombus. (L.)

CHAPTER XXI

COORDINATE GEOMETRY OF THREE DIMENSIONS

21.1 The coordinate system

Let XOX', YOY', ZOZ' be three mutually perpendicular straight lines intersecting at the origin O, (Fig. 174). These lines are called the *coordinate axes* and determine three mutually perpendicular planes (OY, OZ), (OZ, OX), (OX, OY) called the *coordinate planes*.

The position of any point P in space is determined uniquely by its directed distances from these planes. Let L, M, N be the feet of the perpendiculars from P to the coordinate planes and let A, B, C be the other vertices (lying on the coordinate axes) of the cuboid with PL, PM, PN as adjacent edges (Fig. 174). Again with reference to this figure, displacements in the direction $\overrightarrow{OX}$ are defined as positive and displacements in the direction $\overrightarrow{OX'}$ as negative with similar definitions

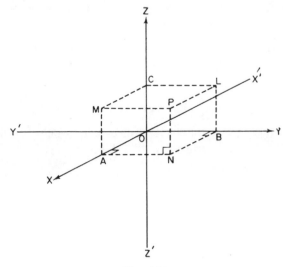

FIG. 174.

in relation to the other coordinate axes. Then the *coordinates* (x, y, z) of P referred to the coordinate axes are defined as follows:

$x = OA(= LP)$ or $-OA$ according as the displacement $\overrightarrow{OA}$ is positive or negative.

$y = OB(= MP)$ or $-OB$ according as the displacement $\overrightarrow{OB}$ is positive or negative.

$z = OC(= NP)$ or $-OC$ according as the displacement $\overrightarrow{OC}$ is positive or negative.

The coordinate planes divide space into eight octants; in the octant $OXYZ$, x, y, and z are all positive; in the octant $OX'YZ'$, x and z are negative and y is positive, etc. The coordinate planes OYZ, OZX, OXY are usually called the yz, zx, xy—planes respectively and the coordinate axes $X'OX$, $Y'OY$, $Z'OZ$, the x, y, z-axes respectively.

Direction of the axes. The directions of the coordinate axes are usually defined as follows:

Rotation through one right angle

(i) from $\overrightarrow{OX}$ to $\overrightarrow{OY}$ is *anticlockwise* in the xy-plane when viewed from the z-positive side of that plane,

(ii) from $\overrightarrow{OY}$ to $\overrightarrow{OZ}$ is *anticlockwise* in the yz-plane when viewed from the x-positive side of that plane,

(iii) from $\overrightarrow{OZ}$ to $\overrightarrow{OX}$ is *anticlockwise* in the zx-plane when viewed from the y-positive side of that plane.

Exercises 21.1

1. Show in a figure the positions of the points $(2, 1, 3)$, $(3, -1, -2)$, $(-4, 0, 0)$, $(-2, -1, -3)$.

2. State the locus of the point
   (i) whose x-coordinate is 2,
   (ii) whose x-coordinate is 2 and whose y-coordinate is 3.

3. With reference to Fig. 174, write down the coordinates of each of the points A, B, C, L, M, N and the point of concurrency of
   (i) the diagonals of the rectangle $MANP$,
   (ii) the diagonals of the cuboid $OANBLCMP$.

21.2 The distance between two points

In Fig. 175, P is the point (x_1, y_1, z_1) and Q the point (x_2, y_2, z_2). If the cuboid with diagonal PQ and faces parallel to the coordinate planes is completed, then

$$PS = (x_2 - x_1), ST = y_2 - y_1, TQ = z_2 - z_1.$$

But $$PQ^2 = PT^2 + TQ^2 = PS^2 + ST^2 + TQ^2$$

$$\therefore PQ^2 = (x_2 - x_1)^2 + (y_2 - y_1)^2 + (z_2 - z_1)^2.$$

$$\therefore PQ = \sqrt{\{(x_2 - x_1)^2 + (y_2 - y_1)^2 + (z_2 - z_1)^2\}}. \tag{21.1}$$

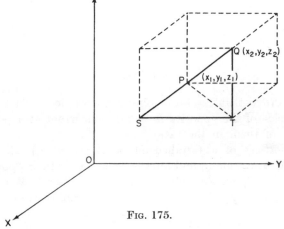

FIG. 175.

21.3 The coordinates of a point which divides the line joining two given points in a given ratio

In Fig. 176, P is the point (x_1, y_1, z_1), Q is the point (x_2, y_2, z_2) and T is the point which divides PQ internally in the ratio $l : m$. Also P_1, Q_1, T_1 are the orthogonal projections of P, Q, T respectively on the xy-plane. Then P_1 is $(x_1, y_1, 0)$, Q_1 is $(x_2, y_2, 0)$ and T_1 divides $P_1 Q_1$ internally in the ratio $l : m$. It follows [Vol. I, § 2.2] that T is $\left(\dfrac{l x_2 + m x_1}{l + m}, \dfrac{l y_2 + m y_1}{l + m}, z' \right)$ where z' is to be determined. But similarly, by projection on either the yz or the zx-plane $z' = \dfrac{l z_2 + m z_1}{l + m}$. Hence T is

$$\left(\frac{l x_2 + m x_1}{l + m}, \frac{l y_2 + m y_1}{l + m}, \frac{l z_2 + m z_1}{l + m} \right). \tag{21.2}$$

If *external* division in the ratio $l : m$ is defined as division in the ratio $l : - m$, the formula (21.2) is true for external division also.

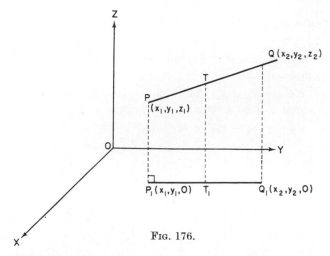

FIG. 176.

Example. Prove that the lines joining the vertices of a tetrahedron to the centroids of the opposite faces are concurrent at a point which divides each of them in the ratio 3 : 1.

Let the vertices of a tetrahedron be $A_1(x_1, y_1, z_1)$, $A_2(x_2, y_2, z_2)$, $A_3(x_3, y_3, z_3)$, $A_4(x_4, y_4, z_4)$ and let the centroids of their opposite faces be G_1, G_2, G_3, G_4 respectively. Then the mid-point M of A_2A_3 is $\{\frac{1}{2}(x_2 + x_3), \frac{1}{2}(y_2 + y_3), \frac{1}{2}(z_2 + z_3)\}$ and the x-coordinate of G_1, which divides A_4M in the ratio $2 : 1$ is $\dfrac{1 \cdot x_4 + 2 \cdot \frac{1}{2}(x_2 + x_3)}{1 + 2} = \dfrac{1}{3}(x_1 + x_2 + x_3)$.

Then the x-coordinate of the point G which divides A_1G_1 in the ratio $3 : 1$ is $\dfrac{1 \cdot x_4 + 3 \cdot \dfrac{1}{3}(x_1 + x_2 + x_3)}{1 + 3} = \dfrac{1}{4}(x_1 + x_2 + x_3 + x_4)$. Similar results are obtained for the y and z-coordinates of G which is therefore $\left(\dfrac{1}{4}\Sigma x_1, \dfrac{1}{4}\Sigma y_1, \dfrac{1}{4}\Sigma z_1\right)$. Since this result is symmetrical in relation to the coordinates of A_1, A_2, A_3 and A_4, the point G also lies on the lines A_2G_2, A_3G_3, A_4G_4 and divides each of them in the ratio $3 : 1$. G is called the centroid of the tetrahedron.

Exercises 21.3

1. Write down, leaving your answer in surd form if necessary, the length of the line joining each of the following pairs of points:

(i) $(3, 1, 2)$, $(-5, 0, 1)$; (ii) $(1, -4, 5)$, $(3, -1, 2)$; (iii) $(-2, 0, 4)$, $(-5, -1, 2)$; (iv) $(-1, -2, 0)$, $(0, 0, 2)$; (v) (a, b, c), (b, c, a).

2. If O is the origin of coordinates, A is $(2, 0, 0)$ and B is $(0, 2, 0)$, find the co-ordinates of the two possible positions of P such that $PO = PB = PA = BA$.

3. A tetrahedron $ABCD$ is defined by the points $A(5, 1, -1)$, $B(1, 2, 3)$, $C(-4, 2, 0)$, $D(6, 3, -6)$. The lines joining the vertices to the centroids of the opposite faces are concurrent at G. Calculate the length OG.

4. Calculate the cosine of the angle A of the triangle ABC in which A is $(1, -1, 2)$, B is $(1, 2, 6)$ and C is $(6, 11, 2)$.

5. Calculate the area of the triangle ABC in which A is $(2, -1, 1)$, B is $(1, 0, -4)$ and C is $(2, -1, 6)$.

6. Write down the coordinates of the points which divide the line joining each of the following pairs of points in the stated ratio.

   (i) $(3, 4, 2)$, $(-3, -2, 8)$; $2 : 1$ internally.

   (ii) $(0, 4, -2)$, $(8, 0, -2)$; $1 : 3$ internally.

   (iii) $(-2, 1, -4)$, $(-6, 5, 0)$; $4 : 1$ externally.

   (iv) $(2, 1, 5)$, $(3, 2, 4)$; $2 : 1$ externally.

   (v) $(2, -1, 4)$, $(7, 0, -1)$; $3 : 2$ internally.

7. Calculate the ratio in which the point $(3, -3, -4)$ divides the line joining $(-1, -1, -5)$ to $(11, -7, -2)$.

8. Prove that the lines joining the mid-points of opposite edges of a tetrahedron are concurrent at a point which is the mid-point of each of them and which is the centroid of the tetrahedron.

9. In each of the following cases find an equation or equations which must be satisfied by the coordinates of the point described and state the nature of the locus.

   (i) The point is equidistant from the three coordinate planes.

   (ii) The distance of the point from the origin is constant and equal to r.

   (iii) The distance of the point from the origin is equal to its distance from the point $A(a, b, c)$.

21.4 The equation of a plane

The first degree equation

$$ax + by + cz + d = 0 \qquad (21.3)$$

represents a plane. For if $P_1(x_1, y_1, z_1)$ and $P_2(x_2, y_2, z_2)$ lie on the locus in space represented by equation (21.3) and if $P_3(x_3, y_3, z_3)$ is the point which divides P_1P_2 in the ratio $l : m$, then

$$x_3 = \frac{lx_2 + mx_1}{l + m}, \quad y_3 = \frac{ly_2 + my_1}{l + m}, \quad z_3 = \frac{lz_2 + mz_1}{l + m}.$$

$$\therefore ax_3 + by_3 + cz_3 + d = \frac{l(ax_2 + by_2 + cz_2 + d) + m(ax_1 + by_1 + cz_1 + d)}{l + m} = 0.$$

Hence every point of any straight line joining two points on the locus represented by equation (21.3) lies on the locus. This locus is therefore a plane.

The equation of the plane through three given points

The equation of the plane through the points (x_1, y_1, z_1), (x_2, y_2, z_2), (x_3, y_3, z_3) is

$$\begin{vmatrix} x & y & z & 1 \\ x_1 & y_1 & z_1 & 1 \\ x_2 & y_2 & z_2 & 1 \\ x_3 & y_3 & z_3 & 1 \end{vmatrix} = 0 \tag{21.4}$$

because

(i) it is linear and therefore represents a plane, and

(ii) it is satisfied by the coordinates of each of the three given points.

If the points (x_1, y_1, z_1), (x_2, y_2, z_2), (x_3, y_3, z_3) are collinear, the left hand side of equation (21.4) vanishes identically. In this case, of course, there is an infinite number of planes through the three points.

The equations of the yz, zx, and xy-planes are $x = 0$, $y = 0$, $z = 0$ respectively.

21.5 The equations of a line—Direction Cosines

In general two planes intersect in a straight line. A straight line is therefore determined by two linear equations of the form

$$a_1 x + b_1 y + c_1 z + d_1 = 0,$$
$$a_2 x + b_2 y + c_2 z + d_2 = 0.$$

The line joining two points

Let A_1 be (x_1, y_1, z_1), A_2 be (x_2, y_2, z_2) and P be any point (x, y, z) on the line $A_1 A_2$, (Fig. 177). Let A_1', P', A_2' be the orthogonal projections of A_1, P, A_2 respectively on the xy-plane and Q_1, Q, Q_2 the orthogonal projections of A_1', P', A_2' respectively on the x-axis. Similarly R_1, R, R_2 are the orthogonal projections of A_1', P', A_2' on the y-axis. Then

$$\frac{A_1 P}{A_1 A_2} = \frac{A_1' P'}{A_1' A_2'} = \frac{Q_1 Q}{Q_1 Q_2} = \frac{R_1 R}{R_1 R_2}.$$

But $Q_1 Q = x - x_1$, $Q_1 Q_2 = x_2 - x_1$, $R_1 R = y - y_1$, $R_1 R_2 = y_2 - y_1$.

$$\therefore \frac{x - x_1}{x_2 - x_1} = \frac{y - y_1}{y_2 - y_1}.$$

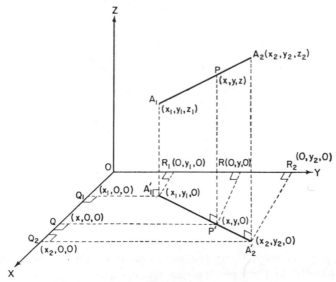

F<small>IG.</small> 177.

Similarly each of these ratios is equal to $\dfrac{z - z_1}{z_2 - z_1}$ and hence the equations of the line $A_1 A_2$ are

$$\frac{x - x_1}{x_2 - x_1} = \frac{y - y_1}{y_2 - y_1} = \frac{z - z_1}{z_2 - z_1}. \tag{21.5}$$

Direction Cosines

If $P(x, y, z)$ is such that the (scalar) distance OP is r and OP makes angles α, β, γ respectively with the *positive* directions of the coordinate axes (i.e., with OX, OY and OZ), then, (Fig. 178),

$$x = r \cos\alpha, \; y = r \cos\beta, \; z = r \cos\gamma.$$

$\cos\alpha$, $\cos\beta$, $\cos\gamma$ are called the *direction cosines* of the line OP, whose equations are

$$\frac{x}{\cos \alpha} = \frac{y}{\cos \beta} = \frac{z}{\cos \gamma}. \tag{21.6}$$

Since $x^2 + y^2 + z^2 = r^2$ we have

$$\cos^2\alpha + \cos^2\beta + \cos^2\gamma = 1. \tag{21.7}$$

Similarly the line through (x_1, y_1, z_1) and with direction cosines $\cos\alpha$, $\cos\beta$, $\cos\gamma$ has equations

$$\frac{x - x_1}{\cos \alpha} = \frac{y - y_1}{\cos \beta} = \frac{z - z_1}{\cos \gamma}. \tag{21.8}$$

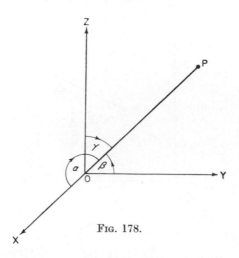

FIG. 178.

Here α, β, γ are respectively the angles made by the line with the lines drawn parallel to the axes through (x_1, y_1, z_1).

Direction Ratios

If $\cos\alpha$, $\cos\beta$, $\cos\gamma$ are the direction cosines of a line and if $\cos\alpha : \cos\beta : \cos\gamma = p : q : r$, then $p : q : r$ are called *direction ratios* of the line. If $p = k\cos\alpha$, $q = k\cos\beta$, $r = k\cos\gamma$, where $k > 0$, then

$$\cos^2\alpha + \cos^2\beta + \cos^2\gamma = 1.$$

$$\therefore \frac{p^2 + q^2 + r^2}{k^2} = 1.$$

$$\therefore \cos\alpha = \frac{p}{\sqrt{(p^2 + q^2 + r^2)}}, \quad \cos\beta = \frac{q}{\sqrt{(p^2 + q^2 + r^2)}},$$

$$\cos\gamma = \frac{r}{\sqrt{(p^2 + q^2 + r^2)}}.$$

The equations

$$\frac{x - a}{p} = \frac{y - b}{q} = \frac{z - c}{r} \qquad (21.9)$$

represent a straight line which passes through the point (a, b, c) in the direction with ratios $p : q : r$. This is the *standard form* of the equations of the line. If we set each of the ratios in equation (21.9) equal to the parameter λ we obtain the *parametric equations* of the line in the form

$$x = a + p\lambda, \quad y = b + q\lambda, \quad z = c + r\lambda. \qquad (21.10)$$

If $p = 0$, then $x = a$ and the line lies in the plane $x - a = 0$. Similar results hold when $q = 0$ or $r = 0$.

Examples. (i) Find the direction cosines of the line joining $(2, -1, 4)$ to $(3, -2, -1)$.

The line is

$$\frac{x-2}{3-2} = \frac{y-(-1)}{-2-(-1)} = \frac{z-4}{-1-4},$$

i.e.,

$$\frac{x-2}{1} = \frac{y+1}{-1} = \frac{z-4}{-5}.$$

Direction ratios of the line are therefore $1 : -1 : -5$. Therefore the line has direction cosines

$$\frac{1}{\sqrt{27}}, \quad \frac{-1}{\sqrt{27}}, \quad \frac{-5}{\sqrt{27}}.$$

(ii) Show that the points $P(2, 1, 1)$ and $Q(-1, -2, -2)$ are on opposite sides of the plane

$$x + y + z + 1 = 0$$

and find the ratio in which PQ is divided by the plane.

If PQ is divided by the plane in the ratio $k : 1$, a positive value of k would indicate internal division of PQ, with P and Q on opposite sides of the plane and a negative value of k would indicate external division of PQ with P and Q on the same side of the plane.

The coordinates of the point in which PQ meets the plane are

$$\left(\frac{2-k}{1+k}, \quad \frac{1-2k}{1+k}, \quad \frac{1-2k}{1+k} \right).$$

$$\therefore \frac{2-k}{1+k} + \frac{1-2k}{1+k} + \frac{1-2k}{1+k} + 1 = 0,$$

whence

$$k = \frac{5}{4}.$$

Hence P and Q are on opposite sides of the plane and PQ is divided internally in the ratio $5 : 4$.

Alternatively, we find the equation of PQ as

$$x - 2 = y - 1 = z - 1 = \lambda,$$

i.e., $\quad x = \lambda + 2, y = \lambda + 1, z = \lambda + 1.$

The line meets the plane $x + y + z + 1 = 0$ where

$$(\lambda + 2) + (\lambda + 1) + (\lambda + 1) + 1 = 0,$$

i.e., where $\qquad \lambda = -5/3.$

Then the point of intersection of the plane and the line is $\left(\dfrac{1}{3}, -\dfrac{2}{3}, -\dfrac{2}{3}\right)$ and this divides PQ internally in the ratio $5:4$.

(iii) Find in standard form the equation of the line of intersection of the planes
$$2x + y - z + 1 = 0, \quad x + 3y + z - 5 = 0.$$

The line of intersection cuts the yz-plane, i.e. the plane $x = 0$, where the equations
$$y - z + 1 = 0, \quad 3y + z - 5 = 0$$

are simultaneously satisfied, i.e., where $y = 1$, $z = 2$. Similarly the line cuts the plane $z = 0$ where $x = -8/5$, $y = 11/5$. The line of intersection is therefore the line joining $(0, 1, 2)$ to $(-8/5, 11/5, 0)$,

i.e.,
$$\frac{x}{4} = \frac{y-1}{-3} = \frac{z-2}{5}. \tag{1}$$

Note that the *equations* of a straight line are not unique. By taking a from each of the ratios in equation (1) we have as an equivalent form for the equations of this line
$$\frac{x - 4a}{4} = \frac{y + (3a - 1)}{-3} = \frac{z - (5a + 2)}{5}.$$

(iv) Show that the lines
$$\frac{x-1}{1} = \frac{y-2}{-3} = \frac{z-1}{-10}, \tag{1}$$

$$\frac{x+1}{1} = \frac{y-3}{1} = \frac{z-1}{6}, \tag{2}$$

intersect. Find the coordinates of their point of intersection. Find also the equation of the plane defined by these lines.

The parametric forms of the lines (1) and (2) are
$$x = 1 + \lambda, \quad y = 2 - 3\lambda, \quad z = 1 - 10\lambda, \tag{1a}$$
$$x = -1 + \mu, \quad y = 3 + \mu, \quad z = 1 + 6\mu. \tag{2a}$$

The lines intersect if there exist values of λ and μ for which the values of x, y, z given by equations (1a), (2a) are respectively equal. The x and y coordinates are equal if
$$1 + \lambda = -1 + \mu, \quad 2 - 3\lambda = 3 + \mu,$$

i.e., if $\quad \lambda = -\dfrac{3}{4}, \mu = \dfrac{5}{4}.\quad$ But, for these values of λ, μ,
$$1 - 10\lambda = \frac{17}{2} = 1 + 6\mu.$$

Hence the lines (1) and (2) intersect at the point $(\frac{1}{4}, 4\frac{1}{4}, 8\frac{1}{2})$.

The equations of the line (1) can be written

$$-3(x - 1) = y - 2, \quad -10(x - 1) = (z - 1),$$

i.e., $\qquad 3x + y - 5 = 0, \quad 10x + z - 11 = 0.$ \hfill (3)

Equations (3) separately represent two planes passing through the line (1). Then the equation

$$(3x + y - 5) + k(10x + z - 11) = 0, \hfill (4)$$

where k is a variable parameter, is the equation of a plane which passes through the line of intersection of the two planes given by equations (3). The plane (4) passes through $(-1, 3, 1)$, a point of the second line (2) other than the point of intersection of the lines (1) and (2), if

$$(-3 + 3 - 5) + k(-10 + 1 - 11) = 0,$$

i.e., if $k = -\dfrac{1}{4}$. The equation of the plane containing the lines is therefore

$$(3x + y - 5) - \frac{1}{4}(10x + z - 11) = 0$$

which reduces to

$$2x + 4y - z - 9 = 0.$$

In general, the lines

$$\frac{x - x_1}{l_1} = \frac{y - y_1}{m_1} = \frac{z - z_1}{n_1} = \lambda_1$$

and $\qquad \dfrac{x - x_2}{l_2} = \dfrac{y - y_2}{m_2} = \dfrac{z - z_2}{n_2} = \lambda_2,$

i.e., $\qquad x = x_1 + l_1\lambda_1, \quad y = y_1 + m_1\lambda_1, \quad z = z_1 + n_1\lambda_1$

and $\qquad x = x_2 + l_2\lambda_2, \quad y = y_2 + m_2\lambda_2, \quad z = z_2 + n_2\lambda_2,$

are coplanar if the equations

$$x_1 - x_2 + l_1\lambda_1 - l_2\lambda_2 = 0,$$
$$y_1 - y_2 + m_1\lambda_1 - m_2\lambda_2 = 0,$$
$$z_1 - z_2 + n_1\lambda_1 - n_2\lambda_2 = 0$$

have a consistent solution for λ_1, λ_2; the condition for this is

$$\begin{vmatrix} x_1 - x_2 & l_1 & l_2 \\ y_1 - y_2 & m_1 & m_2 \\ z_1 - z_2 & n_1 & n_2 \end{vmatrix} = 0.$$

Exercises 21.5

1. Find the equation of the plane through each of the following sets of points:

  (i) $(1, 1, 5)$, $(0, -2, -2)$, $(3, 2, 4)$;

  (ii) $(2, 0, 2)$, $(-5, 1, 1)$, $(4, 4, 4)$;

  (iii) $(1, 0, 6)$, $(-2, -4, 11)$, $(0, 0, 9)$;

  (iv) $(0, 0, a/b)$, $(1, 1, 1)$, $(1, a/b, a/b)$;

  (v) $(2, 0, 7)$, $(0, 5, 2)$, $(-2, 1, -3)$.

2. In each of the following cases find the equation of the plane defined by the given line and the given point.

  (i) $\dfrac{x}{2} = \dfrac{y}{1} = \dfrac{z-1}{1}$; $(0, 1, 6)$.

  (ii) $\dfrac{x+1}{3} = \dfrac{y+3}{6} = \dfrac{z+6}{5}$; $(2, 0, -3)$.

  (iii) $\dfrac{x-1}{1} = \dfrac{y-5}{-7} = \dfrac{z-2}{-2}$; $(3, 3, 4)$.

  (iv) $\dfrac{x+3}{6} = \dfrac{y+4}{8} = \dfrac{z+5}{10}$; $(5, 0, 3)$.

  (v) $\dfrac{x-2}{2} = \dfrac{y-5}{5} = \dfrac{z-2}{-2}$; $(-3, -10, 8)$.

3. Find in standard form the equations of the lines of intersection of the following pairs of planes:

  (i) $x + y + z - 8 = 0$; $2x - 3y + z - 1 = 0$.

  (ii) $x - 3y - z + 14 = 0$; $3x + y + z + 2 = 0$.

  (iii) $2x - y + 3z - 1 = 0$; $3x + y - z - 4 = 0$.

  (iv) $2x - y - 2z - 5 = 0$; $x + 3y - z + 1 = 0$.

  (v) $x + 3y - z + 4 = 0$; $x - y + z = 0$.

4. In each of the following cases find whether the given pair of lines intersects and, where appropriate, find the equation of the plane it defines.

  (i) $\dfrac{x+1}{5} = \dfrac{y+1}{2} = \dfrac{z+6}{11}$; $\dfrac{x}{2} = \dfrac{y-1}{-1} = \dfrac{z-1}{-1}$.

  (ii) $\dfrac{x+1}{2} = \dfrac{y-4}{-3} = \dfrac{z}{1}$; $\dfrac{x+2}{4} = \dfrac{y+2}{2} = \dfrac{z-2}{-1}$.

  (iii) $\dfrac{x}{3} = \dfrac{y-1}{-1} = \dfrac{z-1}{7}$; $\dfrac{x+2}{4} = \dfrac{y+2}{4} = \dfrac{z}{4}$.

  (iv) $\dfrac{x+2}{5} = \dfrac{y+2}{3} = \dfrac{z-2}{-6}$; $\dfrac{x+1}{5} = \dfrac{y-3}{-5} = \dfrac{z-3}{-10}$.

  (v) $\dfrac{x+1}{1} = \dfrac{y+3}{4} = \dfrac{z+1}{1}$; $\dfrac{x+2}{5} = \dfrac{y+1}{2} = \dfrac{z+2}{-2}$.

5. Find whether the points $P(1, -2, 0)$ and $Q(-1, 2, -1)$ are on the same or opposite sides of the plane $x + 3y - z + 4 = 0$ and calculate the ratio in which PQ is divided by the plane.

6. Find the direction cosines of the line joining $(2, -1, 4)$ to $(0, 2, 3)$.

7. Calculate the remaining direction cosines of a line given that $\cos\alpha = \frac{1}{2}$ and that $\cos\beta$, $\cos\gamma$ are equal and positive.

8. Find the coordinates of the point in which the line

$$\frac{x-1}{2} = \frac{y-4}{3} = \frac{z-1}{4}$$

meets the plane

$$x + 2y + z - 5 = 0.$$

9. Calculate the coordinates of the point in which the line of intersection of the planes

$$x - 3y + z + 4 = 0, \quad 2x + y + 2z - 6 = 0$$

meets the xy-plane.

10. Find the equation of the plane which passes through the points $(a, 0, 0)$, $(0, b, 0)$, $(0, 0, c)$.

21.6 The angle between two directions

Suppose that the direction cosines of two lines L_1 and L_2 are $\cos\alpha_1$, $\cos\beta_1$, $\cos\gamma_1$ and $\cos\alpha_2$, $\cos\beta_2$, $\cos\gamma_2$ respectively. Let OP_1, OP_2 be straight lines parallel to L_1 and L_2 and let $OP_1 = OP_2 = r$, $\angle P_1OP_2 = \theta$, (Fig. 179). Then P_1 is the point $(r\cos\alpha_1, r\cos\beta_1, r\cos\gamma_1)$ and P_2 is the point $(r\cos\alpha_2, r\cos\beta_2, r\cos\gamma_2)$. But

$$OF_1^2 + OF_2^2 - P_1F_2^2 = 2OP_1.OP_2\cos\theta.$$

$$\therefore 2r^2 - \{(r\cos\alpha_1 - r\cos\alpha_2)^2 + (r\cos\beta_1 - r\cos\beta_2)^2 +$$
$$+ (r\cos\gamma_1 - r\cos\gamma_2)^2\} = 2r^2\cos\theta.$$

$$\therefore 2\cos\theta = 2 - \{\cos^2\alpha_1 + \cos^2\beta_1 + \cos^2\gamma_1 + \cos^2\alpha_2 + \cos^2\beta_2 + \cos^2\gamma_2$$
$$- 2(\cos\alpha_1\cos\alpha_2 + \cos\beta_1\cos\beta_2 + \cos\gamma_1\cos\gamma_2)\}.$$

Hence, after using equation (21.7),

$$\cos\theta = \cos\alpha_1\cos\alpha_2 + \cos\beta_1\cos\beta_2 + \cos\gamma_1\cos\gamma_2. \tag{21.11}$$

It is customary to use the symbols l, m, n to denote the direction cosines of a line. In this notation, therefore, the angle θ between the lines with direction cosines l_1, m_1, n_1 and l_2, m_2, n_2 is given by

$$\cos\theta = l_1l_2 + m_1m_2 + n_1n_2. \tag{21.11a}$$

The two lines are therefore at right angles if

$$l_1 l_2 + m_1 m_2 + n_1 n_2 = 0$$

and conversely.

Note that the angle between two *skew* lines, i.e., two non-parallel lines which do not meet, is defined as the angle between two lines which are coplanar and respectively parallel to the skew lines.

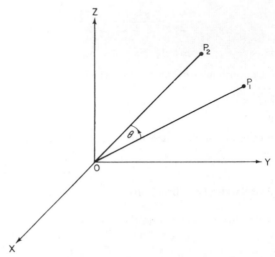

FIG. 179.

The direction of the normal to a plane

In Fig. 180 the normal to a plane from the origin O meets the plane at N where $ON = p$ and the direction cosines of ON are $\cos\alpha$, $\cos\beta$, $\cos\gamma$. If $A(x, y, z)$ is any point in the plane so that $OA = r$, then OA has direction cosines $\dfrac{x}{r}, \dfrac{y}{r}, \dfrac{z}{r}$ and, since ONA is a right angle, $OA \cos \theta = p$ where $\angle AON = \theta$. But using equation (21.11)

$$\cos \theta = \frac{x}{r} \cos \alpha + \frac{y}{r} \cos \beta + \frac{z}{r} \cos \gamma.$$

$$\therefore \frac{x}{r} \cos \alpha + \frac{y}{r} \cos \beta + \frac{z}{r} \cos \gamma = \frac{p}{r};$$

i.e., $x \cos\alpha + y \cos\beta + z \cos\gamma = p$ (21.12)

which is the equation of the plane.

Conversely a normal to the plane

$$ax + by + cz + d = 0$$

has direction ratios $a : b : c$ and direction cosines

$$\frac{a}{\sqrt{(a^2 + b^2 + c^2)}}, \quad \frac{b}{\sqrt{(a^2 + b^2 + c^2)}}, \quad \frac{c}{\sqrt{(a^2 + b^2 + c^2)}} .$$

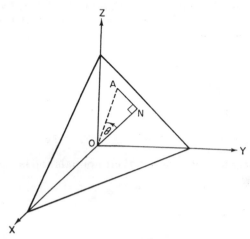

FIG. 180.

The perpendicular distance of a point from a plane

The normal from the origin to the plane

$$x \cos\alpha + y \cos\beta + z \cos\gamma - p = 0$$

is of length p, and the plane through (x_1, y_1, z_1) parallel to this plane has equation

$$x \cos\alpha + y \cos\beta + z \cos\gamma - p' = 0$$

where p' is the length of the normal from the origin to this second plane.

$$\therefore \ x_1 \cos\alpha + y_1 \cos\beta + z_1 \cos\gamma - p' = 0$$

whence the length of the perpendicular from (x_1, y_1, z_1) to the plane $x \cos\alpha + y \cos\beta + z \cos\gamma - p = 0$ is

$$|p' - p| = |x_1 \cos\alpha + y_1 \cos\beta + z_1 \cos\gamma - p|. \qquad (21.13)$$

Then the length of the perpendicular from (x_1, y_1, z_1) to

$$ax + by + cz + d = 0 \text{ is}$$

$$\left| \frac{a x_1 + b y_1 + c z_1 + d}{\sqrt{(a^2 + b^2 + c^2)}} \right|. \tag{21.14}$$

The function $ax + by + cz + d$ can be shown to be positive for all points on one side of the plane and negative for all points on the other side. The perpendicular is defined as positive when it is drawn from a point on the same side of the plane as the origin and negative otherwise.

The angle between two planes

The acute angle between the planes

$$x \cos\alpha_1 + y \cos\beta_1 + z \cos\gamma_1 - p_1 = 0,$$

$$x \cos\alpha_2 + y \cos\beta_2 + z \cos\gamma_2 - p_2 = 0$$

is equal to the acute angle between their normals and is therefore θ where

$$\cos\theta = \left| \cos\alpha_1 \cos\alpha_2 + \cos\beta_1 \cos\beta_2 + \cos\gamma_1 \cos\gamma_2 \right|. \tag{21.15}$$

The acute angle between the planes

$$a_1 x + b_1 y + c_1 z + d_1 = 0, \; a_2 x + b_2 y + c_2 z + d_2 = 0,$$

the direction ratios of whose normals are $a_1 : b_1 : c_1$ and $a_2 : b_2 : c_2$ respectively, is therefore θ where

$$\cos\theta = \left| \frac{a_1 a_2 + b_1 b_2 + c_1 c_2}{\sqrt{(a_1^2 + b_1^2 + c_1^2)} \cdot \sqrt{(a_2^2 + b_2^2 + c_2^2)}} \right|. \tag{21.16}$$

The angle between a line and a plane

The acute angle between the line

$$\frac{x - x_1}{\cos\alpha_1} = \frac{y - y_1}{\cos\beta_1} = \frac{z - z_1}{\cos\gamma_1}$$

and the plane

$$x \cos\alpha_2 + y \cos\beta_2 + z \cos\gamma_2 - p = 0$$

is equal to the complement of the angle between the line and the normal to the plane and is therefore $\frac{1}{2}\pi - \psi$ where

$$\cos\psi = \left| \cos\alpha_1 \cos\alpha_2 + \cos\beta_1 \cos\beta_2 + \cos\gamma_1 \cos\gamma_2 \right|. \tag{21.17}$$

The bisectors of the angles between two planes

The points equidistant from the two planes

$$a_1x + b_1y + c_1z + d_1 = 0, \tag{1}$$

$$a_2x + b_2y + c_2z + d_2 = 0 \tag{2}$$

lie on the planes bisecting the angles between the planes (1) and (2). Hence these bisectors have equations

$$\frac{a_1x + b_1y + c_1z + d_1}{\sqrt{(a_1^2 + b_1^2 + c_1^2)}} = \pm \frac{a_2x + b_2y + c_2z + d_2}{\sqrt{(a_2^2 + b_2^2 + c_2^2)}}. \tag{21.18}$$

Examples. (i) Find the equations of the line which (1) passes through the origin, (2) lies in the xy-plane and (3) is at right angles to the line

$$\frac{x}{7} = \frac{y}{24} = \frac{z}{60}.$$

The direction cosines of the given line are $\dfrac{7}{65}, \dfrac{24}{65}, \dfrac{60}{65}$. If the direction cosines of the required line are $\cos\alpha$, $\cos\beta$, $\cos\gamma$, then

$$\frac{7}{65}\cos\alpha + \frac{24}{65}\cos\beta + \frac{60}{65}\cos\gamma = 0. \tag{1}$$

But since the line lies in the plane $z = 0$ it is perpendicular to the z-axis $\left(\dfrac{x}{0} = \dfrac{y}{0} = \dfrac{z}{1}\right)$ and hence

$$\cos\gamma = 0. \tag{2}$$

Also $$\cos^2\alpha + \cos^2\beta + \cos^2\gamma = 1. \tag{3}$$

The solutions of equations (1), (2), (3) are

$$\cos\alpha = \pm\frac{24}{25}, \quad \cos\beta = \mp\frac{7}{25}, \quad \cos\gamma = 0.$$

Hence the required line is

$$7x - 24y = 0, \quad z = 0$$

or, expressed in standard form,

$$\frac{x}{24} = \frac{y}{-7} = \frac{z}{0}.$$

$\left[\text{Note that taking }\cos\alpha = -\dfrac{24}{25}\text{ and }\cos\beta = \dfrac{7}{25}\text{ gives the same line.}\right]$

(ii) Find the equation of the line through the point of intersection of the lines

$$\frac{x+1}{3} = \frac{y+9}{12} = \frac{z+1}{2}, \quad \frac{x-2}{1} = \frac{y-3}{-1} = \frac{z-1}{1}$$

which is at right angles to both of them.

Expressing the lines in parametric form

$$x = 3\lambda - 1, \quad y = 12\lambda - 9, \quad z = 2\lambda - 1,$$
$$x = \mu + 2, \quad y = -\mu + 3, \quad z = \mu + 1,$$

we see that their point of intersection is $(2, 3, 1)$ corresponding to $\lambda = 1, \mu = 0$.

If $p:q:r$ are the direction ratios of the required line, then

$$3p + 12q + 2r = 0,$$
$$p - q + r = 0$$

whence

$$\frac{p}{14} = \frac{q}{-1} = \frac{r}{-15}.$$

Hence the required line is

$$\frac{x-2}{14} = \frac{y-3}{-1} = \frac{z-1}{-15}.$$

(iii) Find the equation of the plane which perpendicularly bisects the line joining $P(2, -3, 0)$ to $Q(-4, 5, 2)$.

The coordinates of the mid point of PQ are $(-1, 1, 1)$ and the direction ratios of PQ, which is a normal to the required plane, are $3 : -4 : -1$. The equation of the required plane is therefore of the form

$$3x - 4y - z + d = 0$$

and, since it passes through $(-1, 1, 1)$, $d = 8$ and the required equation is

$$3x - 4y - z + 8 = 0.$$

(iv) Find the equations of the perpendicular drawn from $A(2, -1, 1)$ to the line

$$\frac{x-1}{3} = \frac{y-2}{4} = \frac{z+1}{-1}.$$

Find also the coordinates of the point B in which the perpendicular meets the line and the length of the perpendicular.

Let B be the point $(3\lambda + 1, 4\lambda + 2, -\lambda - 1)$ on the given line. Then the direction ratios of AB are $(3\lambda - 1) : (4\lambda + 3) : -(\lambda + 2)$. Thus AB is perpendicular to the given line if

$$3(3\lambda - 1) + 4(4\lambda + 3) + (\lambda + 2) = 0,$$

i.e., if $\hspace{4cm} \lambda = -11/26.$

Thus the direction ratios of the required perpendicular are $59 : -34 : 41$ and its equations are

$$\frac{x - 2}{59} = \frac{y + 1}{-34} = \frac{z - 1}{41}.$$

Also B is the point $\left(-\dfrac{7}{26}, \dfrac{8}{26}, -\dfrac{15}{26}\right)$ and the length of the perpendicular is

$$\sqrt{\left\{\left(2 + \frac{7}{26}\right)^2 + \left(-1 - \frac{8}{26}\right)^2 + \left(1 + \frac{15}{26}\right)^2\right\}} = \frac{3\sqrt{702}}{26}.$$

(v) Find the equation of the plane which passes through the line of intersection of the planes

$$x - y + z + 1 = 0, \quad 2x + y - 3z + 2 = 0$$

and is parallel to the straight line

$$\frac{x - 1}{4} = \frac{y - 2}{3} = \frac{z - 1}{1}.$$

Find, also, the distance between this line and the plane. $\hspace{1cm}$ (L.)

The plane

$$(x - y + z + 1) + k(2x + y - 3z + 2) = 0,$$

where k varies, passes through the line of intersection of the given planes. The direction ratios of the normal to this plane are $(1 + 2k) : (-1 + k) : (1 - 3k)$ and so, if the plane is parallel to the given straight line which has direction ratios $4 : 3 : 1$,

$$4(1 + 2k) + 3(-1 + k) + 1(1 - 3k) = 0,$$

i.e., $\hspace{4cm} k = -\dfrac{1}{4}.$

Hence the equation of the required plane is

$$(x - y + z + 1) - \frac{1}{4}(2x + y - 3z + 2) = 0$$

which reduces to

$$2x - 5y + 7z + 2 = 0.$$

Since the line is parallel to the plane, the distance between the line and the plane is equal to the distance of any point on the line from the plane. Clearly the point $(1, 2, 1)$ lies on the line and hence by (21.14) the required distance is

$$\left| \frac{2.1 - 5.2 + 7.1 + 2}{\sqrt{(2^2 + 5^2 + 7^2)}} \right| = \frac{1}{\sqrt{78}}.$$

(vi) Find the coordinates of the mirror image of the point $(-2, 1, 7)$ in the plane Π, whose equation is $2x - y - 3z = 2$.

Find the equations of the mirror image in Π of the line

$$\frac{x + 2}{3} = y - 1 = \frac{z - 7}{-3}. \tag{L.}$$

The mirror image of the point $A\,(-2, 1, 7)$ in the plane Π is the point A' which is such that AA' is perpendicularly bisected by the plane. The direction ratios of AA' are the same as those of the normal to Π. Hence A' is the point $(-2 + 2\lambda, 1 - \lambda, 7 - 3\lambda)$ where λ is to be chosen so that the mid-point $(-2 + \lambda, 1 - \frac{1}{2}\lambda, 7 - 3\lambda)$ of AA' lies on the plane Π. This is so when $\lambda = 4$ and hence A' is $(6, -3, -5)$.

The point $B(1, 2, 4)$ lies on the given line and, obtained in a similar manner, its mirror image B' in Π is the point $(5, 0, -2)$. But A lies on the given line and hence the mirror image of the given line is $A'B'$ which has equations

$$\frac{x - 6}{1} = \frac{y + 3}{-3} = \frac{z + 5}{-3}.$$

Exercises 21.6

1. Find the equations of the lines through $(1, 2, -1)$ which make equal angles with the three coordinate axes.

2. Calculate each of the angles between the line

$$\frac{x - 3}{3} = \frac{y + 1}{4} = \frac{z - 2}{12}$$

and the coordinate axes.

3. Find the equations of the line through the origin which is parallel to the line of intersection of the planes

$$x + y + z - 3 = 0, \quad 2x - y - 2z + 6 = 0.$$

4. Calculate the cosines of the acute angles between the lines whose equations are given.

(i) $\dfrac{x+1}{7} = \dfrac{y+2}{24} = \dfrac{z+1}{60}$; $\dfrac{x+1}{-12} = \dfrac{y-2}{-3} = \dfrac{z-1}{4}$.

(ii) $\dfrac{x-1}{3} = \dfrac{y+2}{5} = \dfrac{z-3}{2}$; $\dfrac{x}{5} = \dfrac{y-1}{2} = \dfrac{z+2}{3}$.

(iii) $\dfrac{x-1}{3} = \dfrac{y+3}{-4} = \dfrac{z-2}{12}$; $\dfrac{x+1}{4} = \dfrac{y-5}{12} = \dfrac{z-2}{3}$.

(iv) $\dfrac{x-1}{5} = \dfrac{y-2}{-2} = \dfrac{z+1}{4}$; $\dfrac{x-1}{2} = \dfrac{y+1}{3} = \dfrac{z}{-1}$.

5. Find the equations of the line in the xy-plane which passes through the origin, and which is at right angles to the line

$$\frac{x}{l} = \frac{y}{m} = \frac{z}{n}.$$

6. Prove that there are two lines in the xy-plane which pass through the origin and which make angles of 60° with the line

$$\frac{x}{12} = \frac{y}{3} = \frac{z}{4}.$$

Prove also that the cosines of the angles made by these lines with the x-axis are the roots of the equation

$$612\,t^2 - 624\,t + 133 = 0.$$

7. Calculate the angles of the triangle whose sides are the lines

$$\frac{x-4}{1} = \frac{y-1}{-2} = \frac{z+1}{-1}, \quad \frac{x-4}{-5} = \frac{y-1}{1} = \frac{z+1}{2},$$

$$\frac{x+1}{2} = \frac{y-2}{-1} = \frac{z-1}{-1}.$$

8. Prove that the line

$$\frac{x-a}{l} = \frac{y-b}{m} = \frac{z-c}{n}$$

is parallel to the plane

$$(m+n)x - (n+l)y - (l-m)z = d.$$

9. Use direction ratios to show that the points $(1, -1, 4)$, $(-2, 1, -3)$, and $(4, -3, 11)$ are collinear.

10. A tetrahedron $OABC$ has $OA = OB = OC = AC = BC = 2$ units unit and angle $ACB = 90°$. Make a suitable choice of axes and calculate the angle between OC and AB by coordinate methods.

11. Find the equations of the perpendicular drawn from the origin to the line

$$\frac{x-2}{1} = \frac{y-3}{2} = \frac{z+1}{-1}$$

and find the length of this perpendicular.

12. Find the equations of the perpendicular drawn from $(3, 1, -1)$ to the line

$$\frac{x}{3} = \frac{y}{4} = \frac{z}{-12}.$$

13. Find the length of the perpendicular drawn from $A(2, 0, 0)$ to the line

$$\frac{x-1}{2} = \frac{y+1}{-1} = \frac{z+2}{1},$$

by expressing the coordinates of an arbitrary point P of the line in parametric form and finding the minimum value of AP^2.

14. In each of the following cases write down in surd form the perpendicular distance of the given point from the given plane.

(i) $x + y + z - 3 = 0$, $(1, -2, 0)$;

(ii) $2x - 3y + 4z + 1 = 0$, $(0, -1, 1)$;

(iii) $3x - y + z + 2 = 0$, $(1, 4, 1)$;

(iv) $2x + 5y - 3z - 1 = 0$, $(0, 2, 4)$;

(v) $3x - 7y + 9z + 1 = 0$, $(1, 1, -3)$.

15. In each of the following cases calculate the cosine of the acute angle between the planes.

(i) $3x + 4y + 12z - 1 = 0$, $4x - 3y - 12z + 4 = 0$;

(ii) $7x - 24y + 60z - 3 = 0$, $12x - 3y + 4z + 1 = 0$;

(iii) $3x - 4y - 12z - 2 = 0$, $24x + 7y + 60z = 0$;

(iv) $x + y + z - 4 = 0$, $2x - y - z + 6 = 0$.

16. In each case calculate the acute angle between the given line and the given plane.

(i) $\dfrac{x}{3} = \dfrac{y}{4} = \dfrac{z}{12}$, $\quad 7x - 24y + 60z = 0$;

(ii) $\dfrac{x-1}{12} = \dfrac{y+2}{-4} = \dfrac{z-1}{3}$, $\quad 3x + 4y - 12z + 2 = 0$;

(iii) $\dfrac{x+2}{7} = \dfrac{y-1}{60} = \dfrac{z+2}{24}$, $\quad 12x - 3y + 4z = 0$;

(iv) $x = y = z$, $x - y - z - 4 = 0$.

17. Find the equation of the plane through the line

$$\frac{x-1}{3} = \frac{y+4}{4} = \frac{z-1}{1}$$

which is parallel to the line

$$\frac{x}{4} = \frac{y}{3} = \frac{z}{12}.$$

18. Find the equation of the plane which passes through the point (a, b, c), and which is perpendicular to the line

$$\frac{x}{l} = \frac{y}{m} = \frac{z}{n}.$$

19. Find the equations of
(i) the plane through the origin and through the line

$$\frac{x-1}{1} = \frac{y+4}{2} = \frac{z-1}{-1},$$

(ii) the plane through the origin perpendicular to this line.
Hence find the equation of the line through the origin perpendicular to

$$\frac{x-1}{1} = \frac{y+4}{2} = \frac{z-1}{-1}.$$

20. Find the angle between the lines

$$3x + 2y + z - 5 = 0 = x + y - 2z - 3,$$
$$8x - 4y - 4z = 0 = 7x + 10y - 8z. \qquad \text{(L.)}$$

21. Find the equation of the plane passing through the points $(0, 1, 3)$, $(1, 2, 5)$, $(-1, -1, 0)$ and show that the distance of the plane from the origin is $2/\sqrt{3}$.   (L.)

22. Show that the equation of the plane which passes through the origin and contains the line

$$\frac{x-1}{2} = \frac{y-2}{1} = \frac{z-3}{-2}$$

is $7x - 8y + 3z = 0$ (L.)

23. Find the equation of the plane through the point $P(2, 2, 1)$ and l, the line of intersection of the two planes

$$2x - y - z = 3, \; 3x - y - 3z = 4.$$

Find also the equation of the plane π through P perpendicular to l.   (L.)

24. Two planes are given by the equations

$$3x + y + 2z = 3,$$
$$-x + 2y + z = 4.$$

Find (i) the equation of the plane through their line of intersection and the origin, (ii) the equations of the line through the feet of the perpendiculars from the point $(1, 1, 1)$ to the two planes.   (L.)

25. The planes p and q have equations $2y - z - 1 = 0$ and $10x + 3y + 6z - 19 = 0$ respectively. Show that p and q are perpendicular and that their line of intersection l has equations $\dfrac{x-1}{3} = \dfrac{y-1}{-2} = \dfrac{z-1}{-4}$.

The point $A(4, 1, 1)$ is on p and $B(-26, 31, 31)$ is on q. Show that $C(3, 2, 2)$ is on AB and find the coordinates of the point D on l where the line through C perpendicular to l meets l. Prove that CD is perpendicular to AB.   (L.)

26. A tetrahedron has its vertices at $A(0, 0, 1)$, $B(3, 0, 1)$, $C(2, 3, 1)$ and $D(1, 1, 2)$. Find the angles between the faces ABC and BCD, between the edges AB, AC, and between the edge BC and the face ADC. (L.)

27. Find the equation of the plane $\varPi$ which passes through the point $(1, 2, 1)$ and the line of intersection of the planes

$$x - 4y + z - 1 = 0, \, 2x + y - z - 2 = 0.$$

Find also the coordinates of the point P where $\varPi$ meets the line L, which passes through the point $(1, 0, 2)$ and is perpendicular to the plane

$$x + 2y + 3z = 0.$$ (L.)

21.7 The intersection of three planes

When the determinant

$$\varDelta = \begin{vmatrix} a_1 & b_1 & c_1 \\ a_2 & b_2 & c_2 \\ a_3 & b_3 & c_3 \end{vmatrix}$$

does not vanish, the planes

$$a_1 x + b_1 y + c_1 z + d_1 = 0,$$
$$a_2 x + b_2 y + c_2 z + d_2 = 0,$$
$$a_3 x + b_3 y + c_3 z + d_3 = 0$$

have just one point which is common to all three of them and which is given by the solution of the three equations (see § 11.12).

If $\varDelta = 0$, the equations

$$a_1 l + b_1 m + c_1 n = 0,$$
$$a_2 l + b_2 m + c_2 n = 0,$$
$$a_3 l + b_3 m + c_3 n = 0$$

have a set of solutions which are not all zero, of the form

$$l : m : n = k_1 : k_2 : k_3$$

where k_1, k_2, k_3 are constants. There exists, therefore, a line with direction ratios $l : m : n$ which is perpendicular to the normal of each of the three planes. This line is parallel to each of the three planes and therefore, if $\varDelta = 0$, the three planes are all parallel to the same line and the following possibilities arise:

(1) Two of the planes are parallel or all three of the planes are parallel.

(2) The three planes have a common line of intersection.

(3) The three planes intersect in pairs along three parallel lines and form a triangular prism.

(We exclude from consideration the trivial case in which one of the given equations is a multiple of another so that two of the planes are coincident.)

In any particular problem the above three cases can be distinguished as follows:

Case (1) arises when there exists a relation of the form

$$\frac{a_1}{a_2} = \frac{b_1}{b_2} = \frac{c_1}{c_2} \neq \frac{d_1}{d_2}$$

so that two of the planes have a common normal and do not coincide.

In the other two cases we can find the line of intersection of two of the planes. If the coordinates of an arbitrary point on this line (expressed parametrically) satisfy identically the equation of the third plane, then the three planes have a common line of intersection [Case (2)]; otherwise the planes intersect along parallel distinct lines.

Examples. (i) Consider

$$x + y + z - 4 = 0,$$
$$2x - 3y - z + 3 = 0,$$
$$x + 2y + 3z - 9 = 0.$$

Here

$$\Delta = \begin{vmatrix} 1 & 1 & 1 \\ 2 & -3 & -1 \\ 1 & 2 & 3 \end{vmatrix} = -7$$

and hence the planes intersect in one point. Solution of the equations gives this point as $(1, 1, 2)$.

(ii) Consider

$$2x - y - z + 4 = 0, \tag{1}$$
$$x + y - z - 1 = 0, \tag{2}$$
$$5x - 4y - 2z + 13 = 0. \tag{3}$$

Here

$$\Delta = \begin{vmatrix} 2 & -1 & -1 \\ 1 & 1 & -1 \\ 5 & -4 & -2 \end{vmatrix} = 0.$$

Since the direction ratios of the normals to the planes are distinct no two of the planes are parallel, and the planes intersect in lines which are parallel or coincident. But the line of intersection of (1) and (2) has equations

$$\frac{x+5}{2} = \frac{y}{1} = \frac{z+6}{3}, \tag{4}$$

i.e.,
$$x = 2\lambda - 5, \quad y = \lambda, \quad z = 3\lambda - 6.$$

These values of (x, y, z) identically satisfy equation (3) and hence the planes have the common line of intersection (4).

(iii) Consider

$$2x - 3y + 5z - 4 = 0,$$
$$x - y + 3z - 4 = 0,$$
$$4x - 6y + 10z - 5 = 0.$$

The first and third of these planes are parallel and thus the planes have no common point. (The second plane is not parallel to the other two.

(iv) Consider

$$x + 4y - z = 0$$
$$2x + 8y - 2z + 5 = 0,$$
$$3x + 12y - 3z + 1 = 0.$$

The three planes are all parallel and have no common point.

(v) Consider

$$4x + y + z - 10 = 0, \tag{1}$$
$$6x + 3y + z - 4 = 0, \tag{2}$$
$$x - 5y + 2z - 3 = 0. \tag{3}$$

Here

$$\Delta = \begin{vmatrix} 4 & 1 & 1 \\ 6 & 3 & 1 \\ 1 & -5 & 2 \end{vmatrix} = 0$$

and hence the planes intersect in lines which are parallel or coincident. The line of intersection of (1) and (2) is

$$\frac{x}{1} = \frac{y+3}{-1} = \frac{z-13}{-3}$$

or
$$x = \lambda, \quad y = -\lambda + 3, \quad z = -3\lambda + 13.$$

These coordinates do not satisfy equation (3). Hence the planes do *not* have a common line of intersection. Thus the three planes intersect in pairs along three parallel lines with direction ratios $1 : -1 : -3$.

Exercises 21.7

In each of the following questions examine the nature of the intersection of the set of planes given.

(i) If the planes have a common point, find the coordinates of that point.

(ii) If the planes have a common line, find equations of the line in the form
$$\frac{x-a}{l} = \frac{y-b}{m} = \frac{z-c}{n}.$$

(iii) If the planes intersect in parallel lines find the direction ratios of those lines.

1. $x + y + z - 5 = 0; 2x + 3y + 4z - 6 = 0; 5x + 6y + 7z + 3 = 0.$

2. $2x - 3y + z - 4 = 0; x + y - z - 10 = 0; 2x + 5y + 2z + 11 = 0.$

3. $3x + y - z - 5 = 0; x - y - z + 1 = 0; 5x + 3y - z - 11 = 0.$

4. $2x - 3y + z - 8 = 0; x + 5y - z + 1 = 0; 4x - 6y + 2z = 0.$

5. $x - 2y + 3z - 4 = 0; 3x - 4y + 5z - 6 = 0; 7x - 8y + 9z - 11 = 0.$

6. $x - 3y - 5z - 10 = 0; 2x + y + z - 2 = 0; x + 3y - z = 0.$

7. $x + 4y - 2z = 0; 2x + 8y - 4z + 5 = 0; 5x + 20y - 10z + 1 = 0.$

8. $x - 4y - 4z + 9 = 0; 5x - 8y - 2z + 21 = 0; 2x - 2y + z + 6 = 0.$

9. $2x + y + z + 5 = 0; x - 2y - 3z + 5 = 0; 2x + 5y + z + 1 = 0.$

10. $7x - 2y + 3z - 5 = 0; 17x - 2y + 13z + 3 = 0; 10x - y + 8z = 0.$

Miscellaneous Exercises XXI

1. Find the direction cosines of a line perpendicular to the two straight lines whose equations are
$$\frac{x}{2} = \frac{y}{-1} = \frac{z}{3}, \quad \frac{x-2}{2} = \frac{y-3}{1} = \frac{z-4}{-1}. \tag{N.}$$

2. The length of each edge of a regular tetrahedron $ABCD$ is $2a$. A perpendicular is drawn from the vertex D to the opposite face ABC, meeting this face in the point O. Show that O is the centroid of the triangle ABC.

Choosing OA as the axis of x, OD as the axis of z and a line through O parallel to CB as the axis of y, find the equation of each of the planes in which the faces of the tetrahedron lie.

Find the equations of the line through B perpendicular to the face ACD, show that it cuts the line OD and find the coordinates of the point of intersection.   (N.)

3. A cube of edge $2a$ has the squares $ABCD$ and $EFGH$ as opposite faces; AE, BF, CG, DH being parallel edges of the cube. Taking the lines AB, AD, AE as coordinate axes, find the equation of the plane passing through the points C, P, Q, where P and Q are the midpoints of FG and GH. Find the coordinates of the foot of the perpendicular from A on to this plane. Show also that the angle between the planes APQ, CPQ is $\cos^{-1}(5\sqrt{17/51})$.   (N.)

N

4. Find the perpendicular distance from the point $(2, 2, 1)$ to the line of intersection of the two planes

$$2x - y - z = 3, \quad 3x - y - 3z = 4.$$
(L.)

5. Show that the lines

$$\frac{x - 6}{3} = \frac{y - 3}{2} = z - 2$$

and

$$5x + 4y + 7z - 26 = 2x + 3y + 2z - 11 = 0$$

are coplanar. Find the coordinates of the common point and the equation of the plane which contains these two lines.
(L.)

6. The feet of the perpendiculars from $P(x_1, y_1, z_1)$ to the line $x = a$, $y = z$ and to the plane $x + y + z = 3a$ are Q and R respectively; find the coordinates of Q and R.

If QR is parallel to the plane $3x - y + z = 0$, prove that P must lie on the plane $x - y = 0$.
(L.)

7. A plane through the line of intersection of the planes $x + 3y + 5z + 1 = 0$, and $2x + y + 4z + 1 = 0$ is perpendicular to the line

$$\frac{x - 2}{1 - a} = \frac{y}{1 + a} = \frac{z - 1}{a}.$$

Show that a must have the value 13/6. Find the equation of the plane and the coordinates of the point where the plane and the line intersect.
(L.)

8. The feet A, B, C of a tripod with legs of equal length are at the points $(0, 0)$, $(1, 3)$, $(4, 0)$ respectively, referred to rectangular axes in a horizontal plane. The legs are concurrent to the apex P, which is at height 2 above the plane. Find the cosines of the angles between (i) the lines AP and BC, (ii) the planes PAC and PBC.
(L.)

9. Find the ratio in which N, the foot of the perpendicular from the origin, divides the line joining the points $A(4, 6, 0)$ and $B(1, 2, -1)$ and prove that N does not lie between A and B.
(L.)

10. Find the equations of the mirror image of the line

$$\frac{x - 2}{1} = \frac{y - 1}{2} = \frac{z + 3}{1}$$

in the plane $x - y - z = 2$, and show that the acute angle between the line and its image is $\cos^{-1}(5/9)$.
(L.)

11. If (a, b, c) are the rectangular Cartesian coordinates of a point P and if X, Y, Z are the feet of the perpendiculars from P to the axes, find the equation of the plane XYZ.

If the line l through P perpendicular to the plane XYZ meets this plane at D and the coordinate planes at A, B, C, prove that

$$\frac{1}{PA} + \frac{1}{PB} + \frac{1}{PC} = \frac{2}{PD}.$$
(L.)

12. Obtain the standard form, i.e.,

$$\frac{x-a}{l} = \frac{y-b}{m} = \frac{z-c}{n},$$

of the equations of the following two lines

(i) the line L through $(1, 0, -2)$ perpendicular to the plane

$$3x + 2y + z = 0,$$

(ii) the line M of intersection of the planes

$$x - 4y + z = 1, \ 2x + y - z = 2.$$

Hence, or otherwise, find the equations of the line through the point $(0, 0, 7)$ meeting both L and M. (L.)

In questions 13–17 it can be assumed that, given two skew lines L_1, L_2, there is one and only one straight line which intersects them both and is perpendicular to them both. Also this line is the shortest distance between these skew lines.

13. Two lines are given by the equations

$$x = 2, \ y - z = 1; \ 6x = 3(y - 1) = 2(z - 2).$$

Find the length of their common perpendicular and the coordinates of its feet. (L.)

14. Find the length and equations of the common perpendicular to the lines

$$x = y - 1 = 4 - z, \quad \text{and} \quad x - 2y + 9 = 0, x + z - 10 = 0. \quad \text{(L.)}$$

15. Show that the shortest distance between the lines

$$\frac{x+1}{3} = \frac{y-1}{-2} = \frac{z+1}{5}$$

and

$$x - 2y - z = 0, \ x - 10y - 3z = -7,$$

is $\frac{1}{2}\sqrt{14}$.

16. The coordinates of the four points A, B, C, D are respectively $(1, 2, 1)$, $(-1, 0, 2)$, $(2, 1, 3)$ and $(3, -1, 1)$. Find the shortest distance between the lines AB and CD. (L.)

17. Two skew lines have the equations

$$\frac{x-x_1}{l_1} = \frac{y-y_1}{m_1} = \frac{z-z_1}{n_1}. \quad \text{and} \quad \frac{x-x_2}{l_2} = \frac{y-y_2}{m_2} = \frac{z-z_2}{n_2}$$

and $\lambda : \mu : \nu$ are direction ratios of their line of shortest distance.

Prove that

$$\begin{vmatrix} x - x_1 & y - y_1 & z - z_1 \\ l_1 & m_1 & n_1 \\ \lambda & \mu & \nu \end{vmatrix} = 0$$

is the equation of the plane containing the first line and the line of shortest distance between the two given lines.

Find the coordinates of the point in which the line of shortest distance between

$$\frac{x+1}{1} = \frac{y-1}{2} = \frac{z-2}{-1} \quad \text{and} \quad \frac{x-1}{2} = \frac{y}{1} = \frac{z+1}{-3}$$

cuts the plane $y = 0$. (L.)

18. Show that the line

$$\frac{x-1}{2} = \frac{y-2}{3} = \frac{z-3}{4}$$

lies in the plane $x + 2y - 2z + 1 = 0$,

and that the line

$$\frac{x-3}{3} = \frac{y-2}{2} = \frac{z-1}{4}$$

lies in the plane $2x + y - 2z - 6 = 0$. If, in addition, these lines are the lines of greatest slope to the horizontal for the planes in which they lie, find the direction cosines of the vertical. (L.)

19. A plane is drawn through the line

$$x + y = 1, \quad z = 0$$

to make an angle $\sin^{-1}\left(\frac{1}{3}\right)$ with the plane

$$x + y + z = 0.$$

Prove that there are two such planes and find their equations. Prove also that the angle between the planes is $\cos^{-1}(7/9)$. (L.)

20. Find the coordinates of the point of intersection of the following three planes when $k = 4$:

$$2x + 7y - kz = 5,$$
$$3x + 2y - 4z = -8,$$
$$-4x + 3y + 2z = 11.$$

Show that if $k = 6$, the three planes form a triangular prism and find the direction ratios of the edges of the prism. (L.)

21. Vertical drillings are made at three points O, A, B on a piece of level ground to reach an underground seam. A is half a mile from O in a northerly direction, B is half a mile from O in a westerly direction. The depth of the seam is 1320 ft at O, 1287 ft at A and 1276 ft at B. Find, in miles, the shortest distance from O to the outcrop of the seam on the surface, assuming that the ground is level and that the seam is a plane. Give also the direction, in degrees E. of N., in which this outcrop runs. (L.)

22. Find the conditions that the line

$$\frac{x-a}{l} = \frac{y-b}{m} = \frac{z-c}{n}$$

lies in the plane $Ax + By + Cz + D = 0$.

Show that there is a line, equally inclined to the axes, which lies in the plane $x + 2y - 3z + 1 = 0$ and passes through the point $(1, 2, 2)$. Obtain the coordinates of the points of intersection of this line with the sphere $x^2 + y^2 + z^2 = 41$.   (L.)

23. If r is any transversal of the three lines

$$y = mx, \quad z = c; \quad y = 0, \quad z = 0; \quad y = -mx, \quad z = -c$$

and s is any transversal of the three lines

$$y = mz, \quad x = c; \quad y = 0, \quad x = 0; \quad y = -mz, \quad x = -c$$

prove that r meets s.   (L.)

In questions 24 and 25 note that by suitable choice of rectangular coordinate axes the equations of any two skew lines in space can be written in the form

$$y = mx, \quad z = c; \quad y = -mx, \quad z = -c.$$

(The z-axis is taken along the common perpendicular $A_1 A_2$ and the origin as the mid-point of $A_1 A_2$).

24. If points P and Q move, one on each line, in such a way that $OP = \lambda OQ$, where O is the origin of coordinates and $\lambda > 1$, prove that the locus of the point of intersection of the line PQ with the plane $z = 0$ is a hyperbola.   (L.)

25. P and P' are the feet of the common perpendicular to two skew straight lines, Q and Q' are two other points on the respective lines and R is the mid-point of QQ'. If $PQ^2 + P'Q'^2$ is constant, show that the locus of R is an ellipse.   (L.)

26. If a variable line intersects both the lines

$$y = mx, \quad z = c; \quad y = -mx, \quad z = -c,$$

and is equally inclined to them, prove that it lies on one of the surfaces

$$yz = mcx, \quad mzx = cy.$$ (L.)

27. Show that for all values of p the line of intersection of the planes

$$p(x + y) + z + k = 0, \quad x - y + p(z - k) = 0$$

is perpendicular to the line of intersection of the planes

$$p(x - y) - z - k = 0, \quad x + y - p(z - k) = 0.$$

Show also that these two lines intersect on the curve

$$z^2 + y^2 = k^2, \quad x = 0.$$ (L.)

28. The points A, B, and C referred to a right-hand orthogonal reference-frame $O(xyz)$ have coordinates $(1, 1, 1)$, $(2, 3, 4)$ and $(3, 2, 2)$ respectively. Find

(i) the area of the triangle ABC,

(ii) the volume of the tetrahedron $OABC$.

Show that the area enclosed by the orthogonal projection of the triangle ABC on to the plane $x + 2y + z = 0$ is $\frac{1}{2}\sqrt{6}$ units.   (L.)

29. The points A, B, C and D have coordinates $(1, 2, -1)$, $(2, 3, 0)$, $(3, 5, 3)$ and $(4, 4, 1)$ respectively referred to a right-handed system of axes $Oxyz$. Find

(i) the equation of the plane BCD,

(ii) the cosine of the angle BAC,

(iii) the area of the triangle ABC.

Show that the area of the orthogonal projection of the triangle ABC on to the plane of BCD is $(7/5)^{1/2}$ square units.   (L.)

30. Referred to a set of right-handed perpendicular axes with origin O the coordinates of three points A, B, C are $(1, 2, -1)$, $(2, 4, 1)$, $(-1, 3, 4)$ respectively. Find

(i) the cosine of the angle between AB and AC,

(ii) the area of the triangle ABC,

(iii) the volume of the tetrahedron $OABC$.   (L.)

ANSWERS TO EXERCISES

Ex. 11.1 (p. 7)

1. $x = 59/14$, $y = 2/7$. 2. $x = -17/23$, $y = -30/23$. 3. $x = 281/337$, $y = 162/337$.
4 If $a \neq \frac{1}{2}$, $x = (4a - 3)ab/(2a - 1)$, $y = -(4a - 1)ab/(2a - 1)$; if $a = \frac{1}{2}$, the equations have no solution. 5. If a and b are not *both* zero, $x = 2b$, $y = a$; if $a = 0 = b$, then the equations are satisfied for all values of x and y. 6. $x = a^2 - ab$, $y = ab + b^2$.
7. If $a \neq 2$ or -3, $x = y = 1/(a + 3)$; if $a = 2$, the eqns are identical, represent coincident straight lines and have solution $x = t$, $y = \dfrac{1}{3}(1 - 2t)$, where t is arbitrary; if $a = -3$, the eqns have no solution and represent parallel straight lines. 8. Two parallel lines if $a = 1$, coincident lines if $a = -2$; when perpendicular $\left(a = -\dfrac{2}{3}\right)$ lines intersect at $(-7/5, 11/5)$. 9. $k = \pm 1$; when $k = +1$, roots are $-5a/2$, $3a$, and $-5a/2, a$; when $k = -1$ roots are $5a/2, -3a$ and $5a/2, -a$. 10. $(x - 2y + 3)^2 = 5(x + 3y - 2)$, i.e. $x^2 - 4xy + 4y^2 + x - 27y + 19 = 0$. 11. $x^2y^2 - x^2 - 1 = 0$.
13. $13x^2 + 2xy + 2y^2 = 25$. 14. $(2x^2 + y^2)^2 = 13x^2 + 6xy + 10y^2$. 15. $\dfrac{1}{x^2} = \dfrac{9}{y^2} + 16$.
16. $(4p_1 - p_2)^2/x^2 + (p_2 - 3p_1)^2/y^2 = 1$ unless $p_1 = 0 = p_2$ or $p_2 = 3p_1$ or $p_2 = 4p_1$; (i) the locus consists of the origin $(0, 0)$ only, all lines passing through this point; (ii) the line $y = 0$ for $|x| \geq p_1$; (iii) the line $x = 0$ for $|y| \geq p_1$.

Ex. 11.3 (p. 10)

1. -48. 2. -144. 3. 194. 4. 36. 5. -153. 6. 14. 7. 0. 9. $(x - y)(b - a)$.
10. $(a + b)(a - b)^2$.

Ex. 11.4 (p. 15)

1. 4. 2. 789. 3. 0. 4. -3. 5. 0. 6. -8. 7. -182. 8. -345. 9. $xy(y - x) - x - 2y$. 10. $-14abc$. 11. $(a + b)(a^2 + b^2) = 2a^2b$. 12. $a = 1$ giving $x = -2$, $y = 1$ and $a = -\frac{1}{2}$ giving $x = -23/13$, $y = -5/13$.

Ex. 11.5 (p. 17)

1. $-(x - y)(y - z)(z - x)$. 2. $(x - 1)(y - 1)(x - y)(x + y + 1)$.
3. $-(x - y)(y - z)(z - x)(xy + yz + zx)$. 4. $-(a - b)(b - c)(c - a)$.
5. $(a + b + c)(a^2 + b^2 + c^2 - ab - bc - ca)$. 6. $xyz(x^2 + y^2 + z^2 - xy - yz - zx)$. 7. $(x + a)(x^2 + ax + cx + 3ac - 2c^2)$. 8. $(x + y + z)(x^2 + y^3 + xy + 3xz + 3yz)$. 9. $-4abc$. 10. $1, -6 \pm \sqrt{67}$. 11. $-10, \frac{1}{2}(-5 \pm \sqrt{13})$. 12. -4.
13. $-5, \frac{1}{2}(3 \pm \sqrt{13})$.

Ex. 11.8 (p. 22)

1. (i) $3x + 2y + 5 = 0$; (ii) $4x + 3y + 17 = 0$; (iii) $x - 2y = 0$; (iv) $x - 2y - 5 = 0$;   (v) $4x - 3y - 15 = 0$.   3. $a = 3$.   8. (i) $x - t_1 y + at_1^2 = 0$; (ii) $3t_1^2 x - y - 2at_1^3 = 0$; (iii) $3t_1^2 x - 2y - at_1^3 = 0$; (iv) $x + t_1^2 y - 2at_1 = 0$.

Ex. 11.9 (p. 23)

1. (i), (iii), (v).

Ex. 11.10 (p. 25)

1. $19x^2 + 19y^2 - 66x + 32y - 220 = 0$.   2. $9x^2 + 9y^2 + 79x + 71y - 88 = 0$. 3. $3x^2 + 3y^2 - 73x + 73y - 76 = 0$.   4. $bx^2 + by^2 - (a + b)^2 x - 2b^2 y + 2a(a^2 + b^2) = 0$.

Ex. 11.11 (p. 26)

1. 12 sq. units.   2. 5 sq. units.   3. 43/2 sq. units.
4. $\frac{1}{2} c^2 \left| (t_1 - t_2)(t_2 - t_3)(t_3 - t_1)/t_1 t_2 t_3 \right|$ sq. units.   5. $9\sqrt{3}/2$ sq. units.
8. $t_1 + t_2 + t_3 = 0$.

Ex. 11.13 (p. 30)

1. $x = 2$, $y = -3$, $z = -4$.   2. $x = 2$, $y = 3$, $z = 4$.   3. $x = 1$, $y = -1$, $z = 2$. 4. $x = 1$, $y = -1$, $z = 3$.   5. $x = 0$, $y = 4$, $z = -2$.   6. $x = 5$, $y = -1$, $z = -1$. 7. $x = -2$, $y = 3$, $z = -1$.   8. $x = 3$, $y = 0$, $z = -1$.   9. $x = y = z = 1/(a + b + c)$. 10. $x = (c - 1)(1 - b)/(c - a)(a - b)$, $y = (1 - c)(a - 1)/(b - c)(a - b)$, $z = (b - 1)(1 - a)/(b - c)(c - a)$.

Ex. 11.15 (p. 31)

1. $\begin{vmatrix} y & xy + 1 & y^2 \\ x + y & x^2 y + xy^2 + 1 & x + y^2 \\ x^2 & y + x^2 y^2 & x^2 \end{vmatrix}$ 2. $\begin{vmatrix} a^2 + b^2 & 3b & 3a \\ bc & b^2 + 3c & ab \\ ca & ab & a^2 + 3c \end{vmatrix}$

3. $\begin{vmatrix} a^2 + 2f - g & 2a + bf & -a + gc \\ ah + 2b - f & 2h + b^2 & -h + fc \\ ga + 2f - c & 2g + bf & -g + c^2 \end{vmatrix}$.

4. 0.   5. $\cos(x - \alpha) + \alpha \cos x - (2 \cos \alpha + 3\alpha \sin x) x + (3 \sin \alpha - \alpha \cos x) x^2$

Misc. Ex. XI (p. 31)

1. (a) 0; (b) determinant $= 1$; (c) $(x^2 + 3)(x^2 + 6)(x^2 + 12)$.   2. 3040.   3. (i) 0; (ii), 1, 1, -5.   4. (a) $\Delta = 2(1 + \sin^2 \theta)$; $\Delta = 2$ when $\theta = 0$, $\Delta = 4$ when $\theta = \frac{1}{2} \pi$; (b) $(aq - pb) x^3 - (ar - pc) x^2 + (br - qc) x$; $3(aq - pb) x^2 - 2(ar - pc) x + (br - qc)$; $y = 12x - 4x^3$ stationary when $x = \pm 1$.   5. (a) -5; (b) $(a - b)(b - c)(c - a)(a + b + c)$.   6. (a) area $= 1$ sq. unit; (b) $a = 1$, $b = 1$. 7. (a) $2 \cos 2\theta$; (b) 2, 3, -5. 8. $(\alpha - \beta)(\beta - \gamma)(\gamma - \alpha)(\alpha^2 + \beta^2 + \gamma^2 + \alpha\beta + \beta\gamma + \gamma\alpha)$; 0, 1, -1.   9. (a) $30°$, $45°$, $60°$; (b) -1, 2.   10. (a) 0, 3, 8; (b) 1, $-3/2$, -3, $7/2$. 11. (a) 3, $-5/3$; (b) 2.   12. a, $2a$, $-3a$.   13. $x^2 + y^2 - 4x - 5y + 4 = 0$;

$9y + 40x = 0$. 15. $a = 3$ or $\dfrac{11}{7}$; when $a = 3$, $x = 2$, $y = -1$; when $a = \dfrac{11}{7}$, $x = 6$, $y = 5$. 16. (a) $a = 2$ when the lines represented by the eqns are coincident, $a = -4$ when the lines represented by the eqns are concurrent at the point $(-2, -1)$; (b) $2(x + y)(x - y - z)(x + z - y)$. 17. (a) If $m \neq n$, the eqns represent lines intersecting at $\{na/(n - m),\ mna/(n - m)\}$; if $m = n \neq 0$, and $a \neq 0$, the eqns have no solution and represent parallel lines; if $m = n \neq 0$, and $a = 0$, the eqns represent coincident lines; if $m = n = 0$, the eqns represent coincident lines. (b) $2(a - b)(b - c)(c - a)$. 18. (a) $-2, 0, 2$. 19. $x = 0$, $y = 0$; $x = 4a$, $y = 4a$; $x = 2a$, $y = 2a$. 21. (i) Yes, $x = (c - 10)/(a - 6)$, $y = (5a - 3c)/(a - 6)$; (ii) yes, by $x = t$, $y = 5 - 3t$ where t takes all values; each of the eqns represents the line $3x + y = 5$; (iii) No, the eqns represent parallel lines. 23. $x = 0$, $y = 2/3$, $z = 7/3$. 24. $x = a$, $y = b$ or $x = a(a - c)/(b - c)$, $y = b(c - b)/(c - a)$. 25. (a) (i) $3abc - a^3 - b^3 - c^3$; (ii) $(a + b + c)(ab + bc + ca - a^2 - b^2 - c^2)$, $x = p, q, \tfrac{1}{2}(p + q)$; (b) $1, 1, -3/4$. 26. (b) $t = 0, 1, -1$. 27. If $a \neq 1$ or 0, the roots are 1, a, $\tfrac{1}{2}\{-a^2 - a \pm \sqrt{(a^4 + 2a^3 + a^2 + 4a)}\}$; if $a = 1$, the eqn is satisfied for all values of x; if $a = 0$, the roots are $0, 1$. 31. (i) $x = bc/p$, $y = ca/p$, $z = ab/p$ where $p = abc + ab + bc + ca$. (ii) $p + q = 0$. 32. $(3, 1.)$ 34. $x = 7/2$, $y = -5/2$, $z = 1/3$; $a + b + c = 0$. 35. $x = 5/3, y = 1/3, z = 2/3$. 36. $z = 5; x = 2, y = -3$. 37. $2xyz(x + y + z)^3$. 38. $x = -3$, $y = 5$, $z = 2$. 39. (i) $x = -1$, 8; (ii) $(x - y)(y - z)(z - x)(x + y + z)(xy + yz + zx)$. 40. (i) $(x - y)(y - z)(z - x)(x + y + z)$; (ii) $-6, \pm\sqrt{3}$. 43. (i) $-(x - y)(y - z)(z - x)$.

Ex. 12.3 (p. 48)

2. (i) $\pi/6$; (ii) $5\pi/6$; (iii) $-\dfrac{1}{4}\pi$; (iv) $\tfrac{1}{2}\pi$; (v) $-\dfrac{1}{4}\pi$; (vi) $\dfrac{3}{4}\pi$. 3. (i) $\mathrm{Tan}^{-1}(-3)$, second; (ii) $\sin^{-1}(63/65)$, first; (iii) $\sin^{-1}(7/25)$, first; (iv) $\tfrac{1}{2}\pi$; (v) $\tan^{-1}1 = \dfrac{1}{4}\pi$, first. 4. $\mathrm{Cot}^{-1}x = (n + \tfrac{1}{2})\pi - \alpha$; $\sin\alpha = \dfrac{(-1)^n x}{\sqrt{(1 + x^2)}}$, $\cos\alpha = \dfrac{(-1)^n}{\sqrt{(1 + x^2)}}$. 5. $x = \dfrac{3}{4}$. 7. $\tfrac{1}{2}\pi + \alpha$. 10. (i) $2/\sqrt{(1 - 4x^2)}$; (ii) $-1/\sqrt{(4 - x^2)}$; (iii) $1/(1 + x^2)$; (iv) $a/(a^2 + x^2)$; (v) $-1/\{x\sqrt{(x^2 - 1)}\}$; (vi) $\cos x/(1 + \sin^2 x)$; (vii) $2\sec^2 x/(1 + 4\tan^2 x)$; (viii) -1; (ix) -2; (x) $2e^{2x}/(e^{4x} + 2e^{2x} + 2)$. 11. (i) $-2/\sqrt{3}$; (ii) $1/8$; (iii) ∞; (iv) $-2/\sqrt{3}$. 12. (i) $\dfrac{1}{x\sqrt{(x^2 - 1)}}$; (ii) $\dfrac{-1}{x\sqrt{(x^2 - 1)}}$; (iii) $-\dfrac{1}{1 + x^2}$. 14. $2x - y\sqrt{3} - 1 + \pi\sqrt{3}/6 = 0$, $2x + y\sqrt{3} - 1 - 5\pi\sqrt{3}/6 = 0$; $\pi^2\sqrt{3}/8$. 15. $+0.064$. 17. 6 radians per min; zero. 18. (i) 1 radian per sec; (ii) $4/17$ radian per sec. 19. $\log\{x + \sqrt{(x^2 - 1)}\} + C$. 20. (i) $x\cos^{-1}x - \sqrt{(1 - x^2)} + C$; (ii) $x\tan^{-1}x - \tfrac{1}{2}\log(1 + x^2) + C$; (iii) $x\sec^{-1}x - \log\{x + \sqrt{(x^2 - 1)}\} + C$; (iv) $x\cot^{-1}x + \tfrac{1}{2}\log(1 + x^2) + C$. 21. (i) $\tfrac{1}{2}\pi a^2$ sq. units; $b = \dfrac{1}{3}a$ or $\tfrac{1}{2}a$. 22. $\dfrac{1}{4}\pi(\pi + 2)a^3$ cubic units. 23. (i) $\pi/6$; (ii) $\pi/6\sqrt{3}$; (iii) $\pi/9$; (iv) $\tan^{-1}(\tfrac{1}{2}) \doteqdot 0.46$; (v) $\pi/6$. 24. (i) $\tan^{-1}(x + 2) + C$; (ii) $\dfrac{2}{\sqrt{15}}\tan^{-1}\left(\dfrac{4x - 3}{\sqrt{15}}\right) + C$; (iii) $\sin^{-1}\left(\dfrac{x + 1}{\sqrt{2}}\right) + C$; (iv) $\dfrac{1}{\sqrt{3}}\sin^{-1}\{\tfrac{1}{2}(x + 1)\sqrt{3}\} + C$; (v) $\dfrac{1}{\sqrt{15}}\tan^{-1}\left\{\dfrac{(x + 1)\sqrt{5}}{\sqrt{3}}\right\} + C$. 26. $\dfrac{1}{4}\pi - \tfrac{1}{2}\log 2$. 27. $\tfrac{1}{2}(x^2 + 1)\tan^{-1}x - \tfrac{1}{2}x + C$. 28. $\log(x - 2) - \log(x + 2) + \tfrac{1}{2}\tan^{-1}(\tfrac{1}{2}x) + C$. 29. $\tfrac{1}{2}\tan^{-1}(15/17)$. 30. $\pi/12$. 31. $\sqrt{2} - 1$ sq. units.

Ex. 12.4 (p. 56)

11. 13/12, 5/13.   12. $\frac{1}{4}\sqrt{7}$, $3/\sqrt{7}$.   13. $\log 2$, $-\log 4$.   14. $\pm\frac{1}{2}\log 3$.   15. $\log 4$.
17. (i) $2\sinh(2x+5)$;   (ii) $x\cosh x+\sinh x$;   (iii) $2\cosh 2x\cosh 3x +$
$3\sinh 2x\sinh 3x$;   (iv) $e^x(\tanh x+\operatorname{sech}^2 x)$;   (v) $\frac{1}{2}(1-1/x^2)$, [note $\cosh\log x \equiv$
$\frac{1}{2}(x+1/x)$]; (vi) $-\{(1+x\tanh x)\operatorname{sech} x\}/x^2$;   (vii) $\cos x.\cosh(\sin x)$;   (viii) $\operatorname{sech} x$;
(ix) $4\operatorname{cosech} 4x$;   (x) $(\cosh x+\sinh x)e^{\sinh x+\cosh x}$;   (xi) $-\operatorname{sech} x$;   (xii) 0.
18. $bx\cosh t_1-ay\sinh t_1=ab$; $ax\sinh t_1+by\cosh t_1=(a^2+b^2)\sinh t_1\cosh t_1$.
19. $PT=\{c\cosh^2(x/c)\}/\sinh(x/c)$; $PG=c\cosh^2(x/c)$; $TN=c\coth(x/c)$;
$NG=c\sinh\left(\dfrac{x}{c}\right)\cosh\left(\dfrac{x}{c}\right)$; (i) $NZ=c$.   20. $A=2$, $B=4$.   21. 328; 1310.
22. 5 (when $x=-\log 5$). 23. When $|n|>1$, a minimum (n^2-1) at $x=$
$\log\{(n-1)/(n+1)\}$; when $|n|<1$, a minimum $(1-n^2)$ at $x=\log\{(1-n)/(1+n)\}$;
when $n=\pm 1$ the function is $2e^{\pm x}$ and has no stationary values. 24. (i) $\frac{1}{2}\sinh 2x$
$+C$; (ii) $\log\cosh x+C$; (iii) $\log\sinh x+C$; (iv) $\tanh x+C$; (v) $-\operatorname{sech} x+C$;
(vi) $-\operatorname{cosech} x+C$; (vii) $\frac{1}{2}(x+\frac{1}{2}\sinh 2x)+C$; (viii) $(\cosh 6x+3\cosh 2x)/12$
$+C$;   (ix) $x\cosh x-\sinh x+C$;   (x) $\frac{1}{2}(\frac{1}{2}e^{2x}+x)+C$.   25. (i) $1-e^{-1}$;
(ii) $(\sinh 3+9\sinh 1)/12$; (iii) $(\sinh 3+3\sinh 1)/12$; (iv) $2\tan^{-1}(e^2)-\frac{1}{2}\pi$;
(v) $\tanh 1-\log\cosh 1$. 26. $e-1-2\tan^{-1}e+\frac{1}{2}\pi$. 27. $1-e^{-1}$. 28. (i) $3\cdot 31$;
(ii) $0\cdot 707$.   29. $2\pi(1-\tanh 1)=4\pi/(1+e^2)$.

Ex. 12.8 (p. 63)

2. $\pm\log(2+\sqrt{3})$.   3. $\pm\log\left\{\dfrac{1}{3}(4+\sqrt{7})\right\}$.   4. $x=-\log 2$, $y=\log(5/2)$ or
$x=\log(5/2)$, $y=-\log 2$.   5. (i) $\frac{1}{2}\log\left(\dfrac{1+x}{1-x}\right)$;   (ii) $\pm\log\left\{\dfrac{1+\sqrt{(1-x^2)}}{x}\right\}$;
(iii) $\log\left\{\dfrac{1+\sqrt{(x^2+1)}}{x}\right\}$; (iv) $\log\left\{\dfrac{1+\sqrt{(1+x^2)}}{x}\right\}$; (v) $\pm\log\{1+x^2+\sqrt{(x^4+2x^2)}\}$.
6. (i) $-1/(x^2-1)$; (ii) $-1/\{x\sqrt{(1-x^2)}\}$; (iii) $-1/\{x\sqrt{(1+x^2)}\}$.   7. (i) $\cosh^{-1}x$
$+x/\sqrt{(x^2-1)}$;   (ii) $1/(2x)$;   (iii) $\sec x$;   (iv) $-1/\{x\sqrt{(1-x^2)}\}$;   (v) $2\cosh 2x/$
$\sqrt{(\sinh^2 2x-1)}$. 8. (i) $9\{\log(1+\sqrt{2})+\sqrt{2}\}/2\fallingdotseq 10\cdot 3$; (ii) $8\{2\sqrt{3}-\log(2+\sqrt{3})\}\fallingdotseq 17.2$.
9. (i) $\frac{1}{2}\{x\sqrt{(1+x^2)}-\sinh^{-1}x\}+C$; (ii) $\sqrt{(x^2-1)}|-\sec^{-1}x+C$.
10. (i) $x\sinh^{-1}x-\sqrt{(x^2+1)}+C$; (ii) $x\cosh^{-1}x-\sqrt{(x^2-1)}+C$; (iii) $x\tanh^{-1}x$
$+\frac{1}{2}\log(1-x^2)+C$;   (iv) $\frac{1}{2}x^2\cosh^{-1}x-\dfrac{1}{4}x\sqrt{(x^2-1)}-\dfrac{1}{4}\cosh^{-1}x+C$.
11. $ab\{2\sqrt{3}-\log(2+\sqrt{3})\}$ sq. units.   12. (i) $\log\{3(1+\sqrt{2})/(2+\sqrt{13})\}$;
(ii) $\frac{1}{2}\log\{(8+\sqrt{63})/(4+\sqrt{15})\}$;   (iii) $\log\{(3+\sqrt{10})/(2+\sqrt{5})\}$;
(iv) $(1/\sqrt{2})\log\{(\sqrt{18}+\sqrt{15})/(\sqrt{8}+\sqrt{5})\}$;   (v) $\log[\{a+b+\sqrt{(a^2+2ab+2b^2)}\}/$
$\{2a+\sqrt{(4a^2+b^2)}\}]$.   13. $2a^2\log(1+\sqrt{2})$ sq. units.   15. $\sinh^{-1}\{(x-1)/\sqrt{2}\}+C$.
16. $\sin^{-1}\{(x+2)/\sqrt{5}\}+C$.   17. $\cosh^{-1}\{(x-1)/\sqrt{6}\}+C$.

18. $\dfrac{1}{\sqrt{17}}\log\left(\dfrac{4x+5-\sqrt{17}}{4x+5+\sqrt{17}}\right)+C$.

19. $\dfrac{1}{\sqrt{13}}\log\left(\dfrac{\sqrt{13}+1+2x}{\sqrt{13}-1-2x}\right)+C$.   20. $\frac{1}{2}\tan^{-1}\{\frac{1}{2}(x+1)\}+C$.

21. $\dfrac{1}{\sqrt{2}}\sin^{-1}\left\{\dfrac{(x+1)\sqrt{2}}{\sqrt{3}}\right\}+C$.   22. $\dfrac{1}{\sqrt{3}}\cosh^{-1}\left\{\dfrac{(x-1)\sqrt{3}}{\sqrt{8}}\right\}+C$.

23. $\dfrac{1}{2\sqrt{14}}\log\left(\dfrac{\sqrt{7}+(x+1)\sqrt{2}}{\sqrt{7}-(x+1)\sqrt{2}}\right)+C$.   24. $\sinh^{-1}\left\{\dfrac{(2x+1)}{\sqrt{3}}\right\}+C$.

Ex. 12.9 (p. 65)

1. $\frac{1}{2}\log(x^2 + 5x + 1) - \frac{1}{2\sqrt{21}}\log\left(\frac{2x + 5 - \sqrt{21}}{2x + 5 + \sqrt{21}}\right) + C.$

2. $\frac{1}{2}\log(2x^2 + x + 3) + \frac{5}{\sqrt{23}}\tan^{-1}\left(\frac{4x + 1}{\sqrt{23}}\right) + C.$

3. $3\sqrt{(x^2 + x + 1)} + \frac{1}{2}\sinh^{-1}\left(\frac{2x + 1}{\sqrt{3}}\right) + C.$

4. $5\sqrt{(x^2 - 3x - 1)} + \frac{15}{2}\cosh^{-1}\left(\frac{2x - 3}{\sqrt{13}}\right) + C.$

5. $\frac{3}{4}\log(2x^2 + x + 5) + \frac{13}{2\sqrt{39}}\tan^{-1}\left(\frac{4x + 1}{\sqrt{39}}\right) + C.$

6. $\frac{1}{8}\log(4x^2 + 2x - 3) - \frac{1}{8\sqrt{13}}\log\left(\frac{4x + 1 - \sqrt{13}}{4x + 1 + \sqrt{13}}\right) + C.$

7. $-\sqrt{(1 - 3x - x^2)} - \frac{1}{2}\sin^{-1}\left(\frac{2x + 3}{\sqrt{13}}\right) + C.$

8. $\frac{6}{\sqrt{5}}\log\left(\frac{\sqrt{5} - 2 + x}{\sqrt{5} + 2 - x}\right) - \frac{5}{2}\log(1 + 4x - x^2) + C.$

9. $\sqrt{(x^2 + 6x + 4)} - \cosh^{-1}\{(x + 3)/\sqrt{5}\} + C.$

10. $3x + \log(x^2 + x + 1) + \frac{4}{\sqrt{3}}\tan^{-1}\left(\frac{2x + 1}{\sqrt{3}}\right) + C.$

Misc. Ex. XII (p. 68)

1. $x = \log 3$, $y = \log 2$. 3. $A = c^2(1 - e^{-\lambda/c})$; $V_1 = \pi c^2 h$.
4. $1 - (x\sin^{-1}x)/\sqrt{(1 - x^2)}$; y. 5. (i) $2\tan^{-1}x$; (ii) $1/[2\sqrt{\{(2 - x)(x - 1)\}}]$; $1 \leq x \leq 2$.
8. $x = y = \pm\frac{1}{2}\log(3 + 2\sqrt{2})$. 10. $\log(7/5)$. 11. $-\operatorname{cosech} t$. 14. $x = \log(3/2)$, $y = -\log 2$.
16. (i) $\frac{1}{\sqrt{5}}\log\left(\frac{\sqrt{5} - 2 + \tan\frac{1}{2}\theta}{\sqrt{5} + 2 - \tan\frac{1}{2}\theta}\right) + C$;   (ii) $x - \log(x - 1) + 4\log(x - 2) + C$;

(iii) $\frac{1}{2}(\sinh x \cos x + \cosh x \sin x) + C$.   17. $\frac{1}{x} + \frac{1}{x^2} - \frac{x}{x^2 + 1}$; $\frac{1}{2}\log\left(\frac{2n^2}{n^2 + 1}\right) + 1$

$-\frac{1}{n}$; $1 + \frac{1}{2}\log 2$. 18. $x - \frac{x^3}{6} + \frac{3x^5}{40} - \frac{5x^7}{112}$. 19. One, $(\log 4)$.

21. (i) $\frac{1}{2}(x + 2)\sqrt{(x^2 + 4x - 5)} - \frac{9}{2}\cosh^{-1}\left\{\frac{1}{3}(x + 2)\right\} + C$; (ii) $(8x^2 + 4x\sin 4x$
$+ \cos 4x)/32 + C$;   (iii) $\sqrt{(5 - 2x + x^2)} + \sinh^{-1}\{\frac{1}{2}(x - 1)\} + C$;

(iv) $\log\left(\frac{x}{1 - x}\right) + \frac{1}{1 - x} + C$; (v) 2. 22. $\frac{1}{4}\pi - \frac{1}{2}\log 2$. 23. $[\Delta = \sinh 2x - 2\sinh x]$;
$2(2^{2n} - 1)/(2n + 1)!$; $-200e^{-x}$. 24. (i) $2 - \sqrt{2}$; (ii) $\pi/4\sqrt{2} + 1/\sqrt{2} - 1$.

26. (a) $\dfrac{\sqrt{5}}{(3 + x)\sqrt{(1 + 2x)}}$; (b) (i) three; (ii) one.

27. $(\sinh 4x)/32 - (\sinh 2x)/4 + 3x/8 + C$. 28. (i) $x = -1, 2$; (ii) 1. 30. (a) $\frac{1}{3}\pi$; (b) 0.

32. $C - \sinh^{-1}[1/\{(x - 3)\sqrt{2}\}]$.   33. Each derivative is $2(x^2 + 2)/(x^4 + 4)$. The functions are equal to one another. 36. (i) $\cosh^{-1}(x - 2) + C$;

(ii) $\frac{1}{3}\log x - \frac{1}{2}\log(x - 1) + \frac{1}{6}\log(x - 3) + C$;   (iii) $(\sin 4x)/32 - (\sin 2x)/4$

$+ 3x/8 + C.$ 37. (i) $\left\{ 3 \log(x + 1) - \dfrac{3}{2} \log(x^2 + 4) + \dfrac{5}{x + 1} + 4 \tan^{-1}\left(\dfrac{x}{2}\right) \right\} \Big/ 25$
$+ C$; (ii) $\pi/\sqrt{7}$; (iii) $\frac{1}{2}(\sinh^{-1}2 - \sinh^{-1}1) = \frac{1}{2}\log\{(2 + \sqrt{5})/(1 + \sqrt{2})\}$.
38. (i) $\sin^{-1}x - \sqrt{(1 - x^2)} + C$; (ii) $\frac{1}{2}(x^2 + 1)\tan^{-1}x - \frac{1}{2}x + C$. 39. (i) $\frac{1}{2}\log(8/5)$;
(ii) $(\pi - 2)/8$; (iii) $\theta + 2/(1 + \tan\frac{1}{2}\theta) + C$. 40. (i) $x + \dfrac{3}{2}\log(2x + 1) -$
$\log(x^2 + 2) - \dfrac{1}{\sqrt{2}}\tan^{-1}(x/\sqrt{2}) + C$; (ii) $C - \cosh^{-1}\{(x + 1)/x\sqrt{2}\}$. 41. (i) $\frac{1}{2}e^x(\cos x +$
$\sin x) + C$; (ii) $\sqrt{(x^2 + 2x - 3)} + \cosh^{-1}\{\frac{1}{2}(x + 1)\} + C$. 42. $\dfrac{2}{\sqrt{3}}\tan^{-1}(e^x\sqrt{3}) + C$.
43. (i) $\dfrac{1}{4}\{\log x - \frac{1}{2}\log(x^2 + 4) + 2\tan^{-1}(\frac{1}{2}x)\} + C$; (ii) $-\sqrt{(4x - x^2)} +$
$2\sin^{-1}\{\frac{1}{2}(x - 2)\} + C$; (iii) $\frac{1}{2}x^2(\log x)^2 - \frac{1}{2}x^2\log x + \dfrac{1}{4}x^2 + C$. 45. (a) $9(\sqrt{2} - 1) +$
$2\sinh^{-1}1 = 9(\sqrt{2} - 1) + 2\log(1 + \sqrt{2})$; $\dfrac{1}{3}\tan^{-1}(e^{3x}) + C$.
46. (i) (a) $\dfrac{x^{n-1}}{n + 1}\left(\log x - \dfrac{1}{n + 1}\right) + C$; (b) $\frac{1}{2}(\log x)^2 + C$; $\dfrac{1}{4}e^2 - 2e^{-1} + 23/4$;
(b) $15/2 + 8\log 2$.

Ex. 13.1 (79)

5. $16\pi/15$. 8. $\frac{1}{2}\pi(\pi - 2)$. 9. 7π. 10. $\displaystyle\int_0^\pi \dfrac{\sin n\theta}{\sin\theta}\,d\theta = 0$ when n is even.
11. 0. 12. (i) $144\sqrt{3}/35$; (ii) 0; (iii) $360, 448/63$; (iv) 0. 14. 0. 16. 0.

Ex. 13.2 (p. 86)

1. $1, 2, 4$ do not converge. 3. 1. 5. $\pi/12$. 6. 2π. 7. π. 8. $-1/9$. 9. 1.
10. $\dfrac{1}{6}\log 4$. 11. $3\pi/2$. 12. $\frac{1}{2}\pi$. 13. $(\sqrt{2} - 1)/a^2$. 14. 2π. 15. $\frac{1}{2}\pi$. 17. $1/(a^2 + 1)$.
20. $\frac{1}{2}\pi$. 21. 4π sq. units. 22. 1 sq. unit. 23. $\frac{1}{2}\pi^2$ cubic units. 24. $2\log(1 + \sqrt{2})$
sq. units. 25. 2π sq. units.

Ex. 13.3 (p. 92)

1. $au_n = x^n e^{ax} - nu_{n-1}$; $\dfrac{1}{3}\left(x^4 - \dfrac{4x^3}{3} + \dfrac{4x^2}{3} - \dfrac{8x}{9} + \dfrac{8}{27}\right)e^{3x} + C$.
2. $u_n = -x^n\cos x + nx^{n-1}\sin x - n(n - 1)u_{n-2}$; $-x^3\cos x + 3x^2\sin x + 6x\cos x$
$- 6\sin x + C$. 3. $u_n = (\tan^{n-1}x)/(n - 1) - u_{n-2}$; $\dfrac{1}{5}\tan^5 x - \dfrac{1}{3}\tan^3 x + \tan x -$
$x + C$. 4. $2u_n = x^2(\log x)^n - nu_{n-1}$; $x^2\{4(\log x)^3 - 6(\log x)^2 + 6\log x - 3\}/8 + C$.
5. $nu_n = \cosh^{n-1}x\sinh x + (n - 1)u_{n-2}$; $\dfrac{1}{6}\cosh^5 x\sinh x + \dfrac{5}{24}\cosh^3 x\sinh x +$
$\dfrac{5}{16}\cosh x\sinh x + \dfrac{5x}{16} + C$. 6. $(n - 1)u_n = \sec^{n-2}x\tan x + (n - 2)u_{n-2}$;
$\dfrac{1}{4}\sec^3 x\tan x + \dfrac{3}{8}\sec x\tan x + \dfrac{3}{8}\log(\sec x + \tan x) + C$. 7. $35\pi/256$. 8. $16/35$.
9. $2/35$. 10. $\dfrac{3}{4}$. 11. $\pi a^6/32$. 12. $5\pi/16a$. 18. $5\pi/256$. 20. $I_2 + J_2 = 1$;
$I_6 + J_6 = \dfrac{11}{17}$.

Ex. 13.4 (p. 99)

1. $0 \cdot 693 (1)$; $\log_e 2 \doteqdot 0 \cdot 6931$.   2. $0 \cdot 168$; $(2 \log_e 2 - 1) \log_{10} e = 2 \log_{10} 2 - \log_{10} e$ $\doteqdot 0 \cdot 167$.   3. $1 \cdot 32$; $\log_e (2 + \sqrt{3}) \doteqdot 1 \cdot 317$.   4. 9.29.   5. 0.882.   6. $1 \cdot 37$.   7. 0.737.
8. $1 \cdot 15$.   9. $1 \cdot 04$.   10. $1 \cdot 060$ (both methods).   11. $0 \cdot 51$ (both methods).
12. (i) $0 \cdot 448$; (ii) $0 \cdot 437$.   13. $0 \cdot 399$ (both methods).

Ex. 13.5 (p. 104)

1. $(2 \log 2)/\pi \doteqdot 0 \cdot 44$.   2. $0 \cdot 091$.   3. $e^{-1} \doteqdot 0 \cdot 37$.   4. $\frac{1}{4} \pi a \doteqdot 0 \cdot 79 a$.   5. (i) $\frac{14}{\pi}$;
(ii) $\frac{1}{28} (25\pi + 48)$.   6. $\frac{1}{3} \varrho a^2$.   7. $l^2/6$.   8. $5/2$.   9. $0 \cdot 619$.

Ex. 13.6 (p. 108)

1. On the axis, $\frac{1}{4} h$ from the base.   2. $3l/2$ from one end.   3. On the radius
of symmetry, $2r/\pi$ from bounding diameter.   4. On the axis $\frac{1}{4} h$ from the base.
5. $(3a/5, 0)$.   6. $[0, (4\pi - 3 \sqrt{3})/\{4\pi - 6 \log(2 + \sqrt{3})\}]$.   7. $(256a/315\pi, 256a/315\pi)$.
8. $(9/5, 9/5)$.   9. $4/5$.   10. $(3e^4 + 1)/\{2(e^4 - 1)\}$.   11. $(e^2 - 1)/\{4(e - 2)\}$ 12. $5a/8$.

Ex. 13.7 (p. 111)

1. (i) $4r/3\pi$ from bounding diameter; (ii) $4r/3\pi$ from each bounding radius.
2. $128\pi a^3/15$ units³.   3. $8 \sqrt{2\pi} \sin \left(\frac{1}{4} \pi + \theta \right) a^3$ units³; $8 \sqrt{2\pi} a^3$ units³, $8\pi a^3$ units³.
4. $36\pi^2$ units³.   5. 6π units³, 9π units³.   6. $(e - 1)^2/(2e)$ units²; $\{2/(e - 1)^2,$
$(e^4 - 4 e^2 - 1)/[8 e(e - 1)^2]\}$; volume $2\pi/e$ units³.   7. $9c/\{2(15 - 16 \log 2)\}$.   8. $7 \cdot 4$.

Ex. 13.8 (p. 120)

1. $\frac{4}{3} M a^2$, $(M = 2\pi\varrho a^3)$.   2. $2Ma^2/7$, $(M = \frac{1}{4} \pi\varrho a^3)$.   3. $2M/9$, $(M = 2)$.
4. $I_{ox} = M/(8 \log 2)$, $I_{oy} = 3M/(2 \log 2)$, $(M = \log 2)$.   5. (i) $3Mr^2/10$;
(ii) $M(3r^2 + 2h^2)/20$.   6. $\sqrt{\left/ \left\{ \frac{2 (a^5 - b^5)}{5 (a^3 - b^3)} \right\} \right.}$.   7. $10ml^2$; (ii) $20ml^2$.   8. $\frac{1}{2}Ml^2$, $(M = \frac{1}{2}\varrho l^2)$.
9. $\frac{1}{2}R \sqrt{7}$.   10. $\frac{1}{4} \sqrt{5}$.

Misc. Ex. XIII (p. 120)

1. (a) 0; (b) $A = 1$, $B = 1$; $\frac{1}{4} \left(x^4 \tan^{-1} x - \frac{1}{3} x^3 + x - \tan^{-1} x \right) + C$; $1/6$ units².
2. (b) $(3a/5, 2a/35)$.   3. (a) (i) $\frac{1}{2}\pi$; (ii) $\frac{1}{4} (\pi - 1)$; (b) $\frac{1}{4}\pi$.   5. (a) $3 \cdot 6114$; (b) (i) $2a/\pi$;
(ii) $\frac{1}{4} \pi a$.   6. (i) $\frac{3}{4} = 0 \cdot 75$ units²; (ii) $\log 2 \doteqdot 0 \cdot 69$ units²; (iii) $\frac{1}{4} \pi \doteqdot 0 \cdot 79$

units^2; nearer to (iii). 8. 3·988; 0·997(1). 9. $v = 32t$; $v = 8\sqrt{s}$; $\frac{1}{2}V$; $\frac{2}{3}V$. 10. $\frac{1}{2}\pi$.

11. $4\pi a^2(b - a)$ units^3. 12. $3\pi/16$. 15. (i) 0·2983; (ii) 0·2983. 16. 32/15.
17. 28 units^2; (26/3, 16/3); $392\pi\sqrt{2}$ units^3. 18. 2; 1. 20. (i) 0; (ii) $\pi(\pi - 2)/8$.
23. $(3\pi a/16, a/5)$; $\pi^2 a^3/8$ units^3. 24. $17a/6$. 25. $2555a^2/558$. 26. $3\sqrt{3}a^2$.
27. (i) $\frac{1}{4}\pi ab$; $(4a/3\pi, 4b/3\pi)$; (ii) $\frac{2}{3}\pi a^3$; $3a/8$ from base. 28. (i) $\{7\sqrt{2} + 3\log(1 + \sqrt{2})\}/8$; (ii) 1; $\frac{1}{3}(8\sqrt{2} - 10)$. 29. $\frac{1}{2}(\pi^3 - 24\pi + 48)$. 30. $1 - \frac{1}{4}\pi$. 31. 8/7.
32. $4\pi ab^2/3$. 33. Distant $3a/8$ from base. 36. $\{7\sqrt{2} + 3\log(1 + \sqrt{2})\}/16$. 37. 2π.
38. $\pi/\{2\,|\sin\theta|\}$; (i) $\pi/\sqrt{2}$; (ii) $\pi/\sqrt{2}$. 39. $b^2C + a^2S = \frac{1}{2}\pi$; $C = \pi/\{2b(a + b)\}$, $S = \pi/\{2a(a + b)\}$ 40. $u_n - u_{n-1} = 0$; $n(n + 1)\pi$. 41. 7·41. 43. $\frac{1}{2}\pi$.
44. (i) $\pi(\pi + 2)/8$ units^2; (ii) $\pi\log 2$ units^2; (iii) $\{(2\log 2)/\pi, \frac{1}{4\pi}(\pi + 2)\}$. 46. 16.
48. 1·541; 1·541. 49. $6\pi a^3\varrho$; $203\pi\varrho a^5/30 = 203 Ma^2/180$. 50. $\sqrt{\{b(4a + b)/6\}}$.
51. $\frac{1}{4}Ma^2$; $Ma^2(15\pi - 32)/12\pi$. 53. $\pi(1 - \log 2)$ units^2; $(3 - 4\log 2)/\{2(1 - \log 2)\}$.
5. π units^3; on Ox distant $\frac{1}{4}(\pi - 2\log 2)$ from O; $\sqrt{\frac{2}{3}}$. 55. (i) 0; (ii) $\frac{1}{2}\pi$.

Ex. 14.1 (p. 134)

1. (i) $y = 7x$; (ii) $y = x$; (iii) $y = x$; (iv) $y = x$; (v) $y = x$. 2. (i) Maximum at $(0, 1)$, minimum at $(8, -255)$, inflexion at $(4, -127)$; (ii) minimum at $(3/2, -27/16)$, inflexions at $(0, 0)$, $(1, -1)$; (iii) no stationary points, inflexion at $(0, 0)$; (iv) minimum at $(0, 0)$, inflexions at $(\pm a/\sqrt{3}, \frac{1}{4}a)$; (v) maxima at $\left\{\left(2n + \frac{1}{4}\right)\pi, \sqrt{2}\right\}$, minima at $\left\{\left(2n + \frac{5}{4}\right)\pi, -\sqrt{2}\right\}$, inflexions at $\left\{\left(n + \frac{3}{4}\right)\pi, 0\right\}$. 3. $\frac{1}{2}\pi$. 4. $(\pm 1/\sqrt{2}, \text{ e}^{-\frac{1}{2}})$.
5. Maximum at $(1, \text{e}^{-1})$, inflexion at $(2, 2\text{e}^{-2})$; $x + \text{e}^2 y - 4 = 0$. 6. $\cot(\frac{1}{2}\theta)$, $-\frac{1}{2a}\csc^2(\frac{1}{2}\theta)\csc\theta$; stationary points at $\{2a, a(2n + 1)\pi\}$. 7. $\{(n + \frac{1}{2}\pi, (n + \frac{1}{2})\pi\}$. 8. (i) Inflexion at $x = 0$; (ii) inflexion at $x = l$; (iii) inflexions at $x = \pm\frac{1}{2}l$; (iv) inflexions at $x = \pm l/\sqrt{12}$. 9. Minima at $\{(2n + \frac{1}{2})\pi, -5\}$, maxima at $\{(2n - \frac{1}{2})\pi, 3\}$, inflexions at $\{(n\pi + (-1)^n \pi/6), -3/2\}$.

Ex. 14.3 (p. 141)

1. $\frac{1}{2}\{\sqrt{2} + \log(1 + \sqrt{2})\}$ units. 2. $(13\sqrt{13} - 8)a/27$. 3. $\{\sqrt{2} + \log(1 + \sqrt{2})\}a$.
4. $\log(1 + \sqrt{2})$ units. 5. $2(\tan^{-1}\text{e} - \frac{1}{4}\pi)$ units. 6. $8/\sqrt{3}$ units.

Ex. 14.4 (p. 145)

1. $\pi(1 + \frac{1}{2}\sinh 2)$ units^2. 2. $64\pi a^2/3$. 3. $6\pi a^2/5$. 4. $8\pi(2\sqrt{2} - 1)/3$ units^2.
5. $384(1 + \sqrt{2})\pi/5$ units^2. 6. $\pi\left[\log\left\{\dfrac{1 + \sqrt{2}}{1 + \sqrt{(1 + \text{e}^2)}}\right\} + 1 + \sqrt{2} - \dfrac{\sqrt{(1 + \text{e}^2)}}{\text{e}^2}\right]$.
7. $2\pi(\text{e} - 1)^2/(\text{e}^2 + 1)$ units^2. 8. $\dfrac{\pi\sqrt{2}}{2}\left\{3\sqrt{10} - \sqrt{2} - \log\left(\dfrac{3 + \sqrt{10}}{1 + \sqrt{2}}\right)\right\}$ units^2.

Ex. 14.5 (p. 147)

1. $4\pi^2 rc$. 2. $48\pi^2$ units². 3. $24\pi a^2$. 4. $12\pi a^2$. 5. $\pi a^2(4\pi - 1)/16$.
7. $2\pi a^2\{2(\cos^3\theta + \sin^3\theta) + \pi\cos\theta\sin\theta(\cos\theta + \sin\theta)\}$; $2\sqrt{2}\pi(1 + \frac{1}{2}\pi)\,a^2$.

Ex. 14.7 (p. 154)

1. $-1/(4a\sqrt{2})$. 2. $-ab\sqrt{\{8/(a^2 + b^2)^3\}}$. 3. $-ab/(a^2\sinh^2 1 + b^2\cosh^2 1)^{3/2}$.
4. $-1/\sqrt{8}$. 5. $1/(6a\sqrt{2})$. 6. $3/(80a\sqrt{10})$. 7. $1/c$. 8. -1. 9. $-2/5\sqrt{5}$.
10. $-1/9\sqrt{2}$. 11. $\frac{1}{2}a$. 12. $\frac{1}{2}a$. 13. $\frac{1}{3}$. 14. $\frac{1}{2}c$. 15. b^2/a. 20. $(2a + 3at^2, -2at^3)$.

Misc. Ex. XIV (p. 155)

1. $\varrho = (1 + \sin^2 t)^{3/2}/\sqrt{2}$. 3. Minimum $-e^{-1}$ at $x = e^{-1}$; $\varkappa = 1/[x\{2 + 2\log x + (\log x)^2\}^{3/2}]$. 5. $(0, -4)$, $(1, 3)$. 6. $\pi(10\sqrt{10} - 1)/27$ units². 7. $8x - y - 6 = 0$, $8x + y + 6 = 0$, $\left(-\frac{1}{3}, -\frac{26}{3}\right)$, $\left(\frac{1}{3}, -\frac{26}{3}\right)$ 9. (i) $y = \pm x$, $x + a = 0$; (ii) $y = \pm x$, $y = 0$; $\frac{1}{4}a$. 10. (i) $\pm\frac{1}{2}$; (ii) $64a^2/3$; (iii) $2\{2\sqrt{5} + \log(2 + \sqrt{5})\}a$.
11. $(t^2 + 2t)/(t + 1)$. 12. Minimum. 13. $s = 4a\sin\psi$. 14. $\frac{1}{8}$. 15. (i) $k = -\frac{1}{3}$; (ii) inflexion at $x = 0$, minimum at $x = \frac{1}{2}\pi$. 16. $22x - y = 7$ at $(1, 15)$ meets curve again at $(7, 147)$; $32x + y = 128$ at $(4, 0)$ meets curve again at $(-2, 192)$.
17. $3a\sin t\cos t$; $\frac{1}{3}\pi$. 18. $y - \sqrt{2}x + 2\sqrt{2} = 0$; $-\frac{1}{2}\sqrt{2}$, $\varrho = 9\sqrt{3}/2$. 19. $-\text{cosech}\,t$; $a\sinh t$. 20. $-xe^{-x}$, $(x - 1)e^{-x}$; turning point, max e^{-1} at $x = 0$; inflexion at $(1, 2e^{-1})$. 21. $\tan t$, $(\sec t)^3/(at)$; $x\cos t + y\sin t - a(1 - \cos t) = 0$. 22. $(2a, 13a)$,
23. $s = c\sinh 1$; $S = \pi c^2(1 + \cosh 1\sinh 1)$; $V = \frac{1}{2}\pi c^3(1 + \cosh 1\sinh 1)$.
24. $t = \log\left(\dfrac{s + \sqrt{(s^2 + 2)}}{\sqrt{2}}\right)$. 25. Inflexions at $(\pm\frac{1}{2}\pi a, a)$, $\left(\pm\frac{3}{2}\pi a, a\right)$. 27. $64\pi a^2/3$;
$5\pi^2 a^3/8$. 29. $\frac{1}{4}17\sqrt{17}$. 31. $(a, 2a)$; $4\sqrt{2}a$, $\frac{1}{4}17\sqrt{17}a$. 32. $\varrho = a\theta$; $(a\cos\theta, a\sin\theta)$.
33. $9/4$ units. 36. $y = \pm x$. 37. $4a\cos\psi$. 39. $8(2\sqrt{2} - 1)\pi a^2/3$; $\dfrac{4(2\sqrt{2} - 1)a}{3\{\sqrt{2} + \log(1 + \sqrt{2})\}}$.
41. (i) $8 + 2\log 3$ units; (ii) $8(19 - 9\log 3)/3$ units².

Ex. 15.4 (p. 171)

1. $r = c\sec\theta$. 2. $r = c\,\text{cosec}\,\theta$. 3. $r = (a/\sqrt{2})\sec\left(\theta - \frac{1}{4}\pi\right)$. 4. $r\cos(\theta + \pi/6) = \frac{1}{2}a$.
5. $r = 2a\cos\theta$. 6. $r^2 = 4a^2\cos^2\theta$. 7. $r = a/(1 - \cos\theta)$. 8. $r = 2a/(2 - \cos\theta)$.
9. $r = 2a(1 + \cos\theta)$. 11. $2a/\sqrt{3}$. 13. $\frac{1}{3}\pi, \frac{2}{3}\pi$. 15. $r = a(\sec\theta - \cos\theta)$. 16. $2k\pi$.
22. $r = k^2(1 - \cos\theta)/a$. 24. Another circle.

Ex. 15.5 (p. 174)

1. (i) $(x^2 + y^2)^2 = a^2(x^2 - y^2)$; (ii) $y^2 = a(a - 2x)$; (iii) $(x^2 + y^2 - 2ax)^2 = a^2(x^2 + y^2)$; (iv) $x^4 - y^4 = 2a^2xy$; (v) $x\cos\alpha + y\sin\alpha = a$. 2. (i) $r = 2a\cos\theta$; (ii) $r = 2a\sin\theta$; (iii) $r^2 = 2\sin 2\theta$; (iv) $r^2\cos 2\theta = a^2$; (v) $r^2\sin 2\theta = 2c^2$.

Ex. 15.6 (p. 177)

1. $3\pi a^2/2$.   2. $\frac{1}{2}a^2$.   3. $\pi a^2/16$.   4. $\frac{1}{2}\pi a^2$ when n is even; $\frac{1}{4}\pi a^2$ when n is odd.
5. $1:7:19:37$.   7. $\frac{1}{4}(e^{8\pi}-1)a^2$.   8. $r = 2a/(1-\cos\theta)$; $64\sqrt{3}a^2/27$.   9. $\frac{1}{4}(3\pi-8)a^2$
10. $\frac{1}{2}(4-\pi)a^2$.

Ex. 15.8 (p. 182)

1. $\{a(e^{2k\pi}-1)\sqrt{(1+k^2)}\}/k$.   2. $8a$.   3. $3\pi a/2$.   4. $\left(\frac{27}{20}+\frac{1}{2}\log 4\right)a$.   5. $8\pi a^3/3$.
6. $(\pi\sqrt{2a})/8$.   7. $4\pi a^3/15$.   8. $2\sqrt{2}\pi(e^{2\pi}+1)/5$ units^2.

Ex. 15.10 (p. 187)

1. (i) $\frac{1}{3}\pi$; (ii) $\tan^{-1}\left(\frac{1}{3}\right)$; (iii) $\tan^{-1}\left(\frac{\sqrt{3}}{5}\right)$; (iv) $\tan^{-1}\alpha$; (v) at $(1,7\pi/12)$, $\varphi = \tan^{-1}\left(\frac{2}{3\sqrt{3}}\right)$; at $(5,7\pi/12)$, $\varphi = \pi - \tan^{-1}\left(\frac{2}{3\sqrt{3}}\right)$. 3. $\frac{1}{2}\pi$.   5. $90°$; $51°50'$ and $308°10'$.
6. $\frac{1}{2}\pi$.   8. $\frac{3}{4}\pi$; the pole and $\left\{\frac{2\sqrt{2}}{3},\ \cos^{-1}\left(\sqrt{\frac{1}{3}}\right)\right\}$, $\left\{\frac{2\sqrt{2}}{3},\ \pi\pm\cos^{-1}\left(\sqrt{\frac{1}{3}}\right)\right\}$, $\left\{\frac{2\sqrt{2}}{3},\ 2\pi-\cos^{-1}\left(\sqrt{\frac{1}{3}}\right)\right\}$.   10. $\varphi = \frac{1}{2}\pi - 2\theta$; $\theta = \frac{1}{4}\pi$, $\frac{3\pi}{4}$ are asymptotes of the curve. 11. $\frac{2}{3}\pi$. 12. $s = 4a\sin\frac{1}{3}(2\psi-\pi)$. 13. $p^2 = r^4/(a^2+r^2)$; $3/(a2\sqrt{2})$.
14. $p^2 = a^2 r^2/(a^2+r^2)$; $1/(a2\sqrt{2})$.   16. $p^2 = a^2 r^2/(2a^2-r^2)$; $2/a$.   18. $\pi/4$, $5\pi/4$, $9\pi/4$, $13\pi/4$.

Misc. Ex. XV (p. 189)

1. (i) Straight line $3x+4y=1$, cuts axes at $\left(\frac{1}{3},0\right)$, $\left(0,\frac{1}{4}\right)$; (ii) circle $x^2+y^2-3x-4y=0$ cuts axes at $(0,0)$, $(3,0)$, $(0,4)$, centre $(3/2,2)$, radius $5/2$.
2. $\cos^{-1}\left(\frac{5}{12}\right)$, $\pi+\cos^{-1}\left(\frac{5}{12}\right)$.   3. $16x^2+25y^2+96x-256=0$; semi-axes of lengths 5 and 4 units; eccentricity $3/5$.   5. $(2n+1)\pi/12$ for $n = 0, 1, 2, \ldots, 11$.
7. 2 units.   10. $(1\cdot618, 51°50')$, $(1\cdot618, 308°10')$.   11. (a) $(4, 30°)$; (b) $y^2 = 4(1-x)$.
13. $98\pi a^3/15$.   15. $\frac{1}{4}\pi a^2$.   17. $\pi a^2/16$.   18. (ii) $\frac{1}{4}\pi$, $5\pi/4$; (iii) $r = e^\theta$.   19. $\pi^2 a^3/4\sqrt{2}$.
20. $20a^2/3$; $8\sqrt{2}a/5$.   21. $r = 2a(1-\cos\theta)$.   23. $(27a^2\sqrt{3})/8$.   24. $3\pi a^2/16$; at $(512a/315\pi, \frac{1}{2}\pi)$.   25. $\pi a^2/12$; $\dfrac{81\sqrt{3}a}{80\pi}$.

Ex. 16.2 (p. 196)

1. (i) $\frac{1}{2}(-1\pm i\sqrt{3})$;   (ii) $\frac{1}{2}(1\pm i\sqrt{11})$;   (iii) $\frac{1}{2}(1\pm i)$;   (iv) $\cos\theta\pm i\sin\theta$;
(v) $-\tan\theta\pm i$; (vi) $\frac{1}{\sqrt{2}}(1\pm i)$, $-\frac{1}{\sqrt{2}}(1\pm i)$.   2. (i) $10+3i$; (ii) $3+7i$; (iii) $16+30i$;
(iv) $27-5i$;   (v) $18+16i$;   (vi) $(7+4i)/13$;   (vii) $(-2+36i)/5$;   (viii) -4;

(ix) $41 - 38i$; (x) $(-7 + 24i)/625$. 3. (i) $x^2 - y^2 + i\,2xy$; (ii) $x/(x^2 + y^2) - iy/(x^2 + y^2)$; (iii) $(x^2 + y^2 - 1)/(x^2 + y^2 - 2x + 1) - i\,2y/(x^2 + y^2 - 2x + 1)$.
4. (i) $\pm(4 - 3i)$; (ii) $\pm(4 - i)$; (iii) $\pm(3 + 2i)$; (iv) $\pm(8 + i)/13$; (v) $\pm(1 + i)$;
(vi) $\pm(1 - i)$. 5. (i) $x^2 + 4 = 0$; (ii) $x^2 - 4x + 13 = 0$; (iii) $x^4 - 2x^3 + x^2 + 2x - 2 = 0$; (iv) $x^4 - 4x^3 + 5x^2 = 0$; (v) $x^4 - 6x^3 + 15x^2 - 18x + 10 = 0$.
6. $2(ac + bd)/(c^2 + d^2)$. 7. (i) $(x + 1 + 2i)(x + 1 - 2i)$; (ii) $(x - 2 + i)(x - 2 - i)$;
(iii) $(2x + 1 + i)(2x + 1 - i)$; (iv) $(x - 1)(x - 1 + i)(x - 1 - i)$; (v)
$(x + a + ib)(x + a - ib)$. 8. $\dfrac{3 - i}{6(x - 1 - 3i)} + \dfrac{3 + i}{6(x - 1 + 3i)}$.
9. $\frac{1}{2}\left(\dfrac{i}{x + i} - \dfrac{i}{x - i}\right)$. 10. $a^2 = 3b^2$. 11. $i(1 \pm \sqrt{2})$. 12. $i(1 \pm \sqrt{2})$. 13. $4 - 2i$,
$-1 + 2i$. 14. $1 + i$, $\frac{1}{2}(1 - i)$. 15. $i(\sec\theta \pm \tan\theta)$.

Ex. 16.4 (p. 200)

1. $-1, -\omega, -\omega^2$. 2. $(x + 2)^3 - 1 = 0$; $x = -1, -2 + \omega, -2 + \omega^2$. 4. $w^3 = 2 \pm 2i$.
10. $x^3 + 1 = 0$. 11. 1. 12. $x = 5/2, \frac{1}{2}(3 \pm 2i)$. 13. $x^4 - 8x^3 + 32x^2 - 80x + 100 = 0$.
14. $-c(1 \pm i)$. 15. -2; $2 - i, -1 + i, -1 - i$. 16. $1, 1, \omega, \omega^2$.

Ex. 16.5 (p. 204)

2. (i) $5 + i$; (ii) $-1 + i$; (iii) $2 + 2i$; (iv) $1 - i$; (v) $-1 - 4i$.

Ex. 16.7 (p. 212)

1. (i) $13(\cos 67° 23' + i \sin 67° 23')$; (ii) $5(\cos 53° 8' + \sin 53° 8')$; (iii) $1(\cos 0 + i \sin 0)$; (iv) $1(\cos 90° + i \sin 90°)$; (v) $1(\cos 120° + i \sin 120°)$; (vi) $1\{\cos(-120°) + i \sin(-120°)\}$; (vii) $5(\cos 53° 8' + i \sin 53° 8')$; (viii) $65\{\cos(-14° 15') + i \sin(-14° 15')\}$; (ix) $1(\cos 90° + i \sin 90°)$; (x) $\dfrac{1}{\sqrt{13}}\{\cos(-56° 19') + i \sin(-56° 19')\}$.
3. (i) A circle centre $(0, -1)$, radius 1; (ii) a circle centre the origin, radius $1/a$;
(iii) the straight line $y = x$; (iv) the straight line $x - y + 1 = 0$ for $y > 0$; (v) the straight line $y = x \tan\alpha$ for $y \geq 0$ if $0 \leq \alpha \leq \pi$, the straight line $y = x \tan\alpha$ for $y \leq 0$ if $0 \geq \alpha \geq -\pi$. 5. A circle centre $(2, 0)$, radius 2. 6. (i) A circle centre O, radius $\frac{1}{2}\sqrt{\{r_1^2 + r_2^2 + 2r_1r_2 \cos(\theta_1 - \theta_2)\}}$ where P is $r_1 \angle \theta_1$, and Q is $r_2 \angle \theta_2$; (ii) a straight line perpendicular to the fixed direction of PQ; (iii) the perpendicular bisector of the fixed line OP. 7. (i) 1; (ii) $6\left(\cos\dfrac{\pi}{12} - i \sin\dfrac{\pi}{12}\right)$; (iii) -27;
(iv) $-2\left(\cos\dfrac{\pi}{12} + i \sin\dfrac{\pi}{12}\right)$; (v) $\cos\dfrac{13\theta}{6} + i \sin\dfrac{13\theta}{6}$. 8. (i) $1 \angle -\theta$; (ii) $2 \cos\theta \angle 0$;
(iii) $2 \cos 2\theta \angle 0$; (iv) $2 \cos 3\theta \angle 0$. 9. $-\frac{1}{2}(1 + 4\sqrt{3}) + \frac{1}{2}i(3\sqrt{3} - 2)$, $\frac{1}{2}(4\sqrt{3} - 1)$
$-\frac{1}{2}i(3\sqrt{3} + 2)$. 10. (i) ellipse, foci $(\pm 1, 0)$, major axis of length 4; (ii) the real axis between $(-1, 0)$ and $(1, 0)$; (iii) the circle $x^2 + y^2 = 1$ for $y > 0$; (iv) the real axis between $(-1, 0)$ and $(1, 0)$ and the positive imaginary axis; (v) the circle (of Apollonius) $3x^2 + 3y^2 + 10x + 3 = 0$. 12. (a) $x = \pm\dfrac{3}{2}$,
$y = \pm 2$; (b) (ii) $(-1, -\frac{1}{2})$. 13. $2i\sqrt{(ac - b^2)}/a$; other diagonal $2\sqrt{(ac - b^2)}/a$; sides $(\pm 1 \pm i)\sqrt{(ac - b^2)}/a$. 15. (i) centre (a, b) radius $\sqrt{(a^2 + b^2 - c)}$.

Ex. 16.8 (p. 220)

1. $\cos 9\theta - i\sin 9\theta$.　2. $\cos\dfrac{3\pi}{4} - i\sin\dfrac{3\pi}{4} = -\dfrac{1}{\sqrt2}(1+i)$　3. $\cos\dfrac{5\pi}{6} - i\sin\dfrac{5\pi}{6}$.　4. (i) $1\left(\cos\dfrac{\pi}{4} + i\sin\dfrac{\pi}{4}\right)$, $1\left(\cos\dfrac{5\pi}{4} + i\sin\dfrac{5\pi}{4}\right)$; (ii) $\sqrt5(\cos 26°\,34' + i\sin 26°\,34')$, $\sqrt5(\cos 206°\,34' + i\sin 206°\,34')$;　(iii) $\sqrt{13}\{\cos(-33°\,41') + i\sin(-33°\,41')\}$, $\sqrt{13}(\cos 146°\,19' + i\sin 146°\,19')$; (iv) $5(\cos 53°\,8' + i\sin 53°\,8')$, $5(\cos 233°\,8' + i\sin 233°\,8')$;　(v) $1\left(\cos\dfrac{\pi}{4} + i\sin\dfrac{\pi}{4}\right)$, $1\left(\cos\dfrac{5\pi}{4} + i\sin\dfrac{5\pi}{4}\right)$.

5. (i) $1(\cos 0 + i\sin 0)$, $1\left(\cos\dfrac{2\pi}{3} + i\sin\dfrac{2\pi}{3}\right)$, $1\left(\cos\dfrac{4\pi}{3} + i\sin\dfrac{4\pi}{3}\right)$;

(ii) $1\left(\cos\dfrac{\pi}{6} + i\sin\dfrac{\pi}{6}\right)$, $1\left(\cos\dfrac{5\pi}{6} + i\sin\dfrac{5\pi}{6}\right)$, $1\left(\cos\dfrac{3\pi}{2} + i\sin\dfrac{3\pi}{2}\right)$;

(iii) $1{\cdot}12\left(\cos\dfrac{\pi}{12} + i\sin\dfrac{\pi}{12}\right)$, $1{\cdot}12\left(\cos\dfrac{3\pi}{4} + i\sin\dfrac{3\pi}{4}\right)$, $1{\cdot}12\left(\cos\dfrac{17\pi}{12} + i\sin\dfrac{17\pi}{12}\right)$;
(iv) $1{\cdot}71(\cos 17°\,43' + i\sin 17°\,43')$, $1{\cdot}71(\cos 137°\,43' + i\sin 137°\,43')$,
$1{\cdot}71(\cos 257°\,43' + i\sin 257°\,43')$. 7. $1{\cdot}27 \pm i(0{\cdot}79)$, $-1{\cdot}27 \pm i(0{\cdot}79)$. 8. $\cos\dfrac{\pi}{6} + i\sin\dfrac{\pi}{6}$, $\cos\dfrac{\pi}{2} + i\sin\dfrac{\pi}{2} = i$, $\cos\dfrac{5\pi}{6} + i\sin\dfrac{5\pi}{6}$, $\cos\dfrac{7\pi}{6} + i\sin\dfrac{7\pi}{6}$, $\cos\dfrac{3\pi}{2} + i\sin\dfrac{3\pi}{2} = -i$, $\cos\dfrac{11\pi}{6} + i\sin\dfrac{11\pi}{6}$. 9. $\cos 5\theta \equiv 16\cos^5\theta - 20\cos^3\theta + 5\cos\theta$;
$\tan 5\theta \equiv \dfrac{5t - 10t^3 + t^5}{1 - 10t^2 + 5t^4}$ where $t = \tan\theta$. 10. $\cos n\theta \equiv c^n - {}_nC_2c^{n-2}s^2 +$
${}_nC_4c^{n-4}s^4 - \ldots$, the expression ending in $(-1)^{n/2}s^n$ or $(-1)^{(n-1)/2}\,{}_nC_{n-1}\,cs^{n-1}$ according as n is even or odd; $\sin n\theta \equiv {}_nC_1c^{n-1}s - {}_nC_3c^{n-3}s^3 + {}_nC_5c^{n-5}s^5 - \ldots$, the expression ending in $(-1)^{(n-2)/2}\,{}_nC_{n-1}cs^{n-1}$ or $(-1)^{(n-1)/2}s^n$ according as n is even or odd. $[c = \cos\theta,\ s = \sin\theta.]$ 11. $\sin^6\theta \equiv (10 - 15\cos 2\theta + 6\cos 4\theta - \cos 6\theta)/32$;
$\cos^6\theta \equiv (10 + 15\cos 2\theta + 6\cos 4\theta + \cos 6\theta)/32$. 12. (i) $\left(10\,\theta - \dfrac{15\sin 2\theta}{2} + \dfrac{3\sin 4\theta}{2}\right.$
$\left. -\dfrac{\sin 6\theta}{6}\right)\Big/32 + C$; (ii) $\left(35\theta + 28\sin 2\theta + 7\sin 4\theta + \dfrac{4\sin 6\theta}{3} - \dfrac{\sin 8\theta}{8}\right)\Big/128 + C$
(iii) $\left(2\theta - \dfrac{\sin 2\theta}{2} - \dfrac{\sin 4\theta}{2} + \dfrac{\sin 6\theta}{6}\right)\Big/32 + C$. 13. $\cos(\alpha + \beta - \gamma - \delta)$
$+ i\sin(\alpha + \beta - \gamma - \delta)$. 14. 0, $\pm i\sqrt3$, $\pm i/\sqrt3$. 15. $\pm(3+i)$, $\pm(1-3i)$. 18. The given expression equals $2^{n+1}\cos\dfrac{2n\pi}{3}$. 19. $\cos 6\theta \equiv 32c^6 - 48c^4 + 18c^2 - 1$,
$\sin 6\theta/\sin\theta \equiv 32c^5 - 32c^3 + 6c$ where $c = \cos\theta$. 20. $\left(z^2 - 2z\cos\dfrac{4\pi}{7} + 1\right)$,
$\left(z^2 - 2z\cos\dfrac{6\pi}{7} + 1\right)$.

Ex. 16.10 (p. 227)

1. (i) $\sqrt2 e^{i\pi/4}$; (ii) $\sqrt2 e^{-i\pi/4}$; (iii) $e^{i\pi/2}$; (iv) $e^{-i\pi/2}$; (v) $2e^{-i\pi/3}$. 2. (i) $\dfrac{1}{\sqrt2} - \dfrac{i}{\sqrt2}$;
(ii) $-1 + i0$; (iii) $-1 + i0$; (iv) $\dfrac{e}{2} + i\dfrac{e\sqrt3}{2}$; (v) $e^x\cos y + ie^x\sin y$. 3. $2\cos\tfrac12\theta e^{i(\theta+4n\pi)/2}$.
4. $2\cos\theta e^{i2n\pi}$　　5. $(\cos x + \sin x)\cosh y - i\,(\cos x + \sin x)\sinh y$.

6. (i) $\sinh 1 \cos 1 - i \cosh 1 \sin 1$; (ii) $\cosh 1 \cos 1 - i \sinh 1 \sin 1$. 8. $\sinh 2x/(\cosh 2x + \cos 2y)$, $\sin 2y/(\cosh 2x + \cos 2y)$.

9. $\cos^6\theta \equiv (e^{i6\theta} + 6e^{i4\theta} + 15e^{i2\theta} + 20 + 15e^{-i2\theta} + 6e^{-i4\theta} + e^{-i6\theta})/64$;

$$\int \cos^6\theta \, d\theta = \left(\frac{\sin 6\theta}{6} + \frac{3\sin 4\theta}{2} + \frac{15\sin 2\theta}{2} + 10\theta \right) \Big/ 32 + C.$$ 10. $(105\pi + 320)/1536$.

12. (i) $(2n + \tfrac{1}{2})\pi \pm i\log(2 + \sqrt{3})$; (ii) $2n\pi \pm i\log(3 + \sqrt{8})$; (iii) $(2n - \tfrac{1}{2})\pi \pm i\log(2 + \sqrt{3})$. 13. $\theta + r\sin\theta = 0$, $\pm\pi$, $\pm 2\pi$, $\ldots$. 17. $z = \tfrac{1}{2}\log 3 + i(\tfrac{1}{2} \pm n)\pi$.

23. $\cosh(x\cos\theta)\sin(x\sin\theta)$.

Misc. Ex. XVI (p. 229)

1. (b) $2^{1/4}$; $\dfrac{11\pi}{24}$, $\dfrac{23\pi}{24}$, $-\dfrac{35\pi}{24}$, $\dfrac{47\pi}{24}$. 4. $(1 - \tfrac{1}{2}i)z_1 + iz_2$.

5. (a) $\tfrac{1}{2}\left\{1 + i\left(\cot\dfrac{2\pi}{n} + \operatorname{cosec}\dfrac{2\pi}{n}\right)\right\}$. 6. (a) $|w| = 2\cos\tfrac{1}{2}\theta$, $\arg w = \tfrac{1}{2}\theta$.

7. (a) 1; (b) within the unit circle $|z| = 1$, i.e. $x^2 + y^2 = 1$. 8. (a) $1 \angle 27°$, $1 \angle 45°$, $1 \angle 99°$, $1 \angle 117°$, $1 \angle 171°$, $1 \angle 189°$, $1 \angle 243°$, $1 \angle 261°$, $1 \angle 315°$, $1 \angle 333°$.

10. $2^{1/3}\left(\cos\dfrac{2\pi}{9} + i\sin\dfrac{2\pi}{9}\right)$, $2^{1/3}\left(\cos\dfrac{4\pi}{9} + i\sin\dfrac{4\pi}{9}\right)$, $2^{1/3}\left(\cos\dfrac{8\pi}{9} + i\sin\dfrac{8\pi}{9}\right)$, $2^{1/3}\left(\cos\dfrac{10\pi}{9} + i\sin\dfrac{10\pi}{9}\right)$, $2^{1/3}\left(\cos\dfrac{14\pi}{9} + i\sin\dfrac{14\pi}{9}\right)$, $2^{1/3}\left(\cos\dfrac{16\pi}{9} + i\sin\dfrac{16\pi}{9}\right)$. 11. (i) $\dfrac{-(\sqrt{3}-1)+i(\sqrt{3}+1)}{2\sqrt{2}}$, $\dfrac{-(\sqrt{3}+1)-i(\sqrt{3}-1)}{2\sqrt{2}}$, $\dfrac{(\sqrt{3}-1)-i(\sqrt{3}+1)}{2\sqrt{2}}$. 12. (i) $x^2 - 2x + 1$, $x^2 - (\sqrt{3}-1)x + 1$, $x^2 + (\sqrt{3}-1)x + 1$, $x^2 + 2x + 1$, $x^2 + (\sqrt{3}+1)x + 1$; (ii) (a) $ABCD$ is a parallelogram; (b) $AC = 2BD$ and AC, BD are perpendicular; a rhombus. 14. (b) The straight line $x = \tfrac{1}{2}$.

15. (a) Each root has modulus 1, arguments are $\pi/6$, $\pi/3$, $7\pi/6$, $4\pi/3$; $z = -1 + 2i$ or $3 - 2i$. 16. $\cos\alpha\cosh\beta - i\sin\alpha\sinh\beta$. 17. $-3 \pm i2$. 18. $(a - ib)(c - id)$; $(a^2 + b^2)(c^2 + d^2)$. 20. (a) (i) $\dfrac{3}{2} \angle -\dfrac{\pi}{3}$; (ii) $1/\sqrt{5} \angle -100°18'$; (b) $\cos\dfrac{1}{3}\pi$ $- i\sin\dfrac{1}{3}\pi$, $z^3 + z^2 - 1 = 0$. 21. $(1 - z^n)/(1 - z)$. 23. (i) $\omega = \pm\sqrt{\left\{\dfrac{(CR_1^2 - L)}{(CR_2^2 - L)LC}\right\}}$;

(ii) $\dfrac{1}{5}(3 \pm i4)$, $\tfrac{1}{2}(1 \pm i3)$. 24. $4r\pi/n(n + 1)$ where r is $= 0$, ± 1, ± 2, $\ldots$. 28. Centre $-1 + i0$, radius 1. 31. 0. 32. (i) $x = -2$, $y = 1$; (ii) $\dfrac{1}{\sqrt{2}}\cosh\dfrac{1}{4}\pi + \dfrac{i}{\sqrt{2}}\sinh\dfrac{1}{4}\pi$, modulus $= \sqrt{(\tfrac{1}{2}\cosh\tfrac{1}{2}\pi)}$. 34. (i) $x = \tfrac{1}{2}$, $y = -\dfrac{3}{2}$; (ii) circle $x^2 + y^2 - 2x - y = 0$ (centre $1 + i\tfrac{1}{2}$; radius $\tfrac{1}{2}\sqrt{5}$). 36. z describes once (anti-clockwise) the circle centre $2 + 0i$, radius 1, starting from $1 + 0i$; w describes the line $x = 2$ from $2 - i\infty$ to $2 + i\infty$. 37. $\sin^{-1}\left\{\sqrt{\dfrac{3 - \sqrt{5}}{2}}\right\} \doteq 38° 10'$. 38. $\sqrt{2}\,e^{i5\pi/12}$; $-\dfrac{5\sqrt{3}}{4} + \dfrac{3i}{4}$.

40. $\sqrt{2}\left(\cos\dfrac{1}{4}\pi + i\sin\dfrac{1}{4}\pi\right)$. 41. $32\cos^6\varphi - 36\cos^3\varphi + 4$; $\varphi = 0°$, $60°$, $300°$, $360°$.

42. (i) $(1 - i)z$, $2z$, $(1 + i)z$; (ii) $\tfrac{1}{2}\left(1 - i\tan\dfrac{r\pi}{5}\right)$, $r = 0, 1, 2, 3, 4$. 45. $n\pi$, $\left(n \pm \dfrac{1}{6}\right)\pi$ for $n = 0, \pm 1, \pm 2, \ldots$.

Ex. 17.1 (p. 239)

1. $y\dfrac{\mathrm{d}y}{\mathrm{d}x} + x = 0$; the tangent at P is perpendicular to OP.   2. $2x\dfrac{\mathrm{d}y}{\mathrm{d}x} = y$; the subtangent is of length $2x$.   3. $\dfrac{\mathrm{d}^2y}{\mathrm{d}x^2} = k^2y$.   4. $x\dfrac{\mathrm{d}^2y}{\mathrm{d}x^2} + 2\dfrac{\mathrm{d}y}{\mathrm{d}x} = 0$.   5. $\dfrac{\mathrm{d}^2y}{\mathrm{d}x^2} = -n^2y$.

6. $x\dfrac{\mathrm{d}^2y}{\mathrm{d}x^2} + \dfrac{\mathrm{d}y}{\mathrm{d}x} = 0$.   7. $(1+x^2)\dfrac{\mathrm{d}^2y}{\mathrm{d}x^2} + 2x\dfrac{\mathrm{d}y}{\mathrm{d}x} = 0$.

8. $\dfrac{\mathrm{d}^2y}{\mathrm{d}x^2} + 2k\dfrac{\mathrm{d}y}{\mathrm{d}x} + k^2y = 0$.   9. $y\dfrac{\mathrm{d}y}{\mathrm{d}x} = k$.   10. $\dfrac{\mathrm{d}^3y}{\mathrm{d}x^3} = 0$.

11. $a^2\left(\dfrac{\mathrm{d}^2y}{\mathrm{d}x^2}\right)^2 = \left\{1 + \left(\dfrac{\mathrm{d}y}{\mathrm{d}x}\right)^2\right\}^3$.   12. $\dfrac{\mathrm{d}y}{\mathrm{d}x} = \lambda y^2$.   13. $x\dfrac{\mathrm{d}y}{\mathrm{d}x} = \mu y$.

14. $(x^2 - 2xy)\left(\dfrac{\mathrm{d}y}{\mathrm{d}x}\right)^2 - 2xy\dfrac{\mathrm{d}y}{\mathrm{d}x} + y^2 - 2xy = 0$.   15. $\dfrac{\mathrm{d}r}{\mathrm{d}\theta} = r\cot\tfrac{1}{2}\theta$.

16. $\dfrac{\mathrm{d}r}{\mathrm{d}\theta} + r\cot 2\theta = 0$.   17. $v\dfrac{\mathrm{d}v}{\mathrm{d}s} = -\dfrac{k}{s^2}$.   18. $\dfrac{\mathrm{d}\theta}{\mathrm{d}t} = -k(\theta - \theta_0)$, where θ is the temperature of the body and θ_0 is that of the surroundings.

19. $\dfrac{\mathrm{d}^2x}{\mathrm{d}t^2} + k\dfrac{\mathrm{d}x}{\mathrm{d}t} + cx = 0$.   20. $m\dfrac{\mathrm{d}v}{\mathrm{d}t} = P$.

Ex. 17.4 (p. 245)

1. $y = x\log x - x + \mathrm{e}$.   2. $y = \dfrac{wx}{24EI}(l-x)(l^2 + lx - x^2)$ 3. $x + y = C(1 - xy)$.

4. $\operatorname{cosec} y + \cot y = C\mathrm{e}^{x^2/2}$ or $\cot\tfrac{1}{2}y = C\mathrm{e}^{x^2/2}$.   5. $y = \mathrm{e}^{Cx}$.   6. $y = C\mathrm{e}^{\sin x}$.

7. $y = -\log(1 + Cx)$.   8. $y + \log(y-1) = \dfrac{1}{3}x^3 + C$.   9. $y^2 = 1 + C(1-x^2)/x^2$.

10. $y = -\tanh(\mathrm{e}^x + C)$.   11. $\sin y = 1 + C\mathrm{e}^{\sin x}$.   12. $y^2 = (3x^2 - 1)/(x^2 + 1)$.

13. $y = 4x$.   14. $y = 1/(\mathrm{e}^x - x)$.   15. $r = a\sin^2\tfrac{1}{2}\theta$.   16. $y^2 = \cot^2 x$.

17. $r^3 = a^3\sin 3\theta$.   18. $\log(\sec y + \tan y) = 2\tan^{-1}\mathrm{e}^x - \tfrac{1}{2}\pi$.   19. $\sin y = \sin x + \cos x$.   20. $y = \tan(x + \tfrac{1}{4}\pi) - x$.   21. $y = -\log(C - \mathrm{e}^x)$.   22. $\tan^{-1}\left(\dfrac{y}{x}\right) = \tfrac{1}{2}\log(x^2 + y^2) + C$.   23. $x^2 + y^2 - 5y = 0$.   24. $y^2 = 2x^2 - 7$.

25. $y = \pm a\cosh(x/a)$.   26. $r = \mathrm{e}^{\theta/k}$.   27. $1 \cdot 7$ units per sec.

Ex. 17.5 (p. 251)

1. (i) After 16 min: (ii) $48°$.   2. (i) $26\cdot8$ units sec^{-1};   (ii) $1\cdot10$ units.

3. $\dfrac{1}{k}\log\left(\dfrac{a}{a-1}\right)$.   4. $26\cdot6$ days.   5. $5\cdot40$ amp.   6. 6.3 ft.   7. $93\cdot6$ gm.

8. $t = \dfrac{1}{k}\log\left(\dfrac{15}{x-15}\right)$; $19\cdot9$ units.   9. (i) $35°$; (ii) $6\cdot8$ min.   10. $\dfrac{\mathrm{d}x}{\mathrm{d}t} = k_1x - k_2$;

$x = \dfrac{k_2}{k_1} + \left(N - \dfrac{k_2}{k_1}\right)\mathrm{e}^{k_1 t}$.

Ex. 17.7 (p. 255)

1. $y = \{(x - 1)e^x + C\}/x$.　2. $y = \frac{1}{2}(1 - \cot x)e^x + C\,\mathrm{cosec}\,x$.　3. $y = 1$ $+ C/\sqrt{(1 + x^2)}$.　4. $y = (x - 2)e^x/(x - 1) + C/(x - 1)$.　5. $y = b/a + C\,e^{-ax}$.
6. $y = \frac{1}{2}(\sin x + \cos x) + Ce^{-x}$.　7. $r = \theta \cos\theta + C \cos\theta$.
8. $y = (Cx + \sin x)/(\mathrm{cosec}\,x + \cot x)$.　9. $y = (\sec x + \tan x)\{C + \log(1 + \sin x)\}$.
10. $y = x \log x + 1 + Cx$.　11. $y = \sin x/(C - x)$.　12. $y = \dfrac{\pm 1}{x\sqrt{(C - 2x)}}$.
13. $\dfrac{5}{4}e^{-2} + \dfrac{1}{4} \doteq 0\cdot42$ units sec^{-1};　$\dfrac{5}{8}(1 - e^{-2}) \doteq 0\cdot54$ units.　14. $0\cdot67$ amp.
15. $y = -a^4\left\{1 - \sin\left(\dfrac{x}{a}\right)\right\}\bigg/x^3$.　16. (a) $y = (C + x)e^{-2x}$;　(b) $\dfrac{dz}{dx} = \dfrac{1}{4} - z^2$; $y = (xe^x - 2x)/(2e^x + 4)$.　17. $y = (x^5 - 1)/5x^2$.　18. $y = x^4 - 2x$.
19. $y = \tan x - \sqrt{2}\sin x$.　20. $y = \mathrm{cosec}\,x - \cot x$.

Ex. 17.8 (p. 259)

1. $y = \sin 2x - 2\cos 2x$.　2. $y = 2\cos 3x$.　3. $y = (\sin 8x)/8$.
4. $y = \dfrac{1}{3}(7\sin 3x - \cos 3x)$.　5. $y = Ae^{nx} + Be^{-nx}$　or　$C\cosh nx + D\sinh nx$.
6. $y = 3e^x + e^{-x}$.　7. $y = A + Bx + \sin x$.　8. $y = \sqrt{(2x^2 + 2x + 1)}$.

Ex. 17.10 (p. 264)

1. $y = Ae^{-x/2} + Be^{3x}$.　2. $y = Ae^{5x/4} + Be^{-x}$.　3. $y = Ae^{3x} + Be^{-3x}$.
4. $y = A + Be^{-5x}$.　5. $y = (Ax + B)e^{2x}$.　6. $y = e^{-x/2}(A\cos\frac{1}{2}x + B\sin\frac{1}{2}x)$.
7. $y = e^{-3x/2}\{A\cos(\frac{1}{2}\sqrt{7}x) + B\sin(\frac{1}{2}\sqrt{7}x)\}$.
8. $y = Ae^{(-5 + \sqrt{21})x/2} + Be^{(-5 - \sqrt{21})x/2}$.　9. $y = (Ax + B)e^{-3x/2}$.
10. $y = Ax + B + Ce^{-x}$.

Ex. 17.11 (p. 268)

1. $y = Ae^{-2x} + Be^{-3x} + \dfrac{2}{3}$.　2. $y = A\cos 3x + B\sin 3x + \dfrac{1}{9}$.　3. $y = Ae^{2x}$ $+ Be^{-x} + (\cos x - 3\sin x)/10$.　4. $y = (A + Bx + x^2)e^x$.　5. $y = Ae^{-2x} + Be^{-x}$ $- xe^{-2x}$.　6. $y = e^{-x/3}\left\{A\cos\left(\dfrac{x\sqrt{2}}{3}\right) + B\sin\left(\dfrac{x\sqrt{2}}{3}\right)\right\} + \dfrac{4\sin 2x - 11\cos 2x}{137}$.
7. $y = e^{3x/2}\left\{A\cos\left(\dfrac{x\sqrt{3}}{2}\right) + B\sin\left(\dfrac{x\sqrt{3}}{2}\right)\right\} + \dfrac{2}{3}$.　8. $y = e^{2x}(A\cos x + B\sin x)$ $+ \dfrac{x^2}{5} + \dfrac{13x}{25} + \dfrac{67}{125}$.　9. $y = Ae^{2x} + Be^{-x} - \dfrac{x}{2} - \dfrac{1}{4}$.　10. $y = (Ax + B)e^{2x}$ $+ \dfrac{x}{4} + \dfrac{1}{4} + e^x$.　11. $y = e^{-2x}(A\cos x + B\sin x) + (\sin 2x - 8\cos 2x)/65$.
12. $y = A + Be^{5x} - \dfrac{4x}{5}$.　13. $y = e^{-x}(A\cos x + B\sin x) - (\sin 2x + 2\cos 2x$ $- 4\sin x - 2\cos x)/10$.　14. $y = A + Bx + Ce^{-x} + \frac{1}{2}x^2 + xe^{-x}$.
15. $y = (A + Bx + \frac{1}{2}x^2)e^{3x}$.　16. $y = Ae^{2x} + Be^x + (3\cos 2x - \sin 2x)/20$.
17. $y = A + Be^{x/p} - \dfrac{ax}{p} + \dfrac{ae^x}{p(p - 1)}$.　18. $y = Ae^{2x} + Be^{-2x} + C\cos 2x +$

$+ D \sin 2x - x/16.$ 19. $x = A e^{3t} + B e^{-t} - \left\{4 \sin\left(2t + \dfrac{1}{4}\pi\right)\right.$

$+ 7 \cos\left(2t + \dfrac{1}{4}\pi\right)\Big)\Big\}\Big/ 65.$ 20. $x = e^t(A \cos t + B \sin t) - (\cos 2t + 2 \sin 2t)/10.$

22. $y = 3e^{-2t} - 3e^{-t} + 3 \sin t + \cos t.$ 23. $x = e^{3t} - 3e^{2t}; \quad t = \log 2.$

24. $s = 7{\cdot}35$ units; $v = 1{\cdot}20$ units per sec. $[s = \tfrac{1}{2}ae^{-t}(1 - e^{-t})^2].$ 26. Min $\dfrac{1}{16}$

when $t = \log 4.$ $(s = e^{-t} - 5e^{-2t} + 8e^{-3t}.)$ 27. $-ka \{\sec(\tfrac{1}{2}na) - 1\}/2n^2.$

28. $y = e^{-3t}(4 \cos 4t - \sin 4t)/68 + (4 \sin t - \cos t)/17.$ 29. $y = \sin\theta + \sin 3\theta.$

30. $x = \dfrac{1}{2k^2} e^{-kt} \cos \{t\sqrt{(n^2 - k^2)}\} + \dfrac{1}{2kn} \sin nt;$

$[\cos 2kt - e^{-kt}\{\sqrt{3} \sin (kt\sqrt{3}) + \cos(kt\sqrt{3})\}]/2k.$

Misc. Ex. XVII. (p. 270)

2. (i) $y = x + 1 + C\sqrt{x}; \quad y = A e^{-x} + B e^{2x} + (\cos x - 3 \sin x)/10.$

3. (a) $y - \sin y \cos y = x^2 + 6x + C;$ (b) $y = \dfrac{1}{3} x e^{2x} - \dfrac{1}{3}(e^9 - e^3) x e^{-x}.$ 4. $94 {\cdot} 2\%;$

$1{\cdot}69 \times 10^7$ units.   6. $y = e^{-x}(A \cos 2x + B \sin 2x) + \dfrac{x^2}{5} - \dfrac{4x}{25} - \dfrac{2}{125}.$ 7. (a) 8;

(b) $x = A e^{-2t} + B e^{-3t} + e^{-t};$ 8. $y = A e^x + B e^{2x} + \tfrac{1}{2} - \tfrac{1}{2}x e^x + \dfrac{1}{12} e^{-x}.$

9. $\dfrac{dy}{dt} = -\dfrac{y}{(1+t)^2}; \quad y = 10 e^{-t/(t+1)}.$ 10. (a) $y = 2(1 + 3x^3)^{3/2}/(27x^3) + 11/(27x^3);$

(b) $y = 2(e^{x^2} - 1)/(2 - e^{x^2}).$ 11. $t = \dfrac{2\pi}{15k}\left\{\left(\dfrac{9V}{\pi}\right)^{5/6} - y^{5/2}\right\}; \; t_1/t_2 = 1 - \dfrac{1}{2^{1/3}} \doteqdot 0{\cdot}794.$

12. $y = e^{-kx}(A \cos kx + B \sin kx) + (2 \sin x + k \cos x)/(k^3 + 4k).$ 13. (i) $y = A e^x$

$+ e^{-x/2}\left\{B \cos\left(\dfrac{x\sqrt{3}}{2}\right) + C \sin\left(\dfrac{x\sqrt{3}}{2}\right)\right\} - x^4 - 24x + \dfrac{e^{2x}}{7};$ (ii) $y = 2 + C/\sqrt{(1 + x^2)}.$

14. (i) $y = (x + 1) \{C + \log(x + 1)\}/(x - 1);$ (ii) $4a/9;$ $[x = 3a(e^{-t} - 2e^{-2t} + e^{-3t})].$

15. $\dfrac{H^2 t}{2A} = -H\sqrt{y} - K \log(K - H\sqrt{y}) +$ constant.   16. (i) 8 units per sec;

(ii) 4 units; (iii) 1 unit per sec.   17. $y = C e^{-Rt/L}; \; x = C e^{-Rt/L} + \dfrac{E}{R}.$ 18. (i) $y = x$

$+ \cot x + C \operatorname{cosec} x.$ 19. (a) $y = C x^2 - x^2/(1 + x);$ 14/3.   20. (a) $y = (x + 1)/(x - 1);$

(b) $y = (C + \sec^2 x)/x^2.$ 21. $\dfrac{dm}{dt} = -am; \; \dfrac{1}{p}\dfrac{dp}{dt} = -\dfrac{b}{m}\dfrac{dm}{dt};$

(i) $m = m_0 e^{-at};$ (ii) $p = p_0 m_0^b/m^b; \; p = p_0 e^{abt}$ 22. (i) $\dfrac{1}{y} = C + \log\left\{\dfrac{xy}{(1+x)(1-y)}\right\};$

(ii) $y = e^{-x}(A \cos 2x + B \sin 2x) + \dfrac{3}{5} - \dfrac{e^{-x}}{2} + \dfrac{(4 \sin 2x + \cos 2x)}{17}$

23. (a) $x^2 \dfrac{d^2 y}{dx^2} - 2x \dfrac{dy}{dx} + (x^2 + 2)y = 0;$ (b) (i) $y = A e^{-x} + B e^{-2x} + \tfrac{1}{2}x - \dfrac{3}{4};$

(ii) $y = \dfrac{(1 - x)}{x}\left[x + 4 \log \{\tfrac{1}{2}(1 - x)\} + \dfrac{3}{1 - x} - \tfrac{1}{2}\right].$ 24. (a) $x^2 y^2 = C(x + y).$

25. (i) $2x^2 - 3xy + y^2 = C;$ (ii) $x = e^{-t} + t e^t.$ 26. (a) $y = C e^{x/k};$ (b) $y = A e^{3x}$

$+ B e^{4x} + 1 - x e^{3x}.$ 27. (a) $z \dfrac{dz}{dx} = (1 + x)(1 + z); \quad y - \log(1 + x + y)$

$- \tfrac{1}{2}x^2 = C;$ (b) $5y^2 = x^2 - 4.$ 28. (i) $y = (x + 1) e^x/\{3 - (x + 1) e^x\};$

(ii) $y = \cos \log x$.   29. $y^{3/2} = 2 - x^{3/2}$.   30. (i) $y - x \log \{(x + y)^2/x\} = Cx$;

(ii) $y = \sin x + C \cos x$. (iii) $y = e^{-x}(A \cos 2x + B \sin 2x) + (5x + 3)/25$.

31. $\pi h^5/80a^2$; $\dfrac{dy}{dt} = -\dfrac{16ka^2}{\pi y^3}$ where k is a positive constant.

32. (a) $y^2 = 1/(1 - 2x^2)$;   (b) $\dfrac{dz}{dx} = \dfrac{2}{1 - 2z}$; $y - x - (y + x)^2 = C$.

33. (i) $(2x - y)^4(x + 3y)^3 = C$;   (ii) $y = \pm 1/\sqrt(Ce^{2x} + x + \tfrac{1}{2})$;

(iii) $y = 3(e^{-x} - e^{-2x})$.   34. (i) $y = (x \sin x + \cos x + C)/x^2$;   (ii) $y = 2e^{-5t} \sin 2t$.

35.   (i) $y = (x^3 + 2)/x^3$;   (ii) $y = A e^{-3x} + B e^{-x} + (\sin x - 2 \cos x)/10 + \dfrac{e^x}{8} + 1$.

36. (i) $x = 2/(1 + \cos y)$; (ii) $y = e^{-2x}(A \cos 2x + B \sin 2x) + \dfrac{e^x}{26} + \dfrac{e^{-x}}{10}$.

37. (i) $\tan y = \log\left(\dfrac{1 + x}{1 - x}\right) + 1$; (ii) $2 \log (3x - y - 3) + x - y - 3 = C$.

38. (i) $y = C \sin^2 x - 3 \sin x \cos x$; (ii) $(x - y)^2 = x + y$.

39. (i) $x = (x + 2) \tan \tfrac{1}{2}(x + y)$.   40. (i) $y = \dfrac{1}{3} \sec^3 x - \sec x + 1$;   (ii) 729, [the

curve is $y = x^{2x}$.]   41. $2/3\sqrt{3}$, $[y = e^{-x} - e^{-3x}.]$ 42. $y = e^{6x} - 8e^x + 6x + 7$.

43. $y = 2 \cos 2x - \cos 4x$.   44. (i) $y = (\tfrac{1}{2}x^3 - 2x^2 + x \log x + Cx)/(x - 1)$;

(ii) $\sin y = (\tfrac{1}{2}x^3 - 2x^2 + x \log x + Cx)/(x - 1)$.   45. $y = x \sin (\log x + C)$.

47. (i) $\dfrac{dv}{dt} = k(V - v)$; (ii) $v \dfrac{dv}{ds} = k(V - v)$; $v = V(1 - e^{-kt})$; $s = \dfrac{V}{k} \log\left(\dfrac{V}{V - v}\right)$

$- \dfrac{v}{k}$.   48. (i) $k = \dfrac{1}{3} \log 2$, (ii) 18/7. $[x = 2a(e^{kat} - 1)/(2e^{kat} - 1).]$

49. $x = x_0 \exp \left[\{(a - a') t + \tfrac{1}{2}(b - b') t^2\}/100\right]$; $x_0 \exp \left[(a - a')^2/\{200(b' - b)\}\right]$.

50. (a) $\dfrac{dy}{dx} + \left(\dfrac{x - 1}{y - 1}\right) = 0$; $(x - 1)^2 + (y - 1)^2 = C$; (b) $\dfrac{dy}{dx} = -\sin x(1 + 4 \cos x)$.

51. $\dfrac{dx}{dt} = \dfrac{1}{3} x(312 - 4x)$; $x = 78e^{104t}/(25 + e^{104t})$.   52. 14,000; 2,110 insects

per day.   53. (a) $\dfrac{1}{3} x$; (b) $y = \pm e^x/\sqrt(C + x^2)$.   54. $\dfrac{dN}{dt} = -kNt$; $N = N_0 e^{-kt^2/2}$;

184 days.   55. $k_1 = \log ((5/2))$, $k_2 = \log (5/4)$.

Ex. 18.1 (p. 283)

1. 0·45.   2. 1·17.   3. 0·70 and 0·73.   4. 0·57.   5. 0·59 and 2·13.

Ex. 18.2 (p. 285)

2. 1 and 2.   4. −3 and −2, 0 and 1.   5. One (between 3 and 4).   6. Two; −2 and −1, 0 and 1.

Ex. 18.3 (p. 287)

1. −0·25.   2. 0·17.   3. 0·20.   4. 2·04.   5. 4·12.   6. 3·11.

Ex. 18.4 (p. 293)

1. 1·21. 2. 2·47. 3. 1·40. 4. 0·35. 5. 1·48. 6. 1·17. 7. $-1·05 \pm i1·14$.
8. $1·09 \pm i\,0·98$. 9. $\pm 1·17$. 10. 1·20. 11. 1·56. 12. 7 roots; 8·42. 14. $-\dfrac{3\alpha}{2}$,
$1 + 2\alpha,\ 2 - \dfrac{\alpha}{2}$. 14. $\pi(2 + \varepsilon)/8$.

Misc. Ex. XVIII (p. 294)

1. 1·53. 2. 3·64. 3. 1·91 and 0·25. 4. -6 and -5, 0 and 1, 1 and 2; 1·2.
5. $-2·67$. 6. $y = ex$; 2·15. 7. Maximum at $(1, 0)$, minimum at $(3, -4)$; (i) 1 (positive) root; (ii) 3 real roots (2 positive); (iii) 1 (negative) root; 6·2. 8. $-2·4$, 1·2,
4·2. 9. 0·44. 10. $-2·69$. 11. 4·33; $-2·7 \pm i4·0$. 12. 0·511. 13. 0·45. 14. 0·2016.
15. 1·139. 16. 3·591. 17. Three, 1 negative, 2 positive; 0·505. 18. 0·813.
19. 1·09. 20. (b) 1 negative, two positive.

Ex. 19.1 (p. 304)

1. $1 < x < 3$. 2. $x < -2$ and $0 < x < 3$. 3. $-5 < x < -1$ and $2 < x < 3$.
4. $x \leqq 1$. 5. $x > -1$. 6. $-2 < x < -1$ and $x > 0$. 7. (i) $-(a + b) < x < -b$
and $x > 0$; (ii) $x < 0$ and $-b < x < -(a + b)$. 8. $x < -2$ and $x > 0$.
9. $-3 < x < 3$. 10. $x \leqq \frac{1}{2}$. 11. $(6n - 1)\,\pi/6 \leqq x \leqq (6n + 1)\,\pi/6$ where $n = 0$,
$\pm 1, \pm 2 \dots$. 12. $x \leqq -3$ or $x \geqq 1$. 13. $2 < x < 3$. 14. $(n - \frac{1}{2})\,\pi \leqq x \leqq n\pi$
for $n = 0, \pm 1, \pm 2, \dots$. 15. $-2 < x < 1$. 16. $-1 \leqq x \leqq 1$. 17. $-\sqrt{8} \leqq x \leqq -2$
and $2 \leqq x \leqq \sqrt{8}$. 18. $-3 \leqq x < 0$ and $1 \leqq x \leqq 2$. 19. $x \leqq -\frac{1}{2}(-7 + \sqrt{65})$
and $0 < x \leqq 1$ and $7 < x \leqq \frac{1}{2}(7 + \sqrt{65})$. 20. $\pi/6 < x < \frac{1}{2}\pi$ and $7\pi/6 < x < 3\pi/2$.
22. $x = 0 = y$. 30. (a) $x < -5$ or $x > \dfrac{1}{3}$.

Ex. 19.3 (p. 310)

10. $x + y = 2$; $x + y = \frac{1}{2}\pi$.

Misc. Ex. XIX (p. 311)

1. $-2 < y < 2$. 6. (a) (i) $\dfrac{1}{4} < x < \dfrac{3}{2}$; (ii) $x < -1$ and $1 < x < 2$. 7. The
fourth ${}_{16}C_3\ (2)^{13}(\frac{1}{2})^3 = 573,\,440$. 10. (a) $\dfrac{1}{4}\,(6 - \sqrt{38}) < x < \dfrac{1}{4}\,(6 - \sqrt{2})$ and
$\dfrac{1}{4}\,(6 + \sqrt{2}) < x < \dfrac{1}{4}\,(6 + \sqrt{38})$. 11. $-1 \leqq x \leqq 1$ and $x \geqq 3$. 12. $3(1 + t^2)/2t$;
$3(t^2 - 1)/4t^3$. 13. (b) $x < -4$ and $x > 2$. 16. $\frac{1}{2}\,\sqrt{(\pi^2 - 4)} + \sin^{-1}\left(\dfrac{2}{\pi}\right)$. 19. Both
tend to zero. 20. (a) $0 < \theta < 60°$, $120° < \theta < 180°$, $180° < \theta < 240°$,
$300° < \theta < 360°$. 21. (a) x and y each > -1 or x and y each < -1; (b) $-1 \leqq a$
$\leqq 5/3$. 23. (a) 28; (b) $-2 < x < -1$ and $0 < x < 1$. 25. $\dfrac{\sqrt{(h^2 - k^2)}}{k^2}$.

Ex. 20.5 (p. 326)

1. (i) $9x - 7y + 5 = 0$, non-intersecting, $x^2 + y^2 + 4x - 6y + 1$
$+ \lambda(9x - 7y + 5) = 0$; (ii) $9x - 2y = 0$, intersecting, $2x^2 + 2y^2 - 5x + 4y$
$+ \lambda(9x - 2y) = 0$; (iii) $16x - 7y - 2 = 0$, non-intersecting, $x^2 + y^2 + 5x - 2y$
$- 1 + \lambda(16x - 7y - 2) = 0$; (iv) $ax + by = 0$, intersecting if $c < 0$, touching
if $c = 0$, non-intersecting if $c > 0$, $x^2 + y^2 + c + \lambda(ax + by) = 0$. 2. $(2, -8)$,
$(-89/29, 111/29)$. 3. $(-1, 4)$, $(11/2, -5/2)$. 4. $(0, 6)$; $x^2 + y^2 - 12y - 1 = 0$.
5. $\left(\dfrac{2}{5}, -\dfrac{6}{5}\right)$; $x^2 + y^2 - 4x + \lambda(3x + y) = 0$. 6. $(2,1)$, $(-3, -2)$; $x^2 + y^2 - 23x$
$+ 41y = 0$. 7. Both $\left(\dfrac{12}{5}, -\dfrac{16}{5}\right)$; the circles touch at $\left(\dfrac{12}{5}, -\dfrac{16}{5}\right)$. 8. $x^2 + y^2$
$- 10x + 10y = 0$, $49x^2 + 49y^2 + 550x - 290y + 650 = 0$. 9. $(\pm 2k, 0)$.
10. $(x - a)^2 + (y - b)^2 = k^2(x^2 + y^2 + 2gx + 2fy + c)$.

$cab + y\left(a^2 + b\right) + b^2 - 2c^2 = 0$

Ex. 20.9 (p. 332)

1. (i) $3x - y + 4 = 0$; (ii) $10x + 2y - 23 = 0$; (iii) $14x + 5y + 9 = 0$;
(iv) $x(2x_1 + y_1 + 2) + y(x_1 + 2y_1 + 1) + 2x_1 + y_1 + 8 = 0$;
(v) $ay_1x + (ax_1 + b)y + by_1 - 2c^2 = 0$. 2. (i) $(9/5, 3)$; (ii) $(-8a, -2)$;
(iii) $(5, -1)$; (iv) $(14, 17)$; (v) $(-3, 1)$. 3. (i) $9x^2 - 24xy + 8y^2 + 6x + 8y - 7 = 0$;
(ii) $xy = 0$; (iii) $31x^2 + 40xy + 10y^2 + 40x + 20y + 10 = 0$; (iv) $22x^2 + 4xy$
$- 3y^2 - 40x - 10y + 15 = 0$; (v) $x^2 - 5xy - y^2 + 20x + 8y - 16 = 0$.
5. $y^2(x + 2a) + 4a^3 = 0$. 6. $b^2x^2 + a^2y^2 = 4a^2b^2$.

Ex. 20.10 (p. 336)

3. $\frac{1}{2}ab$. 4. $4ab$, (axes $2a$, $2b$). 5. $9x^2 \pm 9xy - 4y^2 = 0$. 11. A similar
and similarly situated ellipse with axes $\sqrt{13}$ times the original. $(x^2/a^2 + y^2/b^2 = 13.)$
12. A similar and similarly situated ellipse with axes $\sqrt{2}$ times the original.
$(x^2/a^2 + y^2/b^2 = 2.)$

Ex. 20.11 (p. 341)

1. (i) $y^2 = 4a(a + x)$; (ii) $3x^2 + y^2 + 8ax - 16a^2 = 0$; (iii) $3x^2 - y^2 - 8ax$
$+ 4a^2 = 0$; (iv) $3x^2 - y^2 - 8ax + 4a^2 = 0$. 2. (i) $\dfrac{2a}{r} = (3 - \sqrt{3})\cos\theta + \sin\theta$;
(ii) $\dfrac{l}{r} = \dfrac{1}{\sqrt{3}}\cos\theta + 2\sin\theta$. 3. (i) $\dfrac{4a}{r} = \cos\theta - \sqrt{3}\sin\theta$; (ii) $\dfrac{l}{r} = 2\cos\theta + \sin\theta$;
(iii) $\dfrac{2l}{r} = 2\cos\theta + \sqrt{3}\sin\theta$. 7. $l/\sqrt{2}$; the major axes make angles $\dfrac{1}{4}\pi$ with major
axis of given ellipse. 8. $e^2 + 2e\cos\alpha\cos\beta + \cos 2\beta = 0$.

Misc. Ex. XX (p. 342)

2. $(1, 2)$, 3; $(5, -1)$, 4. 3. $(1, 0)$, $(3, 0)$; (i) $x^2 + y^2 - 4x - 4y + 3 = 0$;
(ii) $x^2 + y^2 - 4x \pm 2\sqrt{3}y + 3 = 0$; (iii) $x^2 + y^2 - 4x - 2y + 3 = 0$ and $x^2 + y^2$
$- 4x + 14y + 3 = 0$. 4. $x^2 + y^2 = 9$; $x^2 + y^2 \pm 8x - 9 = 0$; $x^2 + y^2$
$+ 8y + 9 = 0$. 5. 16 sq. units; $16\sqrt{2}$ units. 6. $p^2y + x = 2cp$; $\{2cpq/(p + q),$

$2c/(p + q)\}$; $ay = 2c^2$.   8.   (0, 0), (1, 2).   10. $x = 0$; $x^2 + y^2 + 4 + \lambda x = 0$;
$(\pm 2, 0)$; (i) $x^2 + y^2 + 14x + 4 = 0$ and $x^2 + y^2 - 6x + 4 = 0$; (ii) $x^2 + y^2 - 6y$
$- 4 = 0$.   11. $x + y - 5 = 0$ ànd $x + 4y - 10 = 0$.   16. $\left(\dfrac{a \cos\alpha}{\cos\beta}, \dfrac{b \sin\alpha}{\cos\beta} \right)$.
18. $xx'/a^2 + yy'/b^2 = 1$, $b^4 x^2 + a^4 y^2 = a^4 b^4/c^2$.   23. $n^2 (x^2 + y^2) -$
$2n(gx + fy)(lx + my) + c(lx + my)^2 = 0$.   25. $\left(\dfrac{a \cos\alpha}{\cos\beta}, \dfrac{b \sin\alpha}{\cos\beta} \right)$.
27. $\left(\dfrac{a \cos\frac{1}{2}(\theta + \varphi)}{\cos\frac{1}{2}(\theta - \varphi)}, \dfrac{b \sin\frac{1}{2}(\theta + \varphi)}{\cos\frac{1}{2}(\theta - \varphi)} \right)$, $\frac{1}{2}ab \, |\sin(\theta - \varphi)|$.
30. $r = \frac{1}{2}l \operatorname{cosec}^3 \frac{1}{2}\alpha \sin\left(\dfrac{3\alpha}{2} - \theta \right)$.
33. $(a \cos^2\alpha - 2h \sin\alpha \cos\alpha + b \sin^2\alpha)(x^2 + y^2) + 2(h \cos\alpha - b \sin\alpha) x$
$+ 2(h \sin\alpha - a \cos\alpha) y + a + b - 1 = 0$.   35. The circle $lr = ec^2 \cos\theta$.
37. $(-3, 0)$.   42. $(1 + 2lg + cl^2) x^2 + 2(gm + fl + clm) xy + (1 + 2fm + cm^2) y^2$
$= 0$; $\{(la - lb + 2mh)/2(2lmh - am^2 - bl^2)$, $(mb - ma + 2lh)/2(2lmh - am^2$
$- bl^2)\}$.   43. $t_1 t_2 y + x = c(t_1 + t_2)$; $t_1 t_2 t_3 t_4 = -1$.   48. $3x^2 - 8xy - 3y^2 = 0$;
$x^2 + y^2 - 2x - 4y = 0$.   52. $2(ap + hq) x + 2(bq + hp) y = ap^2 + 2hpq + bq^2$;
$qx - py = 0$.   54. $x^2 + y^2 - 5x + 5y = 0$.   55. $\{\frac{1}{2}a(1 + 3t^2), \frac{1}{2}at(3 - t^2)\}$.

Ex. 21.1 (p. 352)

2. (i) a plane parallel to the yz-plane; (ii) a line parallel to $Z'OZ$.   3. $A(x, 0, 0)$,
$B(0, y, 0)$, $C(0, 0, z)$, $L(0, y, z)$, $M(x, 0, z)$, $N(x, y, 0)$; (i) $(x, \frac{1}{2}y, \frac{1}{2}z)$;
(ii) $(\frac{1}{2}x, \frac{1}{2}y, \frac{1}{2}z)$.

Ex. 21.3 (p. 354)

1. (i) $\sqrt{66}$; (ii) $\sqrt{22}$; (iii) $\sqrt{14}$; (iv) 3; (v) $\sqrt{\{2(a^2 + b^2 + c^2 - ab - bc - ca)\}}$
units.   2. $(1, 1, \pm\sqrt{6})$.   3. 3 units.   4. 36/65.   5. $\frac{1}{2}5\sqrt{2}$ units².   6. (i) $(-1, 0, 6)$;
(ii) $(2, 3, -2)$; (iii) $(-22/3, 19/3, 4/3)$; (iv) $(4, 3, 3)$; (v) $(5, -2/5, 1)$.   7. $1 : 2$.
9. (i) $x^2 = y^2 = z^2$, four straight lines each equally inclined to the axes; (ii) $x^2 + y^2$
$+ z^2 = r^2$, a sphere centre the origin and radius r; (iii) $2ax + 2by + 2cz = a^2$
$+ b^2 + c^2$, a plane perpendicular to OA and bisecting OA.

Ex. 21.5 (p. 362)

1. (i) $2x - 3y + z - 4 = 0$; (ii) $x + 2y - 5z + 8 = 0$; (iii) $3x - y + z - 9 = 0$;
(iv) $ax - by + bz - a = 0$; (v) $5x - 2z + 4 = 0$.   2. (i) $2x - 5y + z - 1 = 0$;
(ii) $x + 2y - 3z - 11 = 0$; (iii) $3x + y - 2z - 4 = 0$; (iv) $3x + 4y - 5z = 0$;
(v) $2y + 5z - 20 = 0$.   3. (i) $\dfrac{x}{4} = \dfrac{y - 7/4}{1} = \dfrac{z - 25/4}{-5}$; (ii) $\dfrac{x}{1} = \dfrac{y - 8}{2} = \dfrac{z + 10}{-5}$;
(iii) $\dfrac{x}{-2} = \dfrac{y - 13/2}{11} = \dfrac{z - 5/2}{5}$;   (iv) $\dfrac{x}{1} = \dfrac{y + 1}{0} = \dfrac{z + 2}{1}$,   i.e., $x = z + 2$,
$y = -1$; (v) $\dfrac{x}{1} = \dfrac{y + 2}{-1} = \dfrac{z + 2}{-2}$.   4. (i) Yes, $x + 3y - z - 2 = 0$; (ii) no;
(iii) yes, $2x - y - z + 2 = 0$; (iv) yes, $3x - y + 2z = 0$; (v) no.   5. Internally,
$1 : 10$.   6. $\left(\dfrac{-2}{\sqrt{14}}, \dfrac{3}{\sqrt{14}}, \dfrac{-1}{\sqrt{14}} \right)$.   7. $\cos\beta = \cos\gamma = \sqrt{\dfrac{3}{8}}$.   8. $\left(\dfrac{1}{6}, 2\dfrac{3}{4}, -\dfrac{2}{3} \right)$.
9. $(2, 2, 0)$.   10. $\dfrac{x}{a} + \dfrac{y}{b} + \dfrac{z}{c} = 1$.

Ex. 21.6 (p. 370)

1. $x - 1 = \pm(y - 2) = \pm(z + 1)$. 2. $\cos^{-1}\left(\dfrac{3}{13}\right)$, $\cos^{-1}\left(\dfrac{4}{13}\right)$, $\cos^{-1}\left(\dfrac{12}{13}\right)$.

3. $\dfrac{x}{1} = \dfrac{y}{-4} = \dfrac{z}{3}$. 4. (i) $\dfrac{84}{845}$; (ii) $\dfrac{31}{38}$; (iii) 0 (lines are perpendicular;

(iv) 0 (lines are perpendicular). 5. $\dfrac{x}{m} = \dfrac{y}{-l} = \dfrac{z}{0}$. 7. $\cos^{-1}\left(\dfrac{5}{6}\right)$, $\cos^{-1}\left(\dfrac{13}{6\sqrt{5}}\right)$,

$\pi - \cos^{-1}\left(\dfrac{3}{2\sqrt{5}}\right)$. 10. $90°$. 11. $\dfrac{x}{1} = \dfrac{y}{0} = \dfrac{z}{1}$; $\dfrac{1}{\sqrt{2}}$ units.

12. $\dfrac{x - 3}{432} = \dfrac{y - 1}{69} = \dfrac{z + 1}{131}$. 13. $\dfrac{3}{\sqrt{2}}$ units. 14. (i) $\dfrac{4}{\sqrt{3}}$; (ii) $\dfrac{8}{\sqrt{29}}$; (iii) $\dfrac{2}{\sqrt{11}}$;

(iv) $\dfrac{3}{\sqrt{38}}$; (v) $\dfrac{30}{\sqrt{139}}$ units. 15. (i) $\dfrac{144}{169}$; (ii) $\dfrac{396}{845}$; (iii) $\dfrac{4}{5}$; (iv) 0, (the planes

are perpendicular). 16. (i) $\sin^{-1}\left(\dfrac{129}{169}\right)$; (ii) $\sin^{-1}\left(\dfrac{16}{169}\right)$; (iii) $0°$ (the line and

plane are parallel); (iv) $\sin^{-1}\left(\dfrac{1}{3}\right)$. 17. $45x - 32y - 7z - 166 = 0$.

18. $lx + my + nz = la + mb + nc$. 19. (i) $x + y + 3z = 0$; (ii) $x + 2y - z = 0$;

$\dfrac{x}{7} = \dfrac{y}{-4} = \dfrac{z}{-1}$. 20. $90°$. 21. $x + y - z + 2 = 0$. 23. $y - 3z + 1 = 0$;

$2x + 3y + z - 11 = 0$. 24. (i) $15x - 2y + 5z = 0$; (ii) $\dfrac{3x - 2}{13} = \dfrac{3y - 5}{37}$

$= \dfrac{3z - 4}{32}$ or equivalent forms. 25. $(1, 1, 1)$. 26. $\cos^{-1}\left(\dfrac{5}{\sqrt{35}}\right)$; $\cos^{-1}\left(\dfrac{2}{\sqrt{13}}\right)$;

$\sin^{-1}\left(\dfrac{9}{\sqrt{140}}\right)$. 27. $5x + y - 2z - 5 = 0$; $(5, 8, 14)$.

Ex. 21.7 (p. 377)

1. Three parallel lines with direction ratios $1 : -2 : 1$. 2. A common point $(4, -1, -7)$. 3. A common line of intersection $\dfrac{x - 1}{1} = \dfrac{y - 2}{-1} = \dfrac{z}{2}$. 4. Two of the planes (first and third) are parallel. 5. Three parallel lines with direction ratios $1 : 2 : 1$. 6. A common point $(2, -1, -1)$. 7. The planes are all parallel. 8. A common line of intersection $\dfrac{x + 1}{4} = \dfrac{y - 2}{3} = \dfrac{z}{-2}$. 9. A common point $(-3, 1, 0)$. 10. Three parallel lines with direction ratios $1 : 2 : -1$.

Misc. Ex. XXI (p. 377)

1. $\dfrac{1}{\sqrt{21}}$, $\dfrac{-4}{\sqrt{21}}$, $\dfrac{-2}{\sqrt{21}}$. 2. ABC, $z = 0$; BCD, $2\sqrt{6}x - \sqrt{3}z + 2\sqrt{2}a = 0$; CDA, $\sqrt{6}x - 3\sqrt{2}y + \sqrt{3}z - 2\sqrt{2}a = 0$; DAB, $\sqrt{6}x + 3\sqrt{2}y + \sqrt{3}z - 2\sqrt{2}a = 0$; $\dfrac{x + a/\sqrt{3}}{\sqrt{6}} = \dfrac{y - a}{-3\sqrt{2}} = \dfrac{z}{\sqrt{3}}$; $(0, 0, a/\sqrt{6})$. 3. $2x + 2y + z - 8a = 0$, $\left(\dfrac{16a}{9}, \dfrac{16a}{9}, \dfrac{8a}{9}\right)$. 4. $\sqrt{(5/7)}$. 5. $(3, 1, 1)$; $5x - 17y + 19z - 17 = 0$. 6. $\{a, \tfrac{1}{2}(y_1 + z_1), \tfrac{1}{2}(y_1 + z_1)\}$, $\left\{a + \dfrac{1}{3}(2x_1 - y_1 - z_1), \quad a + \dfrac{1}{3}(-x_1 + 2y_1 - z_1), \quad a + \dfrac{1}{3}(-x_1 - y_1 + 2z_1)\right\}$.

7. $7x - 19y - 13z - 1 = 0$; $(2, 0, 1)$.   8. (i) $\dfrac{1}{3\sqrt{2}}$; (ii) $\dfrac{1}{\sqrt{5}}$.   9. $-\dfrac{18}{5}$, N is

$\left(\dfrac{-2}{13}, \dfrac{6}{13}, \dfrac{-18}{13}\right)$.   10. $\dfrac{x - \frac{2}{3}}{7} = \dfrac{y - \frac{7}{3}}{2} = \dfrac{z + \frac{5}{3}}{-1}$ or equivalent forms.

11. $\dfrac{x}{a} + \dfrac{y}{b} + \dfrac{z}{c} = 1$.   12. (i) $\dfrac{x-1}{3} = \dfrac{y}{2} = \dfrac{z+2}{1}$; (ii) $\dfrac{x-1}{1} = \dfrac{y}{1} = \dfrac{z}{3}$; $\dfrac{x}{2} = \dfrac{y}{1}$

$= \dfrac{z-7}{-4}$.   13. $\dfrac{4}{\sqrt{3}}$; $\left(2, \dfrac{11}{3}, \dfrac{8}{3}\right)$, $\left(\dfrac{2}{3}, \dfrac{7}{3}, 4\right)$.   14. $3\sqrt{2}$; $\dfrac{x-13}{1} = \dfrac{y-11}{0} = \dfrac{z+3}{1}$.

16. $\dfrac{7}{3}$.   17. $\left(\dfrac{-46}{5}, 0, -1\right)$.   18. $\dfrac{1}{\sqrt{38}}, \dfrac{1}{\sqrt{38}}, \dfrac{6}{\sqrt{38}}$.   19. $x + y + 2z = 1$, $5x + 5y$

$+ 2z = 5$.   20. $(2, 3, 5)$; $\dfrac{16}{\sqrt{645}}, \dfrac{10}{\sqrt{645}}, \dfrac{17}{\sqrt{645}}$.   21. 12 miles; $36° 52'$.   22. $Aa + Bb$

$+ Cc + D = 0 = lA + mB + nC$; $(3, 4, 4),)$ $\left(\dfrac{-13}{3}, \dfrac{-10}{3}, \dfrac{-10}{3}\right)$.   28. (i) $\frac{1}{2}\sqrt{35}$

units^2; (ii) $\dfrac{1}{6}$ units^3.   29. (i) $x - 5y + 3z + 13 = 0$; (ii) $\dfrac{9}{\sqrt{87}}$; (iii) $\frac{1}{2}\sqrt{6}$ units^2.

30. (i) $\dfrac{5}{3\sqrt{19}}$; (ii) $\frac{1}{2}\sqrt{146}$ units^2; (iii) $3\frac{1}{2}$ units^3.

SUBJECT INDEX